Proceedings of the 30th International Geological Congress

Volume 16

Mineralogy

Proceedings of the 30th International Geological Congress

Volume 1 : Origin and History of the Earth
Volume 2 : Geosciences and Human Survival, Environment, and Natural Hazards
Volume 3 : Global Changes and Future Environment
Volume 4 : Structure of the Lithosphere and Deep Processes
Volume 5 : Contemporary Lithospheric Motion / Seismic Geology
Volume 6 : Global Tectonic Zones / Supercontinent Formation and Disposal
Volume 7 : Orogenic Belts / Geological Mapping
Volume 8 : Basin Analysis / Global Sedimentary Geology / Sedimentology
Volume 9 : Energy and Mineral Resources for the 21st Century / Geology of Mineral Deposits / Mineral Economics
Volume 10 : New Technology for Geosciences
Volume 11 : Stratigraphy
Volume 12 : Paleontology and Historical Geology
Volume 13 : Marine Geology and Palaeoceanography
Volume 14 : Structural Geology and Geomechanics
Volume 15 : Igneous Petrology
Volume 16 : Mineralogy
Volume 17 : Precambrian Geology and Metamorphic Petrology
Volume 18.A : Geology of Fossil Fuels - Oil and Gas
Volume 18.B : Geology of Fossil Fuels - Coal
Volume 19 : Geochemistry
Volume 20 : Geophysics
Volume 21 : Quaternary Geology
Volume 22 : Hydrogeology
Volume 23 : Engineering Geology
Volume 24 : Environmental Geology
Volume 25 : Mathematical Geology and Geoinformatics
Volume 26 : Comparative Planetology / Geological Education / History of Geosciences

PROCEEDINGS OF THE 30TH INTERNATIONAL GEOLOGICAL CONGRESS

BEIJING, CHINA, 4 - 14 AUGUST 1996

VOLUME 16

MINERALOGY

EDITORS:
HUANG YUNHUI
INSTITUTE OF MINERAL DEPOSITS, CAGS, BEIJING, CHINA
CAO YAWEN
INSTITUTE OF MINERAL DEPOSITS, CAGS, BEIJING, CHINA

CRC Press
Taylor & Francis Group
Boca Raton London New York

CRC Press is an imprint of the
Taylor & Francis Group, an **informa** business

First published 1997 by VSP BV Publishing

Published 2019 by CRC Press
Taylor & Francis Group
6000 Broken Sound Parkway NW, Suite 300
Boca Raton, FL 33487-2742

First issued in paperback 2019

ISBN 13: 978-0-367-44795-3 (pbk)
ISBN 13: 978-90-6764-266-8 (hbk)

Visit the Taylor & Francis Web site at
http://www.taylorandfrancis.com

and the CRC Press Web site at
http://www.crcpress.com

CONTENTS

Proc. 30th Int'l. Geol. Congr., Vol. 16, pp. 1-15
Huang Yunhui and Cao Yawen (Eds)
VSP 1997

Defects in Natural Diamonds Depending on Geological Environment

TORSTEN SCHERER[a], S.S. HAFNER[a], S.M. SUKHARZHEVSKIY[b], G.I. SHAFRANOVSKIY[c], AND D.K. HALLBAUER[d]

[a] *Scientific Center of Materials Sciences and Institute of Mineralogy, University of Marburg, 35032 Marburg, Germany;* [b] *Institute of Earth Crust, University of St Petersburg, Russia;* [c] *Karpinskiy Geological Institute, St Petersburg, Russia;* [d] *Department of Geology, University of Stellenbosch, South Africa*

Abstract

Electron paramagnetic resonance of mantle derived IIa type diamonds from Finsch pipe, South Africa, and diamonds from the impact craters of Nördlinger Ries, Germany, and Popigai, Russia, were studied. Generally the crystals from Finsch pipe exhibited a relatively weak fine structure (FS) line as well as weak superhyperfine splittings of the P1 and P2 type, whereas the diamonds from the impact areas revealed a relatively strong FS line only, P1 and P2 lines being apparently absent. The results hint to differences in the amount of N substitution for C and "dangling" bonds depending on speed of crystal growth associated with mantle or shock wave processes.

Keywords: Diamond, Defects, EPR, Finsch, Nördlinger Ries, Popigai

INTRODUCTION

Defects in crystals depend on the conditions during crystallization and history of temperature and pressure in the subsolidus region of the phase. In natural crystals defects may, therefore, be related to geological processes during and after crystallisation. In the present study paramagnetic centers in diamond have been investigated which result from defects such as vacancies or atomic substitutions at the position of the C atoms. Such centers may be studied by use of electron paramagnetic resonance (EPR) with high precision. Synthetic and natural diamonds have been frequent objects of EPR studies in the past [49, 46, 1, 30, 39, 59, 40, and references therein]. However, the interest was primarily in identification, structure and bonding of defects in the crystal structure without considering the geological environment.

The present study concerned natural diamonds from more or less well documented localities. Generally EPR spectra of natural diamonds are superimposed by those of foreign inclusions such as fluids, silicate minerals, etc. in the crystal. In this work, however, only defects were considered which could be identified as such and of which the EPR is known. EPR due to inclusions was ignored. In this connection it should be noted that natural diamonds are still of great economic interest despite the increasing amount of synthetic crystals being produced recently. The aim of the study was to find distinctions between defects in diamonds of different origin. Mainly 5 different ways of natural diamond formation are known: (1) at conditions of the Earth's upper mantle, (2) by chemical vapor deposition, (3) in U-rich carbonates by radiation, (4) interstellar during collapse of supernova, and (5) at conditions after shock [5, 6, 27, 12]. According to [25] diamond formed after shock events may be also of type 2. In the following, EPR spectra of diamond from type 1 and type 5 will be described.

LOCALITIES STUDIED

Diamonds from Finsch kimberlite pipe, South Africa, and the impact craters, Nördlinger Ries, Germany, and Popigai, Russia, were investigated for detecting differences in paramagnetic centers due to the quite different conditions of formation: crystallization in the upper mantle or during shock metamorphism.

Finsch kimberlite pipe

Finsch pipe [11, 45, 22, 48, 42] is the second largest pipe known in South Africa. Its horizontal cross section is roughly circular with a diameter of 450 to 550 m and surface area of about 0.18 km^2, the walls being nearly vertical. Finsch pipe is located about 37 km east of Postmasburg, Griqualand West, Cape Province, South Africa (Figure 1). The kimberlite diatreme is intruded into banded ironstones outside of the Karoo sedimentary basin. Fragments of Karoo rocks such as siltstone, dolerite, lava, and dolomite are present in the kimberlite. Typical xenoliths are garnet lherzolite, garnet websterite, garnet harzburgite, and, more rarely, eclogite, their sizes being mainly < 4 cm. In general they are highly altered.

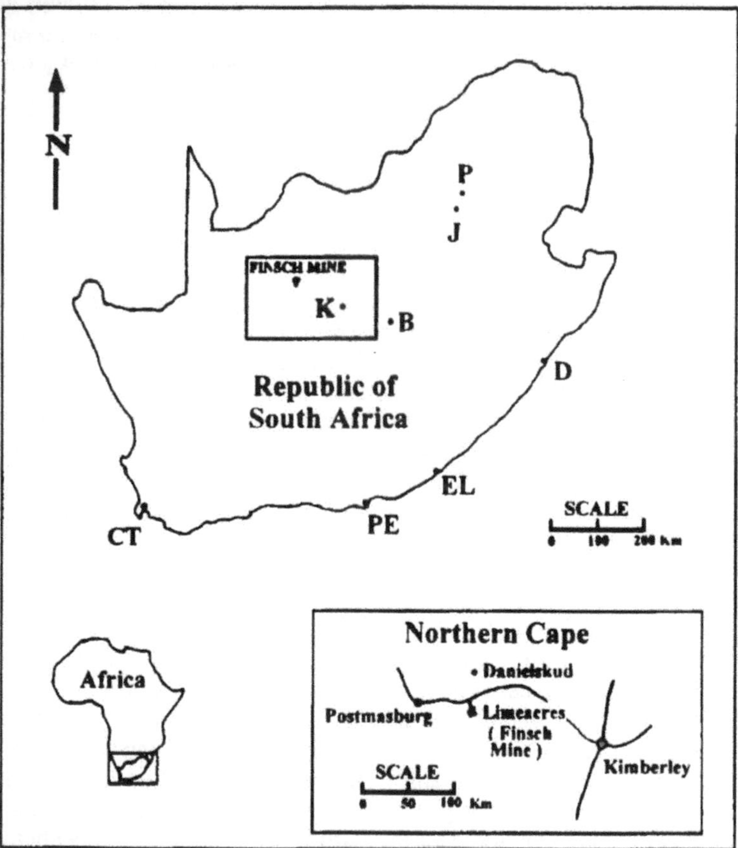

Figure 1. *(from [42], modified).* Location of the Finsch kimberlite pipe. The letters CT, K, PE, B, EL, J, P, and D refer to Capetown, Kimberley, Port Elizabeth, Bloemfontein, East London, Johannesburg, Pretoria, and Durban.

U/Pb radiometric determination of zircon in Finsch kimberlite revealed an age of $93x10^6$ years [13], but ages of Pb isotope ratios in some sulphide inclusions of diamonds were found to be quite incompatible, yielding $2,5x10^9$ years [29]. From this and other observations [43] a syngenetic formation of diamond and host kimberlite can be precluded.

The abundance of diamond in the Finsch kimberlite [43] is generally about 1 ppm or less. Abundances of trace elements were reported in [19]. Studies of inclusions suggested diamond formation at temperatures of approximately 1,100 ± 50 °C and pressures of about 5 GPa [22]. The genesis of diamond with inclusions of peridotitic minerals was probably linked on CO_2-bearing magma [23] formed by the influx of H_2O-CO_2 vapor into a peridotite consisting of clinopyroxene, orthopyroxene, and garnet. Kimberlitic diamonds often contain abundant CO_2, H_2O, N_2, CH_4, Ar, and possibly H_2 [38] which could well be concentrated in the postulated volatile phase. Also CO_3^{2-}- and OH^--ions could be identified [10]. Capture of Ca^{2+} by CO_3^{2-} may account for the calcium depleted nature of the peridotitic mineral inclusions. About 98 % of the diamonds of Finsch pipe were formed during the influx process. The remaining 2 % are considered to be derived from a pre-existing diamond-bearing eclogite horizon. They were sampled by the kimberlite during the ascent through the mantle [22].

Nördlinger Ries

The meteoritic impact crater [2-4,15-18] of the Nördlinger Ries is located in the Swabian Mountains, southern Germany (Figure 2).

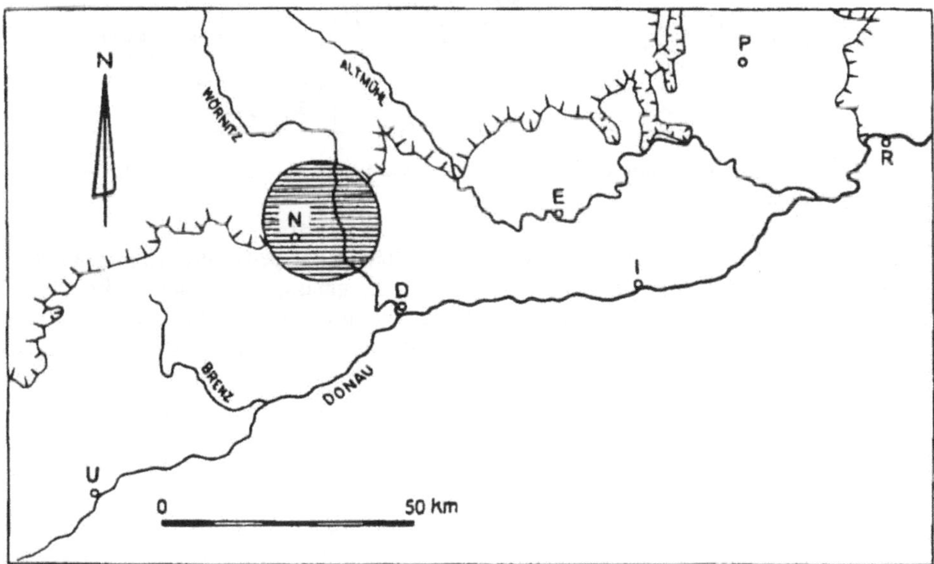

Figure 2. *(from [21], Figure 6, modified).* Location of the Nördlinger Ries. Circle with horizontal lines: impact crater area. The letters U, N, D, E, I, P, and R refer to Ulm, Nördlingen, Donauwörth, Eichstätt, Ingolstadt, Parsberg, and Regensburg. The spinous line indicates the border of the Swabian and Franconian Mountains.

The impact penetrated in a jurassic and triassic arrangement of sandstone, clay, marl, and lime-stone with a thickness of about 600 m, overlaying crystalline rocks.

The crater area (Figure 3) is of annular structure and consists of a flat, circular basin with an outer rim (diameter 25 km) of surrounding hills which are elevated by 100-200 m with respect to the basin.

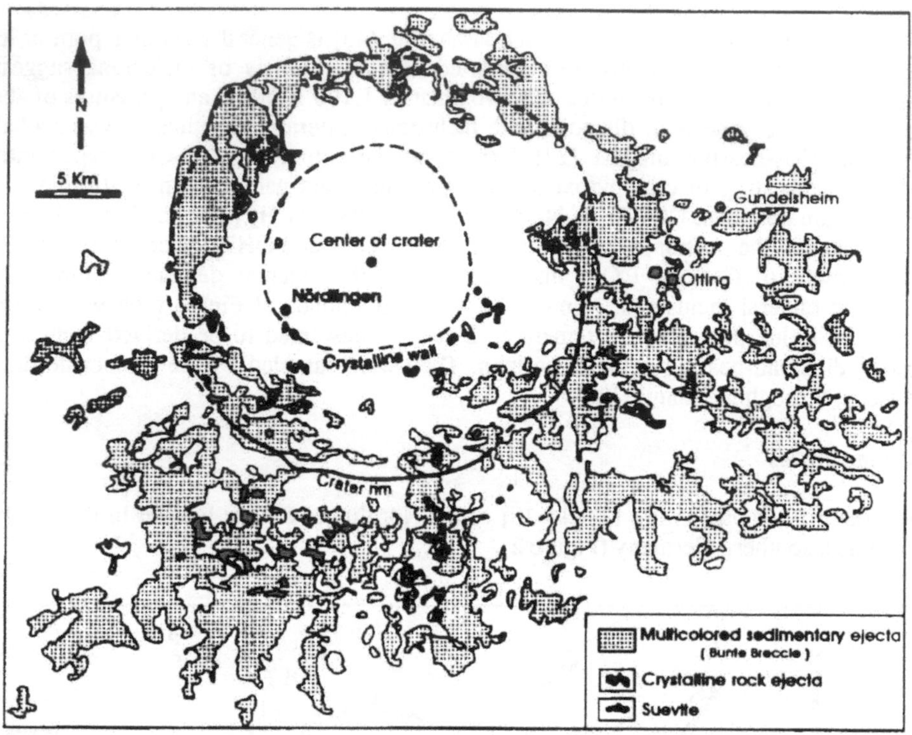

Figure 3. *(from [8], Figure 7, modified).* Map of the impact area, Nördlinger Ries. To the north, the ejecta have already been eroded.

The central crater area is surrounded by an incomplete (U-shaped) inner ring (diameter 12 km) which comprises uplifted and shattered granite, gneiss, and other rocks from the basement. Typical rocks of the impact area are suevite, tektites, breccias, glass bombs, etc. Examples of high-pressure minerals are high-pressure SiO_2 polymorphs [47, 15, 52] such as planar featured quartz, cohesite, stishovite, and high-pressure carbon polymorphs [14, 44]. The various stages of shock metamorphism observed in the Ries area are shown in Table 1. The relative position of the Ries impact energy [52] compared with other events is illustrated in Table 2.

In 1792 Carl von Caspers [7] discovered a rock which he considered to be of volcanic origin: suevite, i.e. a breccia with rock fragments mainly from the crystalline basement, glass bodies, etc. in a glassy matrix which is partly recrystallized or weathered. Suevite often shows all stages of progressive shock metamorphism from 0 to 4 (Table 1), though generally characteristic for stage 4.

While in general the Ries crater was considered to be volcanic since 1792 [7] Shoemaker and Chao in 1961 [47] were able to prove its intrinsic origin by detecting the high-pressure SiO_2 mineral coesite. However, speculations about meteoritic impact had been proposed much earlier [56, 54].

Radiometric K/Ar [20] as well as fission track [53] dating of impact glass and tektite revealed an age of about 14.7x10^6 years (middle Miocene). The first report on diamond was that of Rost et al. [44]. Thus far diamond was found in suevite only.

Table 1. Progressive shock metamorphism in the Nördlinger Ries, from [15], Table 1,completed from [50]

Stage of shock metamorphism	Characteristic deformations and phase transitions	Distance from impact centre [km]	Pressure [GPa]	Residual temp. [°C]
Stage 0	Elastic deformation	> 3.8	-	-
Stage I	Fracturing; plastic deformation (diaplectic quartz and feldspar)	3.8	10	100
Stage II	Phase transitions (diaplectic glasses of quartz and feldspar, high pressure phases of SiO_2)	2.1	35	250
Stage III	Selective melting (glasses of quartz and feldspar, high pressure phases of SiO_2)	1.7	50	1,200
Stage IV	Melting of all main rock forming minerals (inhomogeneous rock melts, "Fladen")	1.5	60	2,200
Stage V	Volatilization (silicate vapor)	1.3	100	5,000

Table 2. Energies of impacts, volcanic eruptions and atomic blasts
Estimated from natural impact events, laboratory experiments [51], and model calculations

For a crater of the size of the Nördlinger Ries an iron meteorite with diameter of 660 m, $\rho = 8$ gcm^{-3}, a stony meteorite with diameter of 915 m, $\rho = 3$ gcm^{-3}, or comet nucleus with diameter of 1,485 m, $\rho = 0.7$ gcm^{-3} would be needed [41]. Impact velocities lie within the range of 11 km/s and 72 km/s.

Event	age [years]	diameter [km]	energy [J]	references
Nördlinger Ries, Germany	14.7 x 10^6	25	10^{20}	[51]
Barringer Crater, Arizona (USA)	50 x 10^3	1.2	10^{17}	[6]
Manicouagan Structure, Canada	212 x 10^6	100	10^{21}	[6]
Popigai crater, Russia	36-39 x 10^6	100	10^{23}	[35]
Tambora volcano, Indonesia	181	-	1.44 x 10^{20}	[24]
20 Mt H-bomb	-	-	10^{17}	-
ten-fold Hiroshima bomb	-	-	8.4 x 10^{14}	[24]

Popigai

Although diamond is well known as meteoritic mineral since the 19th century the study of diamonds from terrestrial impact craters has largely been confined to Russia. Diamond localities in addition to Popigai are e.g. the impact craters of Kara, Khrebet Pay-Khoy Range, Kara sea, and Puchezh-Katunki, 90 km north west of Gorkiy.

The Popigai impact is situated on the northeastern slope of the Anabar shield about 400 km south of the Laptev Sea near the headwaters of the Popigai river in north central Siberia. It was discovered in 1970 [31]. The structure of the impact crater is well preserved though somewhat eroded. K/Ar- [31] and fission-track [28] dating of tagamites and impact glasses revealed an age of about 36-39x10^6 years. The Anabar shield comprises Proterozoic and Cambrian quartzites, dolomites, limestones, Permian sandstones, and argillites, overlaying the Archean basement gneisses. The shield was invaded by Triassic dolerite sills and dikes.

Like the Nördlinger Ries the Popigai crater exhibits a complex inner structure (Figure 4). The overall crater area (diameter about 70 km, depth about 300 m) is surrounded by cliffs made of impact breccias which include giant blocks of country rock.

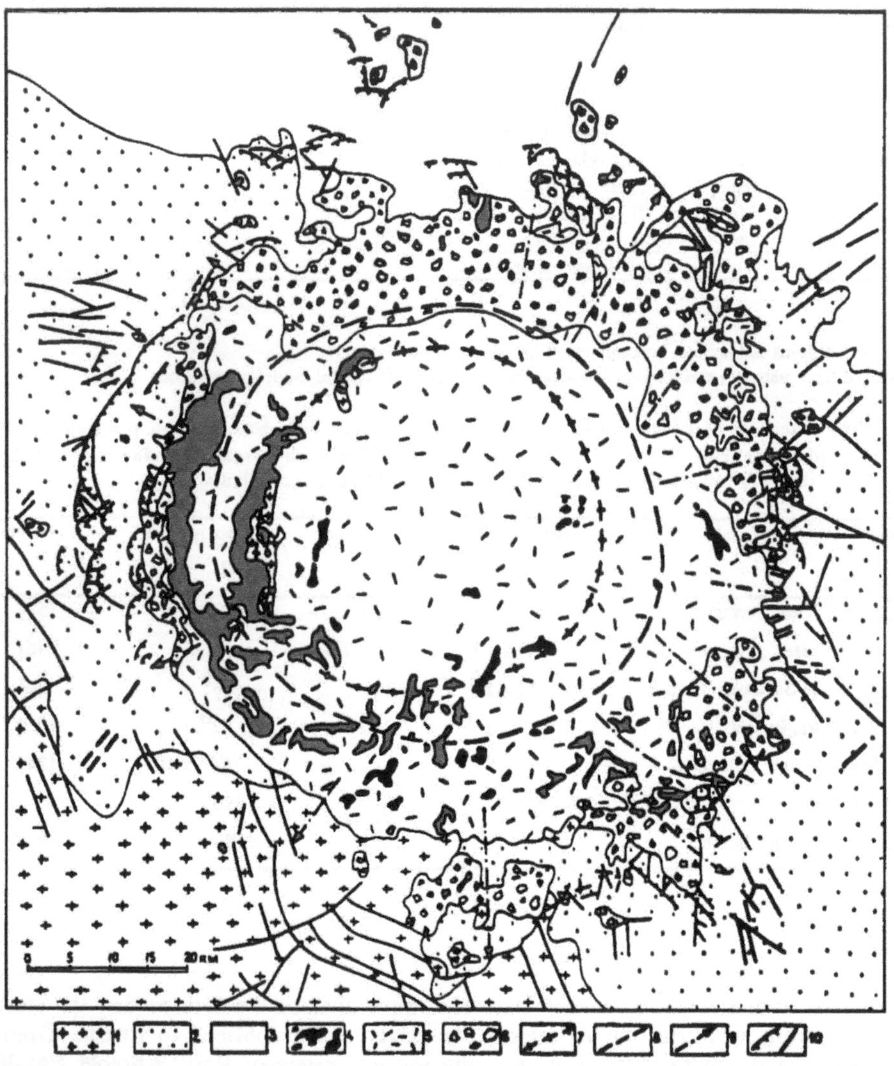

Figure 4. *(from [34], Figure 1).* Map of the Popigai impact crater area: (1) Archean gneisses and schists; (2) Upper Proterozoic and Cambrian quartzites, dolomites, and limestones; (3) Permian sandstones and argillites with sills and dikes of Triassic dolerites; (4) tagamites; (5) suevites with lenses of fine grained allogenic breccia; (6) allogenic breccia; (7) center line of annular uplift; (8) center line of annular trough; (9) axes of radial troughs; (10) thrusts, faults.

The inner crater area consists of a central depression within an annular crystalline uplift made up of shocked and brecciated gneisses, its diameter being about 45 km. Outside this uplift, an annular trough with a width of about 15-18 km is encircled by the crater rim zone as boundary.

Progressive shock metamorphism appears similar to the Nördlinger Ries, but the spatial division into different stages of metamorphosis is larger due to the greater impact size. Gneiss fragments are composed mostly of monomineral impact glass, diaplectic glass, and remnants of diaplectic minerals. The primary texture of the gneiss is more or less preserved.

The Popigai area is known to be especially rich in diamonds formed by the impact event [32]. Diamonds may be found in tagamites, suevites, and even in strongly shocked gneiss fragments included in tagamite or suevite. Graphite in the target rocks must have been the source of the diamonds crystallized after the impact during shock and melting: diamond and graphite exhibit the same isotopic composition [32, 33].

SAMPLES

Finsch pipe
9 single crystals of the IIa type [57] with diameters of about 1.5 mm and weights between 8 mg and 11 mg were studied. The crystals consisted of somewhat distorted octahedra combined with (541) faces or dodecahedra, some of the faces showing etch figures and typical striations. They were apparently free of inclusions except one crystal which exhibited black particles smaller than 10 µm. 3 crystals were lightly yellow, the others colorless.

Nördlinger Ries
Suevite fragments of the Otting quarry 22 km east of the town Nördlingen (Figure 3) were ground and fused with NaOH. After dissolution of the solidified melt with water, the residue was examined under the microscope. 2 apparently polycrystalline diamond grains were separated and identified by X-ray diffraction. Their diameters were about 0.1 mm.

Popigai
About 30 apparently polycrystalline grains with diameters between 0.1 mm and 2 mm separated from tagamites were obtained from a collection [36, 37] and from V. Kvasnitsa and A. Valter in Kiev, Ukraine, their precise localities within the Popigai crater area not being known at this time. The densities varied between 3.28 g/cm^3 and 3.61 g/cm^3. The grains were mainly plate-shaped, a few of them isometric. They were partially rounded, partially with relatively sharp edges. Most grains were translucent with light yellow to wax yellow or brownish color. X-ray diffraction studies revealed that yellowish grains contained lonsdaleite whereas completely transparent, colorless grains were free of lonsdaleite. Brown and black colors were ascribed to inclusions of graphite particles. Optical inspection under the binocular revealed inclusions of black spots with diameters of about 50 µm, sometimes revealing cloudy structures. One grain was opaque black with roughly isometric hexagonal shape and graphite-like planar faces. It was probably the product of progressive shock metamorphism in a carbon bearing target rock.

EXPERIMENTAL

EPR spectra of the diamonds were recorded at frequencies of the X-band (9.5 GHz) between 10 K and 295 K with an E-Line Varian spectrometer equipped with an E-101 microwave bridge and 12 inch magnet in Marburg. Generally microwave powers between $5x10^{-4}$ mW and 2 mW and a modulation frequency of 100 kHz had to be employed in the present study. For the spectra at low temperatures a gas flow cryostat was used. The microwave frequency and applied magnetic field H were calibrated with a frequency counter and a magnetic field control unit for digital recording of the spectra with a computer. Spectra in the Q-band (37 GHz) were recorded with a type RE-1308 spectrometer in St. Petersburg. For the EPR of single crystals in the X-band, final crystallographic orientations with respect to H were obtained by use of a home made two-circle goniometer with a precision of about one degree. For this the maximum splitting of the P1 triplett was employed which occurs when [100] is parallel to H. The goniometer could be operated at low temperatures. Course alignments were made first optically. Crystals were glued to the EPR goniometer with polystyrene dissolved in trichlorethane. This glue was stable down to 4.2 K and did not show any signal in the X-band.

RESULTS

EPR spectra of diamond generally consist of a rather broad quasi-isotropic fine structure (FS) line at g = 2.0025 ± 0.0001 with a width of about $5x10^{-4}$ T and a very large number of superimposed sharp superhyperfine (SHF) split lines between g = 2.0233 and g = 1.9786.

The widths of the SHF lines are $1x10^{-4}$ T or smaller. In this study 3 different groups of lines were considered: (1) the FS line at g = 2.0024, (2) the P1 group of lines, and (3) the P2 group of lines. The FS line was observed in all samples except one from Finsch pipe, whereas P1 and P2 patterns could be detected only in crystals from the Finsch pipe. However, only the colored Finsch pipe crystals revealed all three groups of lines: FS, P1, and P2, the intensity of the P1 and P2 lines being clearly related to the intensity of color. One crystal from Finsch pipe did not reveal any spectrum, neither a FS line nor SHF split lines. This crystal was octahedral with additional (541) faces, completely transparent, colorless, and without apparent inclusions, its weight was 10.1 mg. The FS lines were recorded generally by use of a modulation field with a frequency MF = 100 KHz and amplitude MA = 0.05 mT, the microwave power being MP = 0.01 mW.

The P1 and P2 subspectra were highly anisotropic. They could be observed well with MF = 100 KHz, MA = 0.05 mT, and MP = 0.5 μW. Spectra are shown in Figure 5. The P1 pattern consisted of 9 distinct lines. Its point symmetry was confirmed from special orientations of the crystal with respect to H to be precisely *3m* as shown by the SHF splittings in Figure 5, b, c, d. The hyperfine coupling constants were determined to be $A_{\perp}$ = 27.1 $x10^{-4}$ $cm^{-1}$and $A_{//}$ = $38.1x10^{-4}$ cm^{-1}, respectively. These values agree with those reported in [49]. P2 comprised many lines mainly in the center of the spectrum between g = 2.00835 and g = 1.99442. The correct number of P2 lines could not be determined by inspection of the spectra directly. Typical spectra of a Finsch single crystal are shown in Figure 5. In Figure 6 the dependency of the FS and SHF lines on the MP is plotted.

DISCUSSION

The FS line in diamond has been interpreted generally in terms of a "dangling" bond, i.e. a missing electron in one of the *4* sp^3 lobes of the $2s^22p^2$ valence shell of C. The missing charge in the valence shell must be compensated by a crystallographic defect in the environment apparently without producing SHF splitting.

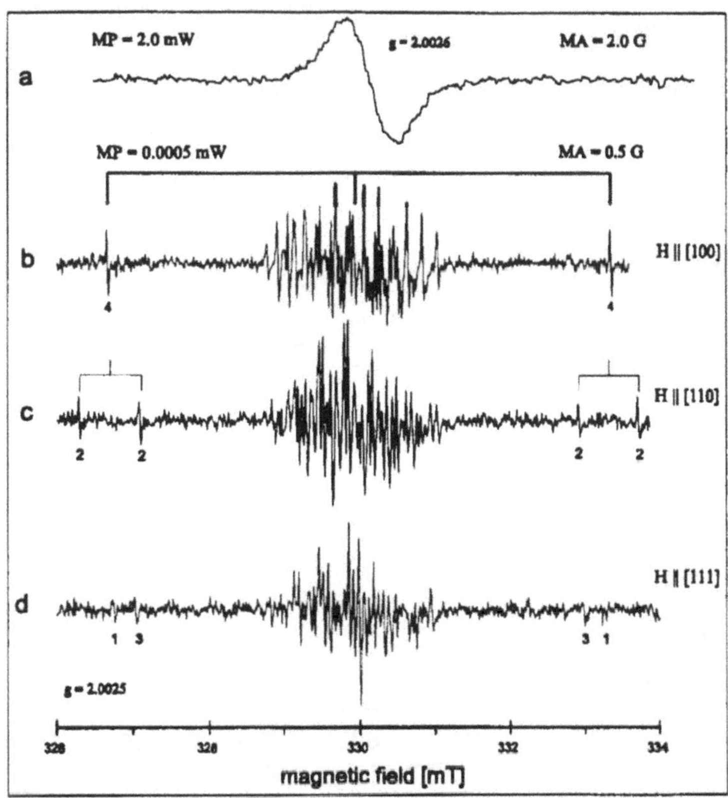

Figure 5. X-band EPR spectra of a single crystal of diamond from Finsch kimberlite pipe. Spectrum (a) shows the FS line ("dangling bond") which is quasi-isotropic and visible only at relatively high microwave power MP. Spectra (b), (c), and (d) refer to the paramagnetic centers P1 and P2 with the applied magnetic field H oriented parallel to [100], [110], and [111] directions, respectively of the crystal. The superimposed FS "dangling bond" line in (b), (c), and (d) cannot be recognized due to the low MP. The arrows mark P1 lines split or degenerated according to the special orientation of the crystal. The numbers designate the multiplicities. The large number of lines of the P2 pattern in the central part of the spectra are not designated.

In the samples we studied the FS line in diamond from impact craters was significantly more intense than that from Finsch kimberlite. This can be concluded from comparison of Figures 7 and 8 with Figure 5. The strong FS line seems to be characteristic for diamond formed by shock metamorphism, at least in case of Nördliger Ries and Popigai. The dangling bond and the crystallographic defect associated with it are probably a result of particular conditions of shock metamorphism, perhaps due to rapid crystal growth. The dependency of FS line intensity on the diamond grain density of samples from the Popigai area is demonstrated in Figure 9.

Diamond specimens from Popigai may include up to 10 vol % [26] domains of lonsdaleite. This is consistent with the birefringence observed in all of our crystals. It is quite possible that part or even most of the FS line intensity measured in the Popigai crystals referred to defects in lonsdaleite which may also show dangling bonds. Lonsdaleite is known to occur also in diamond of the Nördlinger Ries [44]. However, this was not the case for our crystals which were studied in St Petersburg by X-ray diffraction. Whether the very much smaller intensity of the FS line in the Finsch pipe diamonds result from domains or inclusions of lonsdaleite is difficult to conclude. Part of the FS line intensity especially in the Popigai crystals may be due to graphite inclusions. In graphite the dangling bond could be due to an unpaired electron in a planar sp^2 lobe or an empty p_z orbital.

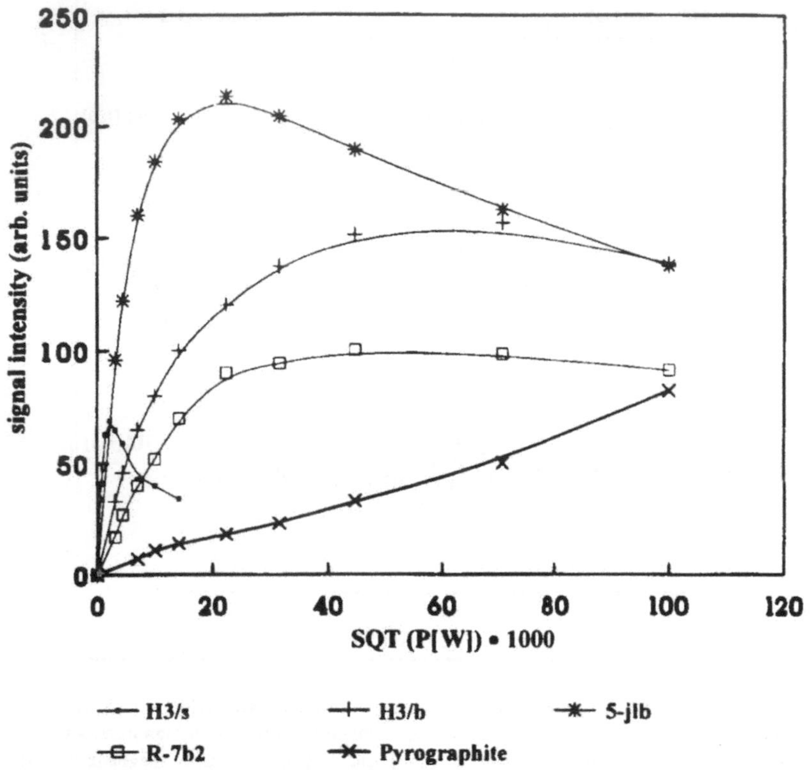

Fig. 6. Intensities of X-band EPR lines of different diamonds and a pyrographite are plotted depending on the microwave power MP. squares: FS line of a grain from Nördlinger Ries; stars: FS line of a grain from Popigai; horizontal crosses: FS line of a Finsch pipe diamond; oblique crosses: FS line of a pyrographite; dots: HSF lines of the P1 and P2 patterns from a Finsch pipe diamond.

Hough *et al.* [25] interpreted the occurrence of intergrown diamond and lonsdaleite plates associated with silicon carbide in impact melts of the Nördlinger Ries in terms of chemical vapour deposition from the ejecta plume of the impact crater. The question whether the mechanism for diamond formation is related to chemical vapor deposition, shock metamorphism, or both is still open.

The HFS split P1 and P2 patterns of the Finsch diamonds show clearly that N atoms are substituted at the crystallographic position of C in diamond. The structural positions of the two centers are illustrated in Figures 10 and 11. The abundance of N in kimberlite diamond is well known, infrared spectra yielding values up to 2,000 atomic ppm generally. Infrared spectra may include also the contribution of foreign inclusions in diamond. A direct correlation with EPR spectra is, therefore, not possible.

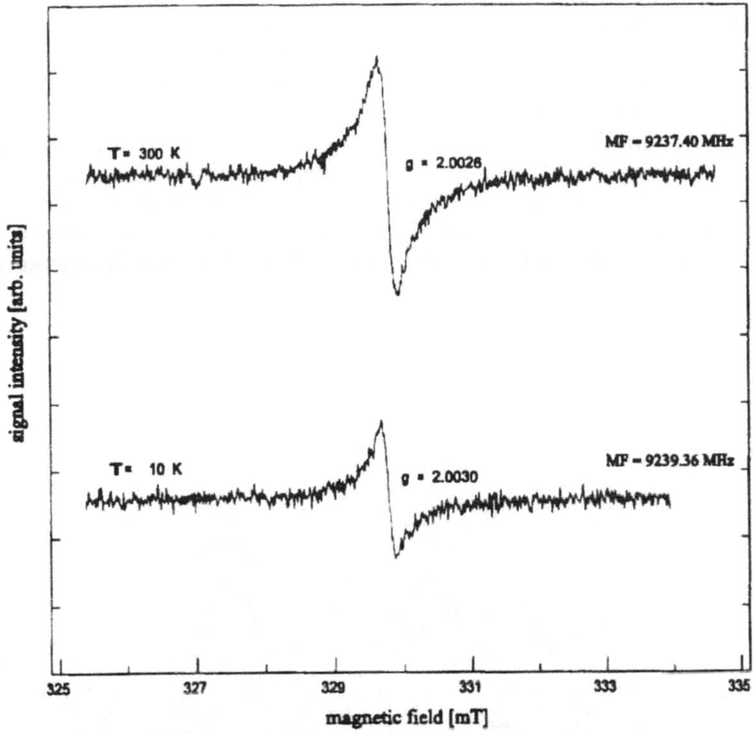

Fig. 7. X-band EPR spectra of a diamond from Nördlinger Ries at 10 K and 295 K, FS line. MF: microwave frequency. The spectra were recorded at a microwave power of 0.01 mW.

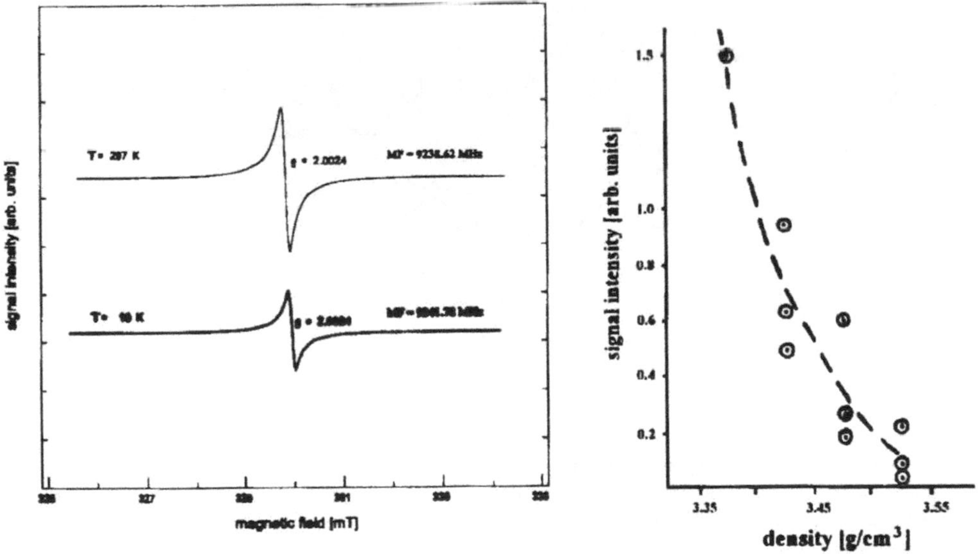

Figure 8, left. X-band EPR spectra of a diamond from Popigai. at 10 K and 295 K, FS line. MF: microwave frequency, MP = 0.01 mW.

Figure 9, right. *(from [54], Figure 3).* EPR FS line intensity (g = 2.0024) of diamond grains from Popigai crater area depending on the density of the grain.

The apparent lack of the P1 as well as P2 patterns in the diamond spectra from Nördlinger Ries and Popigai seems to imply that diamond crystallization at conditions of impact metamorphism occurred under much lower partial N_2 pressures than at subcrustal conditions. It should be mentioned that diamonds of impact origin appear to have less perfect lattices, with a lower degree of long range order. This may produce broadening of the generally very sharp SHF lines which are highly anisotropic. In fact a colorless, transparent diamond with a diameter of 2 mm from Popigai did not show any trace of P1 and P2 patterns. Even if that diamond is assumed to be polycrystalline SHF lines would be expected to be detected in the center part of the spectrum between g = 2.00835 and g = 1.99442, considering the very sharp line widths observed in single crystals from kimberlite.

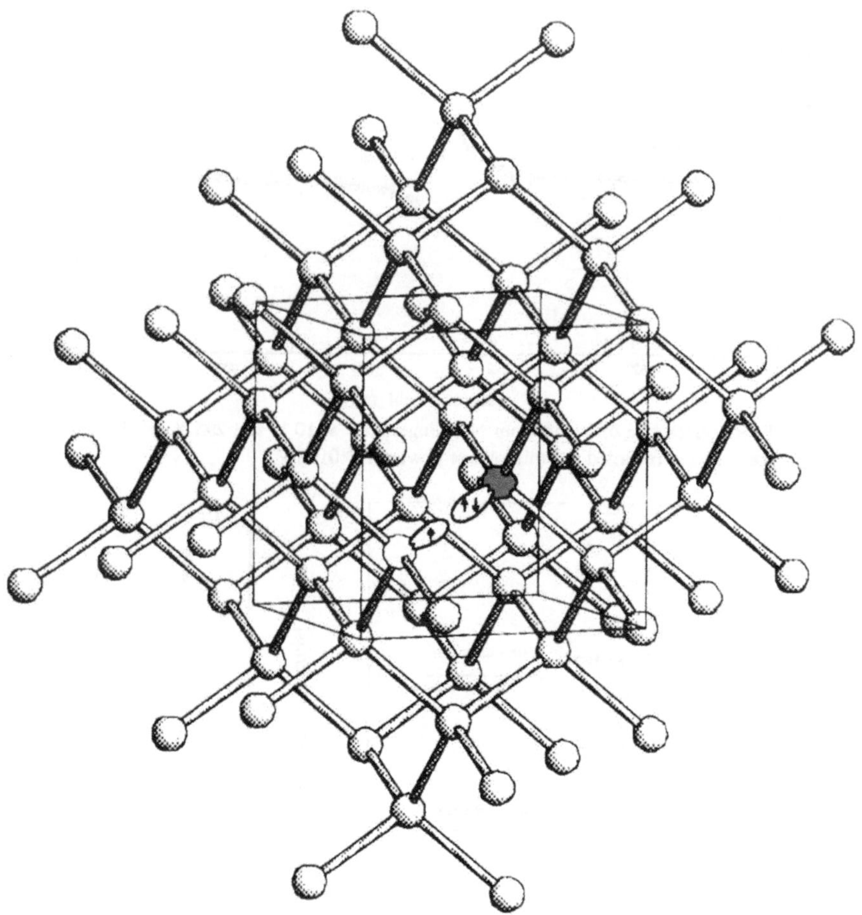

Figure 10. Defect of the paramagnetic center P1 in diamond. The defect consists of an N atom (black sphere) substituted for C. The paramagnetic center is located at the colorless sphere (without shadow) which is a C atom with an unpaired electron pointing to the N. The point symmetry of the void is 3*m*.

Koeberl et al. [26] suggested from the distinction observed in infrared absorption spectra that N may occur in a different form in Popigai impactite diamond compared to mantle derived diamond. In our IIa type crystals from Finsch pipe, however, it is assumed to be < 2 atomic ppm. Here, it should be noted that inclusions with molecular N_2 in diamond cannot be observed in EPR spectra. The question whether crystal structures of diamond formed by impact events include less substitution of N for C as our EPR spectra imply needs further study.

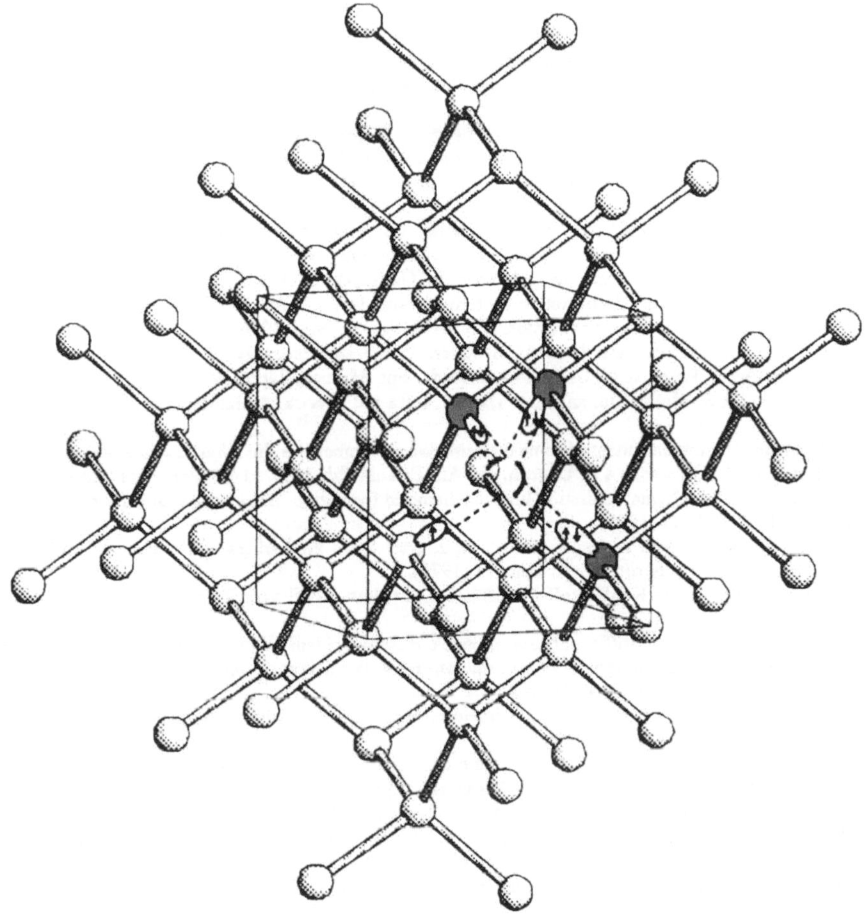

Figure 11. Defect of the paramagnetic center P2 in diamond. The defect consists of a void C position with three adjacent N substituted for C at their crystallographic positions. The paramagnetic center is located at the colorless atom (without shadow) which is a C with an unpaired electron pointing to the void. The point symmetry of the void is $3m$.

Acknowledgements

We thank V. Kvasnitsa and A. Valter, Institute of Geochemistry, Mineralogy, and Ore Formation, National Academy of Sciences, Kiev, Ukraine, for diamonds from the Popigai impact area, D. Robinson, de Beers, Johannesburg, South Africa, for diamonds from Finsch mine, W.C. Tennant, University of Canterbury, Christchurch, New Zealand, for discussion of EPR results, and Dieter Stöffler, Museum für Naturkunde, Humboldt Universität Berlin, Germany, for the information on the Popigai crater area.

REFERENCES

1. R.C. Barklie and J. Guven. ^{13}C hyperfine structure and relaxation times of the P1 centre in diamond. *J. Phys. C: Solid State Phys.* **14**, 3621-3631 (1981).
2. Bayer. Geol. Landesamt (Ed.). Das Ries. Geologie, Geophysik und Genese eines Kraters [in German]. *Geologica Bavarica* **61**, Munich (1969).
3. Bayer. Geol. Landesamt (Ed.). Die Ergebnisse der Ries-Forschungsbohrung 1973: Struktur des Kraters und Entwicklung des Kratersees [in German]. *Geologica Bavarica* **75**, Munich (1977).
4. Bayer. Geol. Landesamt (Ed.). Erläuterungen zur Karte des Rieses 1:50000 [in German]. *Geologica Bavarica* **76**, Munich (1977).
5. D.F. Blake, F. Freund, K.F.M. Krishnan, C.J. Echer, R.Shipp, T.E. Bunch, A.G. Tielens, R.J. Lipari, C.J.D. Hetherington, and S. Chang. The nature and origin of interstellar diamond. *Nature* **332**, 611-613 (1988).
6. D.B. Carlisle. *Dinosaurs, diamonds, and things from outer space*. Stanford University Press, Stanford (1995).
7. C.v. Caspers. Entdeckung des Feuerduftsteins im Herzogthum Pfalz-Neuburg, woraus der zu wassertüchtigen Gebäuden ohnentbehrliche Traß zubereitet wird. Zum Gebrauch der churbajerischen Landen und des ganzen Donaustroms Liebhabern [in German]. Ingolstadt (1792).
8. E.C.T. Chao. Comparison of the cretaceous-tertiary boundary impact events and the 0.77-Ma australasian tektite event: Relevance to mass extinction. *U.S. Geol. Survey Bull.* **2050**,
9. E.C.T. Chao, R. Hüttner, and H. Schmidt-Kaler. *Principal exposures of the Ries meteorite crater in southern Germany*. Bayerisches Geologisches Landesamt, Munich, 84pp (1992).
10. R.M. Chrenko, R.S. McDonald, and K.A. Darrow. Infra-red spectra of diamond coat. *Nature* **213**, 474-476 (1967).
11. C.R. Clement. The emplacement of some diatreme-facies kimberlites. In: *Physics and chemistry of the Earth 9*. L.H. Ahrens, J.B Dawson, A.R. Duncan, and A.J. Erlank (Eds.). pp. 51-60. Pergamon Press (1975).
12. T.L. Daulton and M. Ozima. Radiation-induced diamond formation in uranium-rich carbonaceous materials. *Science* **271**, 1260-1263 (1996).
13. G.L. Davis. The ages and uranium contents of zircons from kimberlites and associated rocks. *Extended Abstr., 2nd. Intern. Kimberlite Conf.*, Santa Fe (1977).
14. A. El Goresy. Eine neue Kohlenstoff-Modifikation aus dem Nördlinger Ries [in German]. *Naturwiss.* **56**, 493-494 (1969).
15. W.v. Engelhardt and D. Stöffler. Stages of shock metamorphism in crystalline rocks of the Ries basin, Germany. In: *Shock metamorphism of natural materials*. B.M. French and M.S. Nicholas (Eds.). pp. 159-168. Mono Book Corp., Baltimore (1968).
16. W.v. Engelhardt, D. Stöffler, and W. Schneider. Petrologische Untersuchungen im Ries [in German]. *Geologica Bavarica* **61**, 229-295 (1969).
17. W.v. Engelhardt and G. Graup. Suevite of the Ries crater, Germany: Source rocks and implications for cratering mechanics. *Geol. Rundschau* **73/2**, 447-481 (1984).
18. W.v. Engelhardt and E. Luft. Origin of moldavites. *Geochim.Cosmochim. Acta* **51**, 1425-1443 (1987).
19. H.W. Fesq, D.M. Bibby, C.S. Erasmus, E.J.D. Kable, and J.P.F. Sellschop. A comparative trace element study of diamonds from Premier, Finsch and Jagersfontein mines, South Africa. In: *Physics and chemistry of the Earth 9*. L.H. Ahrens, J.B Dawson, A.R. Duncan, and A.J. Erlank (Eds.). pp. 817-836. Pergamon Press (1975).
20. W. Gentner and G.A. Wagner. Altersbestimmungen an Riesgläsern und Moldaviten [in German]. *Geologica Bavarica* **61**, 296-303 (1969).
21. P. Groschopf and W. Reiff. Das Steinheimer Becken als Meteorkrater [in German]. In: *Meteorkrater Steinheimer Becken*. 2nd Ed., pp. 19-28. Heimat und Altertumsverein, Heidenheim (1986).
22. J.J. Gurney, J.W. Harris, and R.S. Rickard. Silicate and oxide inclusions in diamonds from the Finsch kimberlite pipe. In: *Kimberlites, Diatremes and Diamonds: Their Geology, Petrology and Geochemistry*. F.R. Boyd and H.O.A. Meyer (Eds.). pp. 1-15. Proc. 2nd. Intern. Kimberlite Conf. **1**, Amer. Geophys. Union, Washington (1979).
23. B. Harte, J.J. Gurney, and J.W. Harris. The formation of peridotitic suite inclusions in diamonds. *Contrib. Mineral. Petrol.* **72**, 181-190 (1980).
24. P. Hédervari. On the energy and magnitude of volcanic eruptions. *Bull. volcanol.* **26**, Naples (1963).
25. R.M. Hough, I. Gilmour, C.T. Pillinger, J.W. Arden, K.W.R. Gilkes, J. Yuan, and H.J. Milledge. Diamond and silicon carbide in impact melt rock from the Ries impact crater. *Nature* **378**, 41-44 (1995).
26. C. Koeberl, V.L. Masaitis, F. Langenhorst, D. Stöffler, M. Schrauder, C. Lengauer, I. Gilmour, and R.M. Hough. Diamonds from the Popigai impact structure, Russia. *Lunar Planet. Sci.* **26**, (1995).
27. C. Koeberl. Diamonds everywhere. *Nature* **378**, 17-18 (1995).
28. A.N. Komarov and A.I. Raikhlin. Comparative study of impactite dating by fission track and K-Ar methods [in Russian]. *Akademiya Nauk SSSR Doklady* **228**, 673-676 (1976).

39. J.D. Kramers. Lead and strontium isotopes in inclusions in diamonds and in mantle-derived xenoliths from southern Africa. *Extend. Abstr., 2nd Internat. Kimberlite Conf.*, Santa Fe (1977).
30. J.H.N. Loubser, J.A. van Wyk, and C.M. Welbourn. Electron spin resonance of a tri-nitrogen centre in cape yellow type Ia diamonds. *J.Phys. C: Solid State Phys.* **15**, 6031-6036 (1982).
31. V.L. Masaitis, M.V. Mikhailov, and T.V. Selivanovskaya. *Popigai meteorite crater* [in Russian]. Nauka Press, Moscow (1975).
32. V.L. Masaitis, G.I. Shafranovsky, V.A. Ezersky, and N.B. Reshetnyak. Impact diamonds from ureilites and impactites [in Russian]. *Meteoritika* **49**, 180-195 (1990).
33. V.L. Masaitis. Diamantiferous impactites, their distribution and petrogenesis [in Russian]. *Regional geology and metallogeny* **1**, 121-134 (1993).
34. V.L. Masaitis. Impactites from the Popigai crater. In: *Large meteorite impacts and planetary evolution*: Boulder, Colorado. B.O. Dressler, R.A.F. Grieve, and V.L. Sharpton (Eds.). pp. 153-162. Geol. Soc. Amer., Special Paper 293 (1994).
35. V.L. Masaitis, M.S. Mashchak, M.V. Naumov, and A.I. Raikhlin. *Large astroblemes of Russia*. Publication No. 020704, VSEGEI St. Petersburg (1994).
36. V.L. Masaitis and G.I. Shafranovsky. Comparative studies of impact diamonds from the Popigai and Ries craters. *Third Int. Workshop „Impact cratering and evolution of the planet Earth"*. Limoges, 19 (1994).
37. V.L. Masaitis, G.I. Shafranovsky, and I.G. Fedorova. The apographitic impact diamonds from astroblemes Ries and Popigai [in Russian]. *Berichte der all-russischen mineralogischen Gesellschaft* **4**, Russian Academy of Sciences, St. Petersburg , 12-19 (1994).
38. C.E. Melton, C.A. Salotti, and A.A. Giardini. The observation of nitrogen, water, carbon dioxide, methane and argon as impurities in natural diamonds. *Amer. Min.* **57**, 1518-1523 (1972).
39. M.E. Newton and J.M. Baker. Models for the di-nitrogen centres found in brown diamond. *J. Phys: Condens. Matter* **3**, 3605-3616 (1991).
40. M.E. Newton. Electron paramagnetic resonance of radiation damage centres in diamond. In: *Properties and growth of diamond*. G. Davis (Ed.). pp. 153-158. EMIS datareviews series **9**, INSPEC, London (1994).
41. G. Pösges and M. Schieber. *Das Rieskrater-Museum Nördlingen* [in German]. Akademiebericht Nr. **253**, Bayerische Akademie für Lehrerfortbildung Dillingen. Verlag Dr. Friedrich Pfeil, Munich (1994).
42. Public Relations Brochure, Finsch Mine (1995).
43. S.H. Richardson, J.J. Gurney, A.J. Erlank and J.W. Harris. Origin of diamonds in old enriched mantle. *Nature* **310**, 198-202 (1984).
44. R. Rost, Y.A. Dolgov, and S.A. Vishnevskiy. Gases in inclusions of impact glass in the Ries crater, west Germany, and finds of high-pressure carbon polymorphs. *Dokl. Acad. Nauk. USSR* **241**, 165-168 (1978).
45. A.P. Ruotsala. Alteration of Finsch kimberlite Pipe, South Africa. *Economic Geology* **70**, 587-590 (1975).
46. M.Y. Shcherbakova, E.V. Sobolev, V.A. Nadolinnyi, and V.K. Aksenov. Defects in plastically-deformed diamonds, as indicated by optical and ESR spectra. *Sov. Phys. Dokl.* **20**, 725-728 (1975).
47. E.M. Shoemaker and E.T.C. Chao. New evidence for the impact origin of the Ries basin, Bavaria, Germany. *J. Geophys. Res.* **66**, 3371-3378 (1961).
48. E.M.W. Skinner and C.R. Clement. Mineralogical classification of southern African kimberlites. In: *Kimberlites, Diatremes and Diamonds: Their Geology, Petrology, and Geochemistry*. F.R. Boyd and H.O.A. Meyer (Eds.). pp. 129-139. Proc. 2nd. Intern. Kimberlite Conf. **1**, Amer. Geophys. Union, Washington (1979).
49. W.V. Smith, P.P. Sorokin, I.L. Gelles, and G.J. Lasher. Electron-spin resonance of nitrogen donors in diamond. *Phys. Rev.* **115**, 1546-1552 (1959).
50. D. Stöffler. Das Nördlinger Ries - Modell für die Bildung der Mondkrater und der Gesteine der Mondoberfläche [in German]. In: *ZEISS Informationen* **19**. pp.54 (1972).
51. D. Stöffler, D.E. Gault, J. Wedekind, and G. Polkowski. Experimental hypervelocity impact into quartz and sand: distribution and shock metamorphism of ejecta. *J. geophys. Res.* **80**, 4062-4077 (1975).
52. D. Stöffler and R. Ostertag. The Ries impact crater. *Fortschr. Mineral.* **61**, Beiheft **2**, 71-116 (1983).
53. D. Storzer and W. Gentner. Spaltspurenalter von Riesgläsern, Moldaviten u. Bentoniten [in German]. *Jber. u. Mitt. oberrh. geol. Ver., N.F.* **52**, 97-111 (1970).
54. O. Stutzer. "Meteor Crater" (Arizona) u. Nördlinger Ries [in German]. *Z. dt. geol. Ges.* **88**, 510-523 (1936).
55. S.M. Sukharzhevskiy, G.I. Shafranovskii and E.L. Balmasov. The EPR study of impact diamonds from astroblemes. *Proc. 23rd Lunar and Planetary Science Conference* , Houston, 1381-1382 (1992).
56. E. Werner. Das Ries in der schwäbisch-fränkischen Alb [in German]. *Bl. Schwäb. Albvereins* **16/5**, 153-168 (1904).
57. G.S. Woods. The 'type' terminology for diamond. In: *Properties and growth of diamond*. G. Davis (Ed.). pp. 83-84. EMIS datareviews series **9**, INSPEC, London (1994).
58. J.A. van Wyk, J.H.N. Loubser, M.E. Newton, and J.M. Baker. ENDOR and high-temperature EPR of the N3 centre in natural type Ib diamonds. *J. Phys.: Condens. Matter* **4**, 2651-2662 (1992).
59. J.A. van Wyk and J.H.N. Loubser. ENDOR of the P2 centre in type-Ia diamonds. *J. Phys.: Condens. Matter* **5**, 3019-3026 (1993).

Proc 30th Int'l. Geol. Congr., Vol. 16, pp. 17-27
Huang Yunhui and Cao Yawen (Eds)

Polysomatic Relationships in Some Titanosilicates Occurring in the Hyperagpaitic Alkaline Rocks of the Kola Peninsula, Russia

GIOVANNI FERRARIS[1], ALEXANDER P. KHOMYAKOV[2], ELENA BELLUSO[1] and SVETLANA V. SOBOLEVA[3]
[1] *Dip. Scienze Mineralogiche e Petrologiche, Univ. Torino, via Valperga Caluso 35, I-10125 Torino, Italy*
[2] *Inst. of Mineralogy, Geochemistry and Crystal Chemistry of Rare Elements, Veresaev st. 15, 121357 Moscow, Russia*
[3] *Inst. of Ore Deposits, Petrography, Mineralogy and Geochemistry, Russian Acad. Sci., Staromonetny per. 35, 109017 Moscow, Russia*

Abstract

The recent description of nafertisite $(Na,K,\square)_4(Fe^{2+},Fe^{3+},\square)_{10}[Ti_2O_3Si_{12}O_{34}](O,OH)_6$ (a = 5.353, b = 16.176, c = 21.95 Å, β = 94.6°) and a review of the titanosilicates from the Khibina-Lovosero complex (Kola Peninsula) show a key role of the polysomatic phenomena in *(i)* modelling unknown crystal structures; *(ii)* correlating phase transformations; *(iii)* explaining mineral associations. The structure of nafertisite has been derived by comparison with bafertisite {a = 5.36, b = 6.80, c = 10.98 Å, β = 94°; $Ba_2(Fe,Mn)_4[Ti_2O_4Si_4O_{14}](O,OH)_2$} and astrophyllite {$a$ = 5.36, b = 11.63, c = 11.76 Å, α = 112.1, β = 103.1, γ = 94.6°; $(K,Na)_3(Fe,Mn)_7[Ti_2O_3Si_8O_{24}](O,OH)_4$}. These titanosilicates have layer structures: two (001) sheets (*H*), based on [100] rows of *Ti* octahedra alternated with different strips of *Si* tetrahedra, sandwich an octahedral sheet (*O*); two such *HOH* layers clamp the large cations. The *heterophyllosilicates polysomatic* series B_mM_n has been defined (*B* bafertisite-like and *M* mica-like modules cut across *HOH*). Bafertisite is not reported in the Khibina-Lovosero complex, but several minerals based on its *H* sheet are widely present. The dimensions of this *H* sheet are fixed by its periodicities along (≈ 5.4 Å) and across (≈ 7 Å) the *Ti* rows. The corresponding *HOH* layer occurs in a group of bafertisite-like structures (*seidozerite derivatives*) which differ only in their interlayer contents. The inclusion among these derivatives of some poorly described species is discussed. Four members of the group (sobolevite, lomonosovite, quadruphite and polyphite) have been observed together in oriented associations indicating close genetic relationships. Corrugated bafertisite-like *H* sheets are tridimensionally connected in the Zr-di-orthosilicates parakeldyshite, keldyshite, an unnamed species and khibinskite (*Zr* plays the role of *Ti*): the first three minerals form, in the order, an evolutionary series where parakeldyshite is the primary phase and the others are its products of alteration. Other (de)hydration/(de)cationization reactions are discussed as well.

Keywords: titanosilicates, polysomatism, crystalchemistry, Kola Peninsula, hyperagpaitic alkaline rocks

INTRODUCTION

The hyperagpaitic alkaline rocks which occur in the Khibina and Lovozero massifs (Kola Peninsula, Russia) represent a unique geological laboratory, where about 500 different minerals are known [17]. More than 50% of these minerals are silicates and about 1/3 of the latter are titanosilicates (including *Zr* and *Nb* silicates). About 120 minerals from the Khibina-Lovosero complex have been described as new species; most of these show high contents of alkalis. Very often the Kola localities are unique;

sometimes they are concurrent (*e.g.*, vitusite [32]), or representative of the second world occurrence for the same species (*e.g.*, hilairite [18]) or closely related species (*e.g.*, altisite [8]). In fact, geochemical features similar to those occurring in the Kola hyperagpaitic rocks are reported, but in smaller scale, only in few other localities as at Mont Saint-Hilaire (Canada) and Ilímaussaq (Greenland) [13, 17]. About 30 rare species occur only at Mont Saint-Hilaire and Khibina-Lovosero complex; few of them have been reported also at Ilímaussaq [13].

The present paper critically reviews results concerning some titanosilicates which occur in the title rocks and have been characterized through the application of the concept of polysomatism. This concept belongs to the wider field of the modular crystallography which describes crystal structures by emphazising the presence of common structural blocks in different compounds [10, 23, 44]. Typical examples are polytypes and polysomes among which the first, and still the most known example, is represented by the biopyriboles [43].

Typically, a group of compounds forms a polysomatic series P_pQ_q if the crystal structure of each member of the group can be described as a linear combination of P and Q modules. In most known cases the two building modules are slices with thickness t_P and t_Q and very close periodicities in the two directions (*e.g.*, **a** and **b**) which define the contact surface; the third periodicity **c** corresponds to the vector stacking the slices. In particular, composition and metric properties of the members linearly depend from those of the building modules P and Q; in other words, the member P_pQ_q shall show: (*i*) $c \approx pt_P + qt_Q$, $b \approx b_P \approx b_Q$ and $a \approx a_P \approx a_Q$; (*ii*) composition = p(composition of P) + q(composition of Q). Actually, the composition can vary by effect of isomorphous substitutions and related minor structural modifications. In describing the seidozerite derivatives an extended concept of polysomatism shall be proposed, where the stoichiometry and the thickness of one module is variable. In particular, the polysomatic theory proves to be very fruitful in *(i)* deriving structural models from geometrical and chemico-physical data; *(ii)* understanding genesis, paragenesis and (oriented) associations; *(iii)* interpreting phase transitions and, in general, formation of secondary phases through solid state reactions.

NAFERTISITE AND HETEROPHYLLOSILICATE SERIES

Since many years it is known [2] that several titanosilicates, *e.g.* bafertisite [12, 26, 31, 42] and astrophyllite [25, 35, 46], show a 2:1 layer structure which differs from that of the *TOT* phyllosilicates only for the presence of rows of Ti-polyhedra in the *T* sheet (Fig. 1, 2).

Nafertisite is a titanosilicate from the Khibina massif recently described [9, 19] with simplified formula $(Na,K,\square)_4(Fe^{2+},Fe^{3+},\square)_{10}[Ti_2O_3Si_{12}O_{34}](O,OH)_6$ and cell parameters a = 5.353, b = 16.176, c = 21.95 Å, β = 94.6(2)° (Z = 2, space group A2/m). The comparison with (*i*) bafertisite $\{Ba_2(Fe,Mn)_4[Ti_2O_4Si_4O_{14}](O,OH)_2$; a = 5.36, b = 6.80, c = 10.98 Å, β = 94°, Z = 1, s.g. $P2_1/m$ [26] (Fig. 1); polytypes have been reported [12, 31]} and (*ii*) astrophyllite

$\{(K,Na)_3(Fe,Mn)_7[Ti_2O_3Si_8O_{24}](O,OH)_4$; $a = 5.36$, $b = 11.63$, $c = 11.76$ Å, $\alpha = 112.1$, $\beta = 103.1$, $\gamma = 94.6°$, $Z = 1$, s.g. $P\bar{1}$ (reduced cell from [46]) (Fig. 2); polytypes are known [25, 35, 47]} allowed to obtain a structural model for nafertisite (Fig. 3) which has been succesfully tested against X-ray diffraction data obtained from a very poor crystal [9]. The characterization of nafertisite has been possible after

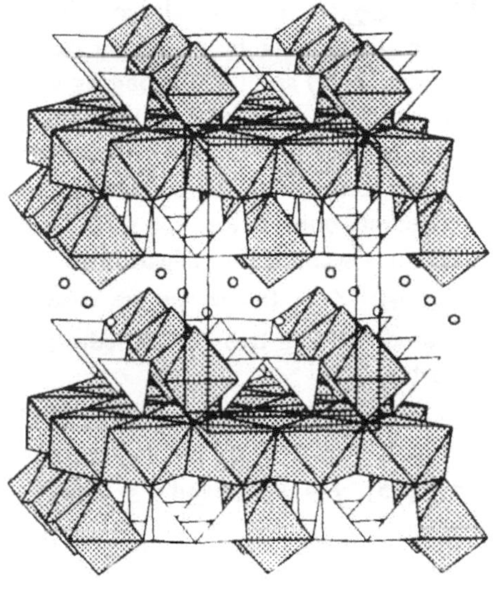

Figure 1. Perspective view along [100] of the crystal structure of bafertisite [26]. The polyhedra of the *O* sheet and the Ti polyhedra of the *H* sheet are dotted; interlayer cations are shown as circles.

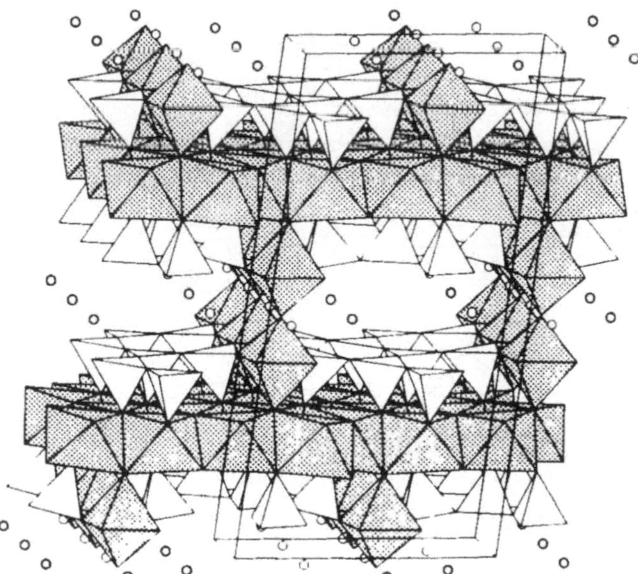

Figure 2. Perspective view along [100] of the crystal structure of astrophyllite [46]. See Fig. 1 for explanation.

it was understood that this mineral, together with bafertisite, astrophyllite and other layer titanosilicates (Table 1) is member of a polysomatic series B_mM_n built up with *B* (bafertisite-like) and *M* (mica-like) modules. Because of its clear relationships with the phyllosilicates, the series has been called [9] ***heterophyllosilicate series*** to stress the extra (***hetero***) presence of Ti-polyhedra in the phyllosilicate-like sheet. The members of the series with $m = 1$ have cell parameters $a \approx 5.4$, $b \approx (6.8 + n \times 4.7)$ and $c \approx k \times 11$ Å, and (with some more generalization with respect to the formula given by [9]) a general formula: $\{A_2R_4[Ti_2(O')_{2+l}\ Si_4O_{14}](O')_2\}_m\{AR_3[Si_4O_{10}](O'')_2\}_n$. For $m = 1$, which is the only case observed, this formula can be usefully written $A_{2+n}R_{4+3n}[Ti_2(O')_{2+l}\ Si_{4+4n}O_{14+10n}](O'')_{2+2n}$; A are large (alkali) interlayer cations and R are octahedral cations. The atoms belonging, even in part, to the H sheet are shown in squared brackets; O' (bonded to *Ti*) and O'' (belonging to the octahedral O sheet only) can be oxygen, *OH*, *F* or H_2O; the *14+10n* oxygens are bonded to *Si*. *Ti* (or crystallochemically equivalent cations, like *Zr* and *Nb*) can be 5- or 6-coordinated forming polyhedra which share one corner with the octahedral sheet and 4 corners with 4 Si-tetrahedra. In case of octahedral coordination, the sixth corner can be unshared ($l = 2$; Fig. 1) or shared with either a second Ti-octahedron ($l = 1$; Fig. 2, 3) or an interlayer anion ($l = 0$; Fig. 4); $l = 0$ also in the cases of coordination number 5 (Fig. 5) and of an edge shared between two Ti-octahedra (Fig. 6). For $m = 1$, the following cases have been reported:

(*a*) $n = 0$ - bafertisite-type structures with $l = 0$ (Table 1); bafertisite [12, 26, 31] and hejtmanite [42, 45] with $l = 2$;

(*b*) $n = 1$ - astrophyllite-type structures with $l = 0$ [35], $l = 1$ [46] and $l = 2$ [47];

(*c*) $n = 2$ - nafertisite with $l = 1$ [9];

(*d*) $n = \infty$ - micas.

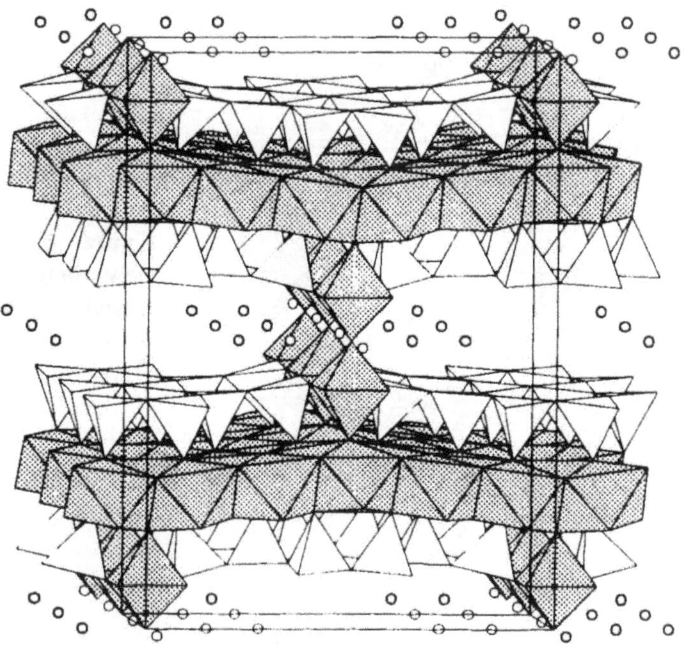

Figure 3. Perspective view along [100] of the crystal structure of nafertisite [9]. See Fig. 1 for explanation.

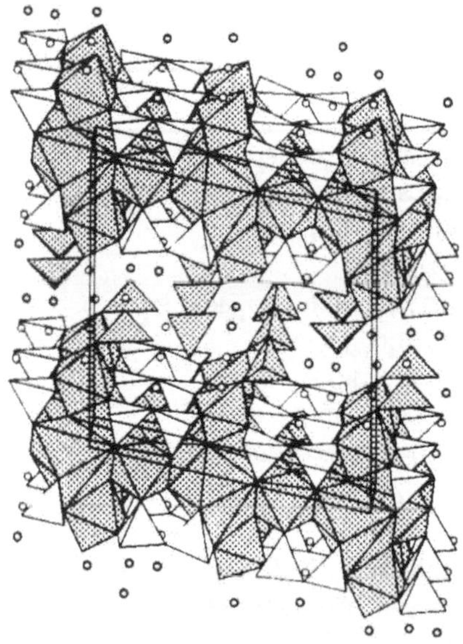

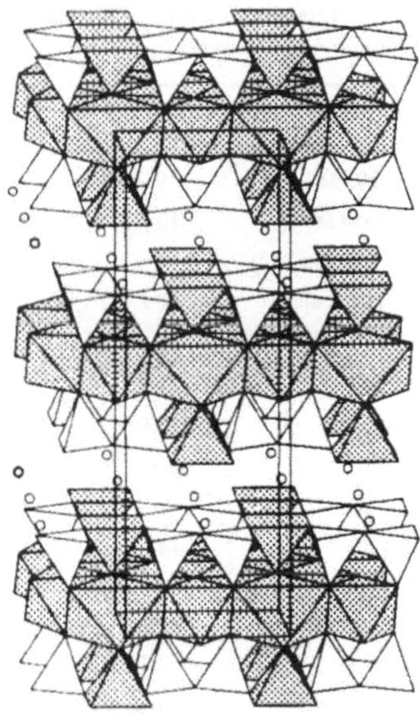

Figure 4 (left side). Perspective view along [100] of the crystal structure of betalomonosovite [27]. See Fig. 1 for explanation; PO_4 tetrahedra are dotted.

Figure 5 (right side). Perspective view along [100] of the crystal structure of K-barytolamprophyllite [30]. See Fig. 1 for explanation.

THE SEIDOZERITE DERIVATIVES

While so far bafertisite has not been reported in the Khibina-Lovosero complex, the bafertisite-type *HOH* layer turns out to be a building module of several titanosilicates occurring there. In Table 1 the chemical formulae of these compounds are written according to the general formula of the heterophyllosilicate series as given above, instead of putting in evidence the di-orthosilicate group Si_2O_7. In particular, the *HOH* layers occur in the structure of seidozerite (Fig. 6, [37, 38]), where they are bridged by Ti-octahedra; pairs of these octahedra share one edge and no interlayer space is left (a special case with $l = 0$). This structure is the parent of a group of minerals (*seidozerite derivatives*) where only the interlayer content varies [7]. These minerals are B_1M_0 members of the heterophyllosilicate series with a variable interlayer content and represented by the formula $A_2R_4[X_2(O')_{2+l}Si_4O_{14}](O'')_2 \cdot Y$, where *A* (large cations) and *Y* (water, tetrahedral anions, cations) are interlayer stuff; *R* and *X* (mainly *Ti*, but also *Zr*, *Nb*, ...) are octahedral cations belonging to the *O* and *H* sheet, respectively. All the seidozerite derivatives are based on a common (sub)cell with $a \approx 5.4$ Å, $b \approx 7$ Å, while the length of the *c* vector stacking the layers depends on the nature and quantity of *A* and *Y*. In Table 1 only the minerals of this type occurring in the Khibina-Lovosero complex and quoted by [17] are reported; they are arranged according to the increasing complexity of their (presumed) interlayer stuff and, consequently, of the length of the stacking *c* vector.

Table 1. Minerals which occur in the Khibina-Lovosero complex and are based on a bafertisite-like *HOH* layer (seidozerite derivatives).

Mineral Name	*c* Parameter (Å)	Formula
Seidozerite [37, 38]	9.2 x 2	$Na_2(Na,Mn,Ti)_4[(Na,Ti,Zr)_2O_2Si_4O_{14}]F_2$
Lamprophyllite [34]	10 x 2	$Na_2(Sr,Ti,Na,Fe)_4[Ti_2O_2Si_4O_{14}](O,F)_2$
Barytolamprophyllite [30]	10 x 2	$(Ba,Na)_2(Na,Ti,Fe,Ba)_4[Ti_2O_2Si_4O_{14}](O,F)_2$
Delindeite [1, 17, a]	10.8 x 2	$Ba_2(Na_2Ti)[Ti_2O_2Si_4O_{14}](O,F)_2$
Shkatulkalite [22, a]	10.4 x 3	$(Na,)_2(Na,Mn,Ca,)_4[(Ti,Nb)_2(O,OH)_2Si_4O_{14}](OH,F)_2\cdot H_2O$
Murmanite [28]	11.94	$Na_2(Ti,Na,)_4[Ti_2(H_2O)_2Si_4O_{14}]O_2\cdot 2H_2O$
Epistolite [16]	12.07	$Na_2(Na_3Ti)[(Nb,Ti)_2O_2Si_4O_{14}](O,F)_2\cdot 4H_2O$
M72 [17, a, b]	-	$BaNa(Na_3Ti)[(Ti,Nb)_2O_2Si_4O_{14}](O,F)_2\cdot 4H_2O$
M73 [17, a, b]	-	$BaNa(Na,Nb,Mn,)_4[Ti_2O_2Si_4O_{14}](O,OH,F)_2\cdot 5H_2O$
Vuonnemite [6]	14.55	$Na_2(Na_3Ti)[(Nb,Ti)_2O_2Si_4O_{14}](O,F)_2\cdot Na_6(PO_4)_2$
Lomonosovite and Betalomosovite [27]	14.6	$Na_2(Na_2Ti_2)[Ti_2O_2Si_4O_{14}](O,F)_2\cdot Na_6(PO_4)_2$
M55 [17, a,c]	-	$(Ba,K,)_2(Ca,Nb,Mn,Fe,)_4[Ti_2O_2Si_4O_{14}]O_2\cdot Na_4Ba(PO_4)_2$
Quadruphite [20, 39]	20.34	$Na_2(NaMgTi_2)[Ti_2O_2Si_4O_{14}]O_2\cdot Na_{11}Ca(PO_4)_4\ F_2$
Polyphite [20, 40]	26.56	$Na_2(Ti,Mn,Mg,Na)_4[Ti_2O_2Si_4O_{14}]F_2\cdot Na_{12}(Ca,Mn,Mg)_5(PO_4)_6F_4$
Sobolevite [41]	40.62	$Na_2(NaMgTi_2)[Ti_2O_2Si_4O_{14}]O_2\cdot Na_{10}CaMg(PO_4)_4F_2$
M74 [17, a]	47.45	$Ba_2(BaFe)_2[Ti_2O_2Si_4O_{14}]O_2\cdot (NaF_2)_4\cdot 4H_2O$
Bornemanite [21, a]	48.2	$Na_2(Na_3Nb)[Ti_2O_2Si_4O_{14}]O_2\cdot Na_2Ba(PO_4)F$

Numbers in brackets indicate reference; [a] structure unknown; [b] cell parameters unknown; [c] only *a* = 5.4 and *b* = 7.0 Å parameters known.

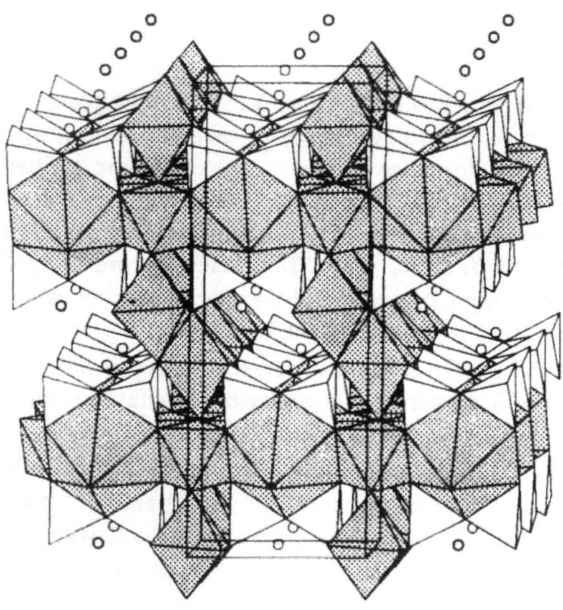

Figure 6. Perspective view along [100] of the crystal structure of calcium seidozerite [38]. See Fig. 1 for explanation.

A bafertistite-like formula is also shown by the following minerals occurring in the hyperlkaline agpaitic rocks [17]: *(i)* gõtzenite, rinkite (also reported by [17] as lovchorrite and mosandrite) and rosenbuschite; *(ii)* lavenite, titanolavenite and wöhlerite. In the structures of minerals *(i)* [4, 11, 29, 36] a *HOH* bafertisite-like layer can be roughly single out and normally the non-tetrahedral polyhedra belonging to the *H* sheets contain cations larger than *Ti* (*e.g.*, *Ca*) and share corners or edges with tetrahedra of the next *HOH* layer. In this way, these non-tetrahedral polyhedra establish a bridge between two *H* sheets belonging to two adjacent layers; that requires a thinner interlayer space and, consequently, a shorter stacking vector. Besides, in the "octahedral" *O* sheet higher coordination polyhedra are present and *Ti* preferentially occupies a small octahedron there. The deviations from an "orthodox" *HOH* layer and mica-like structure increase with the content of large *R* cations; seidozerite [37, 38] (Fig. 6) can be considered at the boundary of the "mica-like" structures [29]. In spite of their bafertisite-like formula, the minerals of group *(ii)* show instead a cuspidine-like structure [23].

UNDER-CHARACTERIZED TITANOSILICATES

The insertion in Table 1 of a mineral with unknown structure is based on its chemical composition and cell parameters as reported by [17]. The chemical formulae are re-arranged according to that given above for the seidozerite derivatives, assuming $\ell = 0$. Such a procedure, based on the concepts of modular crystallography, allows to make the reasonable hypothesis that in the minerals of Table 1 there are *HOH* bafertisite-like layers which clamp the *A* and *Y* part of the corresponding formula. The characterization of the phases labelled M_{**} by [17] is still poor and they have not yet been submitted for approval as new mineral species.

To be noted that, as in other groups of minerals, the presence of cations and anions which have a multiple structural role can prevent a unique choice of the crystallochemical formula. Examples are (Table 1): alkaline and earth-alkaline elements, which can occupy both interlayer and octahedral sites; *Ti* (*Nb*, *Zr*,...), which can stay both in *H* and *O* sheets; *Ca*, which can occur in these two sheets and in the interlayer as well.

SOLID STATE REACTIONS AND ORIENTED GROWTHS

Parakeldyshite [5] ($Na_2ZrSi_2O_7$, a = 9.31, b = 5.42, c = 6.66 Å), keldyshite [15] [$Na_3HZr_2(Si_2O_7)_2$, a = 9.00, b = 5.34, c = 6.96 Å], khibinskite [24] ($K_2ZrSi_2O_7$, a = 9.6 x 2, b = 5.5 x 2, c = 7.0 x 2 Å) and the still unnamed mineral M34 ($NaHZrSi_2O_7 \cdot H_2O$) occur often associated in the Khibina-Lovozero complex [17]. Parakeldyshite is a primary high-temperature phase, while both keldyshite and M34 are secondary phases which are formed by decationization and hydration of parakeldyshite, after which both are pseudomorphs. The known structures of these phases [5, 15, 24] are based on a (010) corrugated bafertisite-like *H* sheet where the tetrahedra occur in two (up and down) orientations; the Ti-octahedra connect both the Si_2O_7 groups within a sheet and pairs of sheets (Fig. 7). The preservation of the *H* sheet through the alteration processes and the observed pseudomorphoses show that the transformations from parakeldyshite to keldyshite and M34 are oriented solid state reactions. Cases where the bafertisite *HOH* layer is completely preserved through (de)hydration and/or (de)cationization processes, which affect only the interlayer stuff, are also known among the Kola titanosilicates as [17]: (*i*) lomonosovite + H_2O ---> murmanite + (Na + P); (*ii*) vuonnemite + H_2O ---> epistolite + (Na + P).

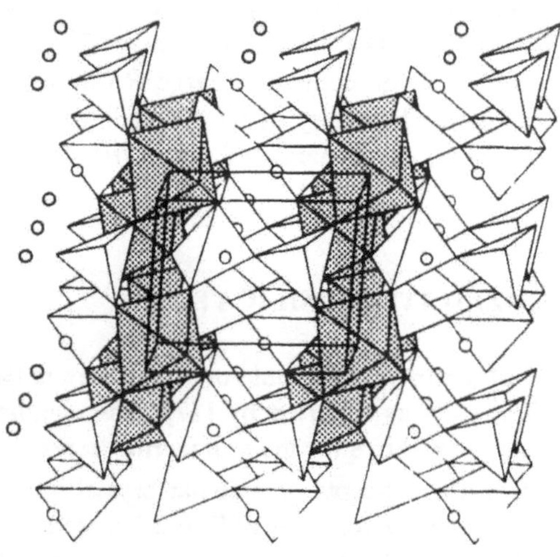

Figure 7. Perspective view along [100] of the crystal structure of keldyshite [15] [(010) is the basal plane]. The Ti-octahedra are dotted.

Polyphyte and quadruphite, which are the two most complex seidozerite derivatives (Table 1), have been reported [17] to occur as intergrowths with the structurally related phases sobolevite and lomonosovite. Taking into account that the chemical and structural differences between these minerals affect almost only their interlayer contents (Table 1), it seems reasonable to make the hypothesis that in a field of stability of the main building block (*i.e.*, the *HOH* layer), a local fluctuation of physico-chemical conditions can favour one more than another of these minerals. Processes of fluctuation in composition, which indicate some kind of unstable equilibrium, occur commonly in phases which are interested by modular phenomena; *e.g.*, such processes originate polysomatic and polytypic defects within the matrix of a parent phase [10, 44] and can be compared with a purely chemical zoning.

The structural interpretation of the discussed processes suggests that on the presence of a derivative phase, also its parent phase can be expected and *viceversa*. Actually, such an idea has been exploited for mineral identification both in the Kola Peninsula and in similar geological situations mentioned above [17]; *e.g.*, lomonosovite [14] and vuonnemite [3, 33] were discovered at Ilímaussaq (Greenland) after realizing that they are, in the order, parent phases of murmanite and epistolite, the latter two species being already known in that locality.

CONCLUSIONS

Titanosilicates based on *HOH* layers are widespread minerals in the hyperagpaitic rocks of the Kola Peninsula. In the known members of the heterophyllosilicate polysomatic series, the similarity between the different *HOH* layers and the *TOT* phyllosilicate layer increases from bafertisite to nafertisite, through astrophyllite; the most frequent layer is that of bafertisite (or seidozerite) whose *H* sheet has formula $[Ti_2O_{2+/}\ Si_4O_{16}]^{(12+2/)-}$. The frequence of this layer and its occurrence, with a variety of interlayer stuff, under different geological conditions prove that it is particularly stable; namely, it maintains through alteration processes involving (de)cationization and (de)hydration. Similar behaviour is shown by the isolated *H* sheet.

The shown examples prove that systematization of the structural properties (*e.g.*, polysomatism) offers keys both to characterize new species (*e.g.*, to model their unknown crystal structures) and to interpret geological features like genesis, (oriented) associations and transformations.

Acknowledgements

Research supported by MURST and CNR (Roma) grants. The use of facilities of the C.S. sulla Geodinamica delle Catene Collisionali (CNR, Torino) is gratefully acknowledged. S.V.S. benefited of MAE (Roma) and University of Torino grants. D.Yu. Pushcharovsky and E.V. Sokolova (Moscow University) kindly helped with some russian references and generously let us know unpublished results.

REFERENCES

1. D.E. Appleman, H.T. Jr. Evans, C.L. Nord, E.J. Dwornik and C.M. Milton. Delindeite and lourenswalsite, two new titanosilicates from the Magnet Cove region, Arkansas. *Min. Mag.* **51**, 417-425 (1987).
2. N.V. Belov. *Crystal Chemistry of Large Cations Silicates*. Consultants Bureau, New York (1961).
3. I.V. Bussen, A.P. Denisov, N.I. Zabavnikova, L.V. Kozyreva, Yu.P. Menshikov and E.A. Lipatova. Vuonnemite, a new mineral. *Zapiski Vses. Mineral. Obshch.* **102**, 423-426 (1973) (in Russian).
4. E. Cannillo, F. Mazzi and G. Rossi. Crystal structure of götzenite. *Sov. Phys. Crystallogr.* **16**, 1026-1030 (1972).
5. A.N. Chernov, B.A. Maksimov, V.V. Ilyukhin and N.V. Belov. Crystalline structure of monoclinic modification of K,Zr-diorthosilicate ($K_2ZrSi_2O_7$). *Sov. Phys. Dokl.* **15**, 711-713 (1971).
6. Yu.N. Drozdov, N.G. Batalieva, A.A. Voronkov and E.A. Kuzmin. Crystal structure of $Na_{11}Nb_2TiSi_4P_2O_{25}F$. *Sov. Phys. Dokl.* **19**, 258-259 (1974).
7. Yu.K. Egorov-Tismenko and E.V. Sokolova. Structural mineralogy of the homologic series seidozerite-nacaphite. *Mineral. Zhurn.* **12**, 40-49 (1990) (in Russian).
8. G. Ferraris, G. Ivaldi and A.P. Khomyakov. Altisite $Na_3K_6Ti_2[Al_2Si_8O_{26}]Cl_3$ a new hyperalkaline aluminosilicate from Kola Peninsula (Russia) related to lemoynite: crystal structure and thermal evolution. *Eur. J. Mineral.* **7**, 537-546 (1995).
9. G. Ferraris, G. Ivaldi, A.P. Khomyakov, S.V. Soboleva, E. Belluso and A. Pavese. Nafertisite, a layer titanosilicate member of a polysomatic series including mica. *Eur. J. Mineral* **8**, 241-249 (1996).
10. G. Ferraris, M. Mellini and S. Merlino. Polysomatism and the classification of minerals. *Rend. Soc. It. Mineral. Petrol.* **41**, 181-192 (1986).
11. E. Galli and A. Alberti. The crystal structure of rinkite. *Acta Cryst.* B27, 1277-1284 (1971).

12. Ya.-S. Guan, V.I. Simonov and N.V. Belov. Crystal structure of bafertisite, $BaFe_2TiO[Si_2O_7](OH)_2$. *Dokl. Acad. Nauk SSSR* **149**, 1416-1419 (1963) (in Russian).
13. L. Horváth and R.A. Gault. The mineralogy of Mont Saint-Hilaire, Quebec. *Min. Record* **21**, 283-362 (1990).
14. S. Karup-Møller. Lomonosovite from the Ilímaussaq intrusion, South Greenland. *N. Jb. Miner. Abh.* **148**, 83-96 (1983).
15. A.D. Khalilov, A.P. Khomyakov and S.A. Makhmudov. Crystal structure of keldyshite $NaZr[Si_2O_6OH]$. *Sov. Phys. Dokl.* **23**, 8-10 (1978).
16. A.D. Khalilov, E.S. Makarov, K.S. Mamedov and L.Ya. Pyansina. Crystal structures of minerals of the murmanite-lomonosovite group. *Dokl. Acad. Nauk SSSR* **162**, 179-182 (1965) (in Russian).
17. A.P. Khomyakov. *Mineralogy of hyperagpaitic alkaline rocks*. Clarendon Press, Oxford (1995).
18. A.P. Khomyakov and N.M. Chernitsova. Hilairite, $Na_2ZrSi_3O_9 \cdot 3H_2O$ - first finds in the USSR. *Mineral. Zhur.* **2**, 95-96 (1980) (in Russian).
19. A.P. Khomyakov, G. Ferraris, G.N. Nechelyustov, G. Ivaldi and S.V. Soboleva. Nafertisite $Na_3(Fe^{2+},Fe^{3+})_6[Ti_2Si_{12}O_{34}]O(OH)_6 \cdot 2H_2O$, a new mineral with a new type of band silicate radical. *Zapiski Vses. Mineral. Obshch.* **124**, 101-108 (1995) (in Russian).
20. A.P. Khomyakov, G.N. Nechelyustov, E.V. Sokolova, G.I. Dorokhova. Quadruphite $Na_{14}CaMgTi_4[Si_2O_7]_2[PO_4]_2O_4F_2$ and polyphite $Na_{17}Ca_3Mg(Ti,Mn)_4[Si_2O_7]_2[PO_4]_6O_2F_6$, two new minerals of the lomonosovite group. *Zapiski Vses. Mineral. Obshch.* **121**, 105-112 (1992) (in Russian).
21. Yu.P. Menshikov, I.V. Bussen, E.A. Goiko, N.I. Zabavnikova, A.N. Merkov and A.P. Khomyakov. Bornemanite, a new silicophosphate of sodium, titanium, niobium and barium. *Zapiski Vses. Mineral. Obshch.* **104**, 322-326 (1975).
22. Yu.P. Menshikov, A.P. Khomyakov, L.I. Polezhaeva and R.K. Rastsvetaeva. Shkatulkalite $Na_{10}MnTi_3Nb_3(Si_2O_7)_6(OH)_2F \cdot 12H_2O$: a new mineral. *Zapiski Vses. Mineral. Obshch. 125, 120-126 (1996).*
23. S. Merlino and N. Perchiazzi. Modular mineralogy in the cuspidine group of minerals. *Can. Mineral.* **26**, 939-943 (1988).
24. N.A. Nosyrev, E.V. Treushnikov, A.A. Voronkov, V.V. Ilyukhin, R.M. Ganiev and N.V. Belov. The crystal structure of synthetic khibinskite $K_2ZrSi_2O_7$ (II). *Sov. Phys. Dokl.* **21**, 696-697 (1976).
25. Z.Z. Pen and Z. Ma. Crystal structure of astrophyllite and a new type of band silicate radical. *Scientia Sinica* **12**, 272-276 (1963) (in Russian).
26. Z.Z. Pen and T.C. Shen. Crystal structure of bafertisite, a new mineral from China. *Scientia Sinica* **12**, 278-280 (1963) (in Russian).
27. R.K. Rastsvetaeva. Crystal structure of betalomonosovite from the Lovozero region. *Sov. Phys. Crystallogr.* **31**, 633 -636 (1986).
28. R.K. Rastsvetaeva and V.I. Andrianov. New data on the crystal structure of murmanite. *Sov. Phys. Crystallogr.* **31**, 44-48 (1986).
29. R.K. Rastsvetaeva, B.E. Borutskii and Z.V. Shlyukova. Crystal chemistry of Hibbing rinkite. *Sov. Phys. Crystallogr.* **36**, 349-351 (1991).
30. R.K. Rastsvetaeva, V.G. Evsyunin and A.A. Konev. Crystal structure of K-barytolamprophyllite. *Cryst. Rep.* **40**, 472-474 (1995).
31. R.K. Rastsvetaeva, R.A. Tamazyan, E.V. Sokolova and D.I. Belakovskii. Crystal structures of two modifications of natural Ba,Mn-titanosilicate. *Sov. Phys. Crystallogr.* **36**, 186-189 (1991).
32. J.C. Rønsbo, A.P. Khomyakov, A.P. Semenov, A.A. Voronkov and V.K. Garanin. Vitusite - a new phosphate of sodium and rare earths from the Lovozero alkaline massif, Kola, and the Ilímaussaq alkaline intrusion, South Greenland. *N. Jb. Miner. Abh.* **137**, 42-53 (1979).
33. J.C. Rønsbo, E.S. Leonardsen, O.V. Petersen and O. Johnsen. Second occurrence of vuonnemite: the Ilímaussaq alkaline intrusion, South West Greenland. *N. Jb. Miner. Mh.*, 451-460 (1983).
34. Yu.N. Safianov, N.O. Vasilieva and V.P. Golovachev. Crystal structure of lamprophyllite. *Dokl. Akad. Nauk. SSSR* **269**, 117-120 (1983) (in Russian).
35. N. Shi, Z. Ma, G. Li, N.A. Yamnova and D. Yu. Pushcharovsky. Refinement of monoclinic astrophyllite crystal structure. *IGC 30 Abstr.* **2**, 445 (1996).
36. R.P. Shibaeva, V.I. Simonov and N.V. Belov. Crystal structure of the Ca,Na,Zr,Ti silicate rosenbuschite, $Ca_{3.5}Na_{2.5}Zr(Ti,Mn,Nb)[Si_2O_7]_2F_2O(F,O)$. *Sov. Phys. Crystallogr.* **8**, 406-413 (1963).
37. V.I. Simonov and N.V. Belov. The determination of the structure of seidozerite. *Sov. Phys. Crystallogr.* **4**, 146-157 (1960).
38. S.M. Skszat and V.I. Simonov. The structure of calcium seidozerite. *Sov. Phys. Crystallogr.* **10**, 505-508 (1966).

39. E.V. Sokolova, Yu.K. Egorov-Tismenko and A.P. Khomyakov. Some features of the crystal structure of $Na_{14}CaMgTi_4[Si_2O_7]_2[PO_4]_4O_4F_2$, a homologue of lomonosovite and sulphohalite. *Mineral. Zhurn.* 9, 28-35 (1987) (in Russian).
40. E.V. Sokolova, Yu.K. Egorov-Tismenko and A.P. Khomyakov. The crystal structure of $Na_{17}Ca_3Mg(Ti,Mn)_4[Si_2O_7]_2[PO_4]_6O_2F_6$, a new member of the layer titanosilicates family. *Dokl. Akad. Nauk. SSSR* **294**, 357-362 (1987) (in Russian).
41. E.V. Sokolova, Yu.K. Egorov-Tismenko and A.P. Khomyakov. The crystal structure of sobolevite. *Dokl. Akad. Nauk. SSSR* **302**, 1112-1118 (1988) (in Russian).
42. E.V. Sokolova, Yu.K. Egorov-Tismenko, L.A. Pautov and D.I. Belakovsky. Structure of the natural Ba-titanosilicate $BaMn_2TiO[Si_2O_7](OH)_2$ member of the seidezorite - nakaphite series. *Zapiski Vses. Mineral. Obshch.* 118, 81-84 (1989) (in Russian).
43. J.B. Thompson. Biopyriboles and polysomatic series. *Am. Mineral.* **63**, 239-249 (1978).
44. D.R. Veblen. Polysomatism and polysomatic series: A review and applications. *Am. Mineral.* **76**, 801-826 (1991).
45. S. Vrana, M. Rieder and M.E. Gunter. Hejtmanite, a manganese-dominant analogue of bafertisite, a new mineral. *Eur. J. Mineral.* **4**, 35-43 (1992).
46. P.J. Woodrow. The crystal structure of astrophyllite. *Acta Cryst.* **22**, 673-678 (1967).
47. B.B. Zvyagin and Z.V. Vrublevskaya. Polytypic forms of astrophyllite. *Sov. Phys. Crystallogr.* **21**, 542-545 (1976).

Proc 30th Int'l. Geol. Congr., Vol. 16, pp. 29-38
Huang Yunhui and Cao Yawen (Eds)
VSP 1997

Fine structures and crystallographic orientations in biogenic magnetite observed byTEM

JUNJI AKAI and AKIHIKO IIDA
Department of Geology, Faculty of Science, Niigata University, Ikarashi-2, Niigata 950-21, Japan

Abstract

Examining manay bacterial magnetite grains, fundamental and characteristic mineralogical features of the biogenic magnetite were described. Varieties of shapes in magnetosome, fine textures and fine strucutres of the magnetite were found: pit-structures were found on the surface of magnetite grains of octahedral and hexagonal prism types. They can be interpreted to be due to growth textures. Twin crystals were also found. Irregularly wavy rough surfaces of the magnetite crystal were described in tear-drop type magnetosomes. Crystallographic directions of elongation in the tear-drop type magnetosomes which may be closely related to crystal growth process under the control by organism were examined. In general, [111] direction in magnetite is the casiest direction for magnetization. However, the observed magnetite grains of tear drop types elongated not only in [111] direction but also in [100] direction in relatively large percentages. This may be due apparently to shape effect.

Keywords : magnetite, bacteria, magnetotactic bacteria, TEM, mineralogy, tear-drop type, magnetosome

INTRODUCTION

Magnetotactic bacterium was first found by Blakemore [2], which contain intracellular magnetite grains arranged in one or more chains. The magnetite grains are called as magnetosomes. Magnetotactic bacteria and contained magnetosomes are closely related to various scopes and topics of earth sciences ; mineralogy, earth histroy, palaeontology and organism evolution, palaeomagnitism, and environmental sciences [12,3,13]. In this study we payed attentions to this magnetite from the stand point of mineralogy. Then, mineralogical characterizations were carried out .The magnetite grains are characteristic in their sizes, compositions, and shapes. Various types of magnetite grains in magnetotactic bacteria are distinguished:such as octahedral, hexagonal prism and tear-drop type [14,9]. Recently, magnetotactic bacteria are considered to be ubiquitous and the occurrences of them in various environments are reported [8] . Biomineralization processes , in general, can be classified into the following two types [14,7] : "Biologically-Induced Mineralization (BIM)" which means by-product mineralization for organism and "Biologically Controlled Mineralization " (BCM) in which minerals are formed under strict biological control. Bacterial magnetite found as fossils are called as "magnetofossils" [11,15]. In general, two types of fossils are known: facies fossils and index (leading) fossils. The former indicates an environment where the organisms have lived, and the latter indicates a specific age of strata. Detailed knowledges are necessary for bacterial fossils to be used for such facies-indicative fossils.

Examining manay magnetite grains, characteristic mineralogical features of the biogenic magnetite grains were found : such as, varieties of magnetosome shapes , fine textures and fine strucutres of the magnetite and crystallographic directions of elongated tear-drop type magnetosomes, which may be closely related to crystal growth process under the control by organisms. The Objective of this paper is to describe these fundamentall and detailed mineralogical data on biogenic magnetite and biomineralization characteristics. Furthermore, some considerations on the problems of magnetite biomineralization will be given.

Experimental methods

Sediments and water samples were collected from the upper sediment layers of aquatic environments (mainly eutrophic freshwater lakes, ponds and so on) in a water depth of 100 - 30cm. Vessels (15 cm in diameter and 8 cm in height) were filled with the sediments and water from some lakes. After several weeks of storage in the laboratory, magnetotactic bacteria naturally enriched just beneath the water-sediment interface. They were further condensed with Sm-Co magnet due to their magnetotaxis. The enriched bacteria were observed without any special treatment under the optical microscope. A few water drops containing bacteria cells were put on carbon-coated, formvar-covered grids or microgrids for electron microscopic observation. They were air-dried and observed by TEM (JEM200CX with energy dispersive spectrometer of TN2000). Minerals were identified based on SAED patterns and the EDS analytical results.

Results and discussions

1.Types of Magnetotacitc bacteria and their environments to live

Sampling localities are the same as those described in the Iida and Akai [5] . Among them many specimens from Lake Sagata were used. A part of the detailed results have already been described in Iida and Akai [5] . More observations were addedd in this study. Then, it was estimated that the localities where magnetotactic bacteria were found were characterized by comparatively eutrophic environments. On the other hand, this method did not detect such bacteria in non-eutrophic environments, where coarse sediments are dominant are characteristic. The pH of the water in which bacteria lived was usually about 6.5-7.2. The bacteria were not found in comparatively acidic lakes. Some magnetotactic bacteria, such as spirillum, were observed on sandy bottom part in Lake Sagata. On humus muddy part, coccoid bacteria were comparatively predominant. Most of the cocci developed remarkably the intracellular inclusions of crystalline sulphur, especially during summer and autumn [6]. Varieties of morphological forms of magnetotactic bacteria were observed although the typical types are the following three : octahedral , hexagonal prism and tear-drop types. The division of magnetotactic bacteria have been found [5] . The same number of magnetosome chains may always belong to the cells of magnetotactic bacteria, although the number of magnetite crystals belonging to a single chain is reduced in half. This fact seems to be common to the magnetotactic bacteria obtained. Major three types of bacterial magnetite forms are known and the schematic figures of the typical three magnetite grains are shown in Iida and Akai [5]. Typical ED pattern obtained from the the bacterial iron minerals was that of magnetite. Size distribution patterns of magnetite grains in each type are found in Butler and Banerjee diagram [4] on which magnetic domain type is distinguished. Size distribution in the tear-drop type is concentrated in narrower single domain region than the other two types, so it may indicate that thistear-drop type is more

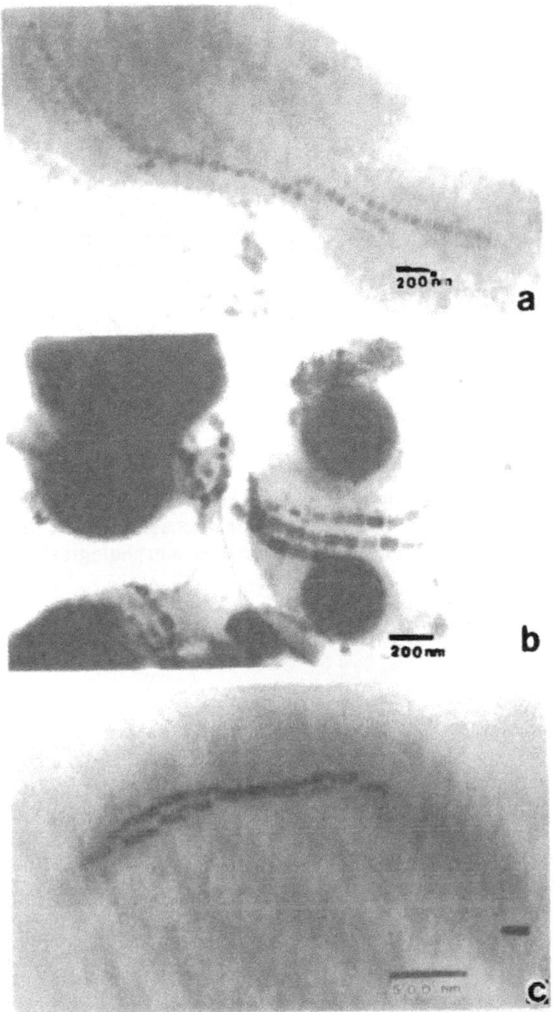

Figure 1. Three types magnetosomes contained in the magnetotactic bacteria. a: Octahedral type magnetite (Lake Fukushimagata, December), b: Hexagonal prism type magnetite (Lake Sagata, September), c:Tear-drop type magnetite (Asahiike pond)

effective in obtaining strong magnetic moment per unit volume of magnetite. In the following, each type of magnetite form is described in detail. Some diversity was also found in the same type of magnetosomes.

2.Fine structures in biogenic magnetites

Mganetite grains of magnetosomes were mineralogically examined in detail.

‹Octahedral type›

Magnetosomes of octahedral type are usually 70-120 nm in length (Fig.1a). It is

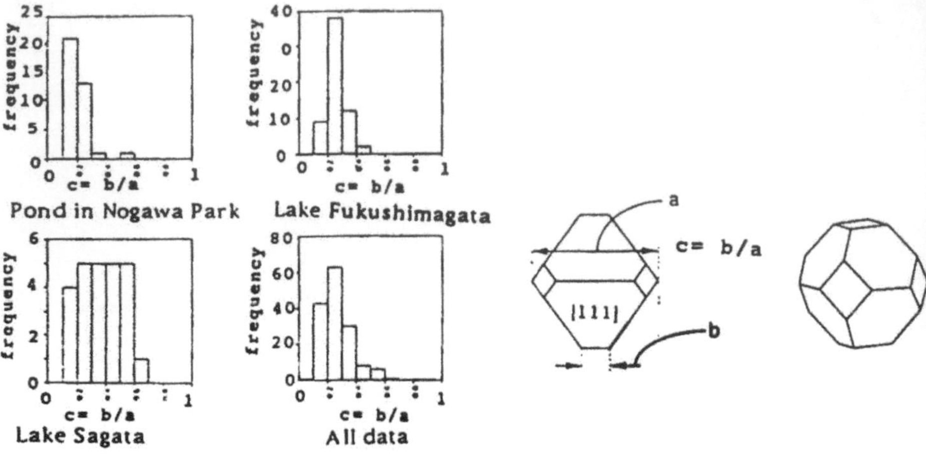

Figure 2. Histograms representing relative size relations of {111} and {100} crystal faces in octahedral type magnetosome grains. Morphological characteristics in crystal forms of octahedral type magnetosomes. Frequencies of various morphological size ratios b/a are represented.

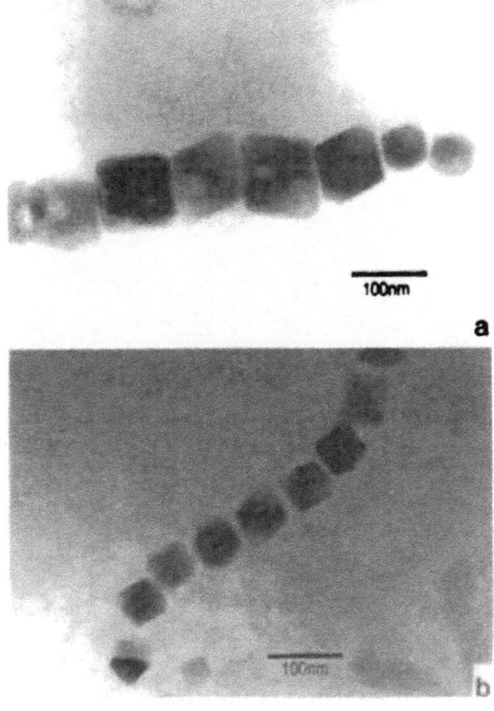

Figure 3. a,b : TEM images of morphological varieties in octahedral type magnetosomes. Pit structure is often found in {111} planes (a). Twin structure is sometimes found (b).

composed of well-defined {111} faces truncated by small {100} faces. Its shape is similar to the inorganic magnetite habit. It is often found that this type of magnetite is contained in large or slender helical bacteria. The crystals were aligned with the [111] direction parallel to the magnetosome chain axis. Mineralogical characteristics in crystal forms were examined. When crystals are growing, the {100} faces tend to become smaller in comparison to the devolopment of the {111} faces (Fig.2). Twinned crystals were relatively often observed in this type. Pit structures on two {111} planes were sometimes found (Fig.3). These may be due to growth texture.Twinning (spinel type) was sometimes found, especially in the octahedral types (Fig. 3)

‹Hexagonal-prism type›

Hexagonal prism type of magnetosome (Fig.1b) is commonly observed in various types of bacteria, while the crystal morphology is composed of {111} planes and some {110} planes which are truncated by the other {111}, {100} and {110} planes [10]. This elongated habit corresponds to uniaxial growth along [111] axes which are well known "easy axis" of magnetization in magnetite. The magnetite grains easily align in the direction along the long axis of magnetosome chain. Two morphological types are sometimes found: dumbell-like and spindle-like in shape. Furthermore, long type, small type, normal type and short type are also found. Fine structures and fine textures are also examinedd in this type of magnetite grains by TEM observation: pit structures on two {111} planes (Fig.4)

were found. These may be interpreted to be due to growth texture. Twinned crystals were sometimes found. Smaller crystals, 10-30 nm in size, are similar to cube-like and /or comprised of ill-defined planes.

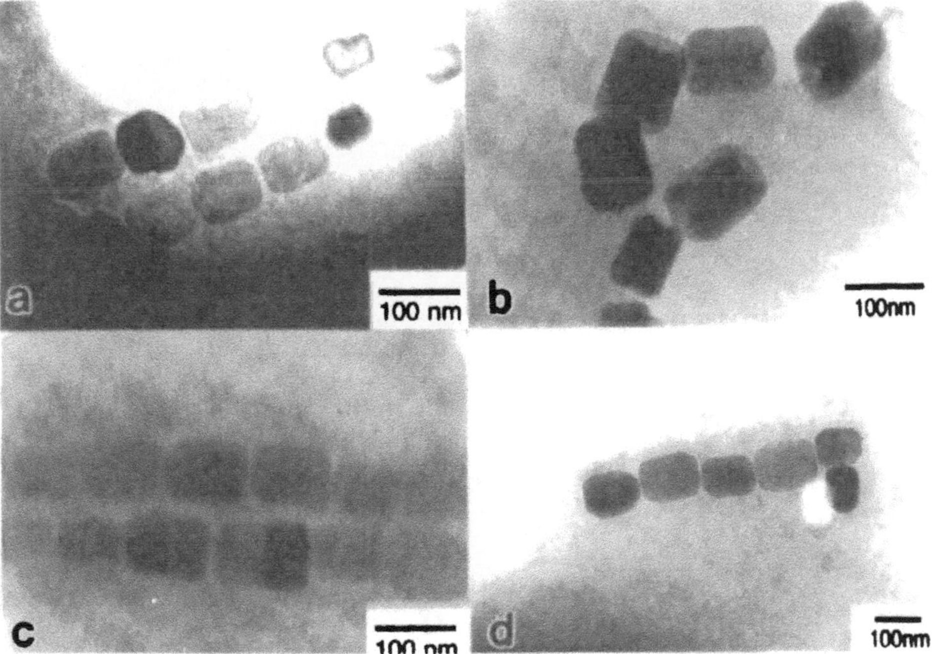

Figure 4. a,b,c,d : TEM images of morphological varieties in hexagonal prism type magnetosomes. Pit structure is sometimes found (b). Twinning is rarely found (c).

‹Tear-drop type›

Generally, tear-drop type of magnetosomes consists of characteristic tear-drop shaped crystals (Fig.1c). They can be easily distinguished morphologically from any other types and inorganic one. Measuring crystal length and width it is found that the crystal width are generally becoming constant ranging within 30-50 nm. Mineralogical characteristics in crystal forms were examined. Well-ordered {111} faces are sometimes found in the rear part. The long axis of the crystal usually elongates up to about 300 nm at maximum. Morphological varieties of this tear-drop types were found as shown in Fig.5. This type of magnetosome sometimes indicates the three patterns of the arrangement; straight and a zigzag type against the long axis of grains and a bundle type with remarkable kinked and curved crystals (Fig.5). Irregularities in surface shape and surface topography in the magnetite crystals were found. There are, for examples, band structure oblique to elongated direction. This oblique band contrasts in HRTEM image of the specimen from the Japan Sea sediments [1] . may be interpreted to be due to surface roughness (Fig. 6). Because the resutls of image analysis on HRTEM photograph of Fig.6a using software of image 1.47 showed enhanced image indicating surface roughness. Bending in shape was also found in long elongated tear-drop type magnetosomes.

3. Crystallographic directions of elongation in tear-drop type magnetosome

Crystallographic direction of the elongation in tear drop types is often [111] direction or [100] direction. Same tendency has also been known in the magnetite

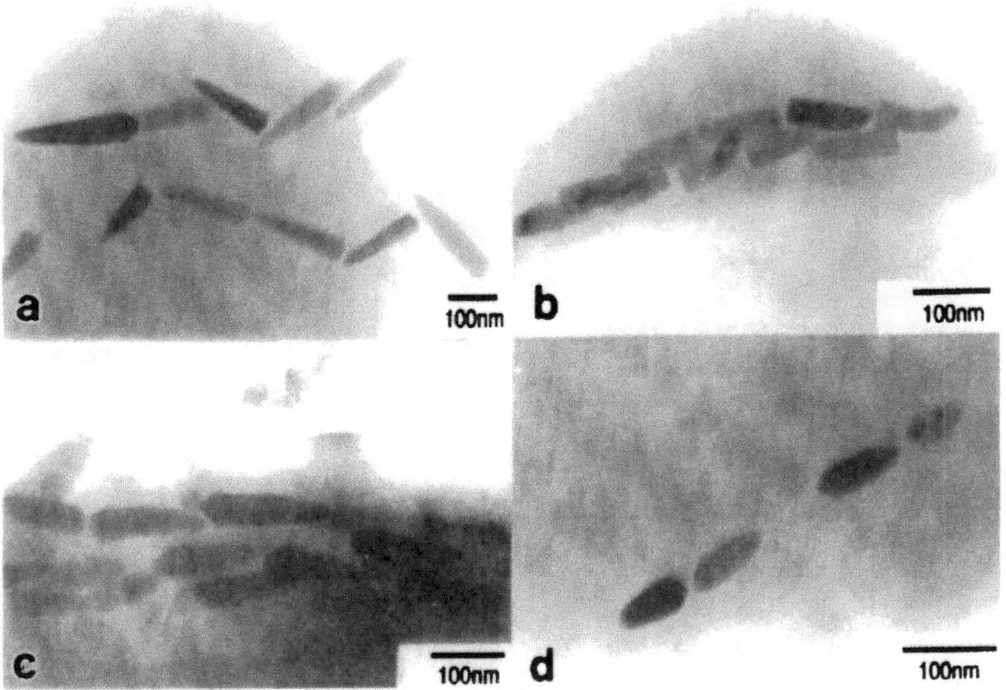

Figure 5. a,b,c,d : Some morphological and arrangement varieties in tear-drop type magnetosome grains.

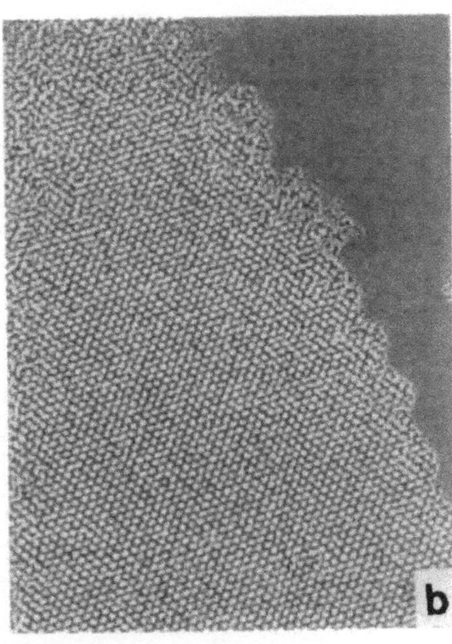

Figure 6. HRTEM image of tear-drop type magnetosome (a) and the results of image analysisof the HRTEM photograph by image 1.47 (b). Outline of the magnetosome is enhanced and sharply found : irregular outline is characteristic. The oblique band contrast found in (a) may be due to surface roughness.

grains from the Japan Sea [1]. More detailed examination in this study showed that tear-drop type oftén elongates along both |100| and |111| , and also rarely elongates along the directions |110| (Fig.7). The elongations along |111| direction are typically characteristic in the tear-drop type with slightly rounded anterior part. In general, [111] direction in magnetite is the easiest direction for magnetization. However, the observed magnetite grains of tear drop types elongated not only in [111] direction but also in [100] direction in relatively large percentages. The ratio diagram based on examined 60 grains is shown in Fig.8. This may be related to shape effect in magnetization.

Some considerations and summary

(1)There is a fundamental question about the evolutional relations: Which is the oldest type among the major three different magnetosomes (or oldest magnetotactic bacterium ?) and so on. Octahedral type is the most similar to the inorganic magnetite in their crystal shapes. The tear-drop type is well designed and is the farthest from the inorganic type. Combining the features observed , there is a possibility that the tear-drop type is the most recent type among the three types. Octahedral type is the simplest and may be the oldest type of biomineralization in these magnetosomes.

(2) More detailed classification in types of the magnetosome shapes could be possible : for example, slightly different modifications especially in tear-drop type

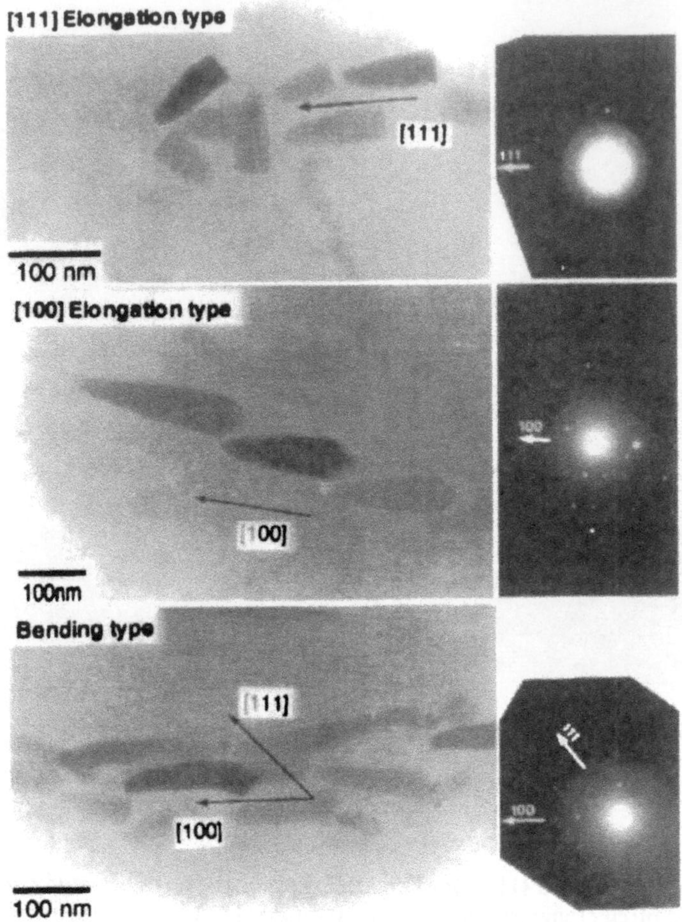

Figure 7. Varieties of crystallographic directions of elongations in tear-drop type magnetosome.

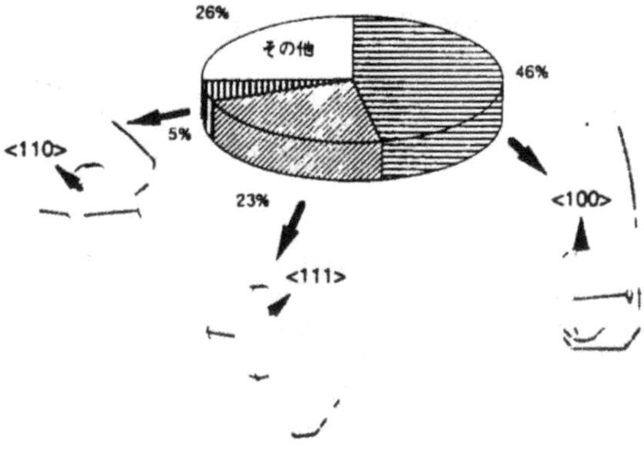

Figure 8. Histogram representing the frequencies of the crystallographic directions of elongations in tear-drop magnetosome. Total number of grains examined is more than 60.

magnetosomes were recognized. However, it is not yet clarified at present whether they are due to species differences or due to some diversity within the same species.

(3) Fine textures and fine strucutres of the magnetite crystals were found:

1) Pit-structures were found on the surfaces of magnetite grains of octahedral and hexagonal prism types. However, pit shapes were not found in the tear-drop type.
2) Twin crystals were also found. Twinning was not observed in tear-drop type.
3) Irregularly wavy roughnesses of the magnetite crystal surfaces in TEM image were described in tear-drop type magnetosomes applying image anlysis method. The wavy pattern is obliquely arranged to the elongated direction of the tear-drop type magnetite grains. It was clearly clarified to be due to surface roughness.
4) Bending in shape was also found in long elongated tear-drop type magnetosomes.

(4) Crystallographic direction of the elongated tear-drop type magnetosomes were examined. In general, [111] direction in magnetite is the easiest direction for magnetization. However, the observed magnetite grains of tear drop type elongated not only in [111] direction but also in [100] direction in relatively large percentages. This may be due apparently to shape effect, but its meaning in evolution is not known yet.

(5) Mangetosome in magnetotactic bacteria has been considered to be one of typical BCM (Biologically Controled Mineralization) products. This study demonstrated its detail and concrete characteristics of the controlled biomineralization with fundamental data. These data will become some important clues for further detailed examination of BCM type biomineralization of biogenic magnetite.

Acknowledgements.

We greatly thank Prof. H. Fukuhara, Prof. T. Hirokawa, Dr. K. Sekiya and Prof. H. Takeda of Niigata University fortheir helpful discussions and many technical advices. Especially, Prof.H. Fukuhara has kindly helped us in biological treatment of the bacteria. We acknowledges Prof. Matsunaga of Tokyo Agr. Techn. Univ. for his advices and suggestions for sampling of the bacteria. We also thank to Prof. T. Yoshimura of Niigata Univ. for his encouragement in this study.

References

1. J. Akai, T. Sato,and S. Okusa. TEM study on biogenic magnetite in deep-sea sediments from the Japan Sea and the western Pacific Ocean. *Journ. Electron Microsc.* 40, 110 117 (1991).
2. R.P.Blakemore. Magnetotactic Bacteria. *Science* 190, 377-379 (1975).
3. R.P. Blakemore and N.A.Blakemore. Magnetotactic magnetogens, In: Frankel, R. B., Blakemore, R. P. (Eds) *Iron Biominerals*. New York: Plenum Press, 51-67. (1991)
4. Butler and Banerjee. Theoretical single-domain grain size range in magnetite and titanomagnetite. *Journ. Geophys. Res.* 80, 29, 4049-4058 (1975).
5. A. Iida and J. Akai. TEM study on magnetotactic bacteria and contained magnetite grains as biogenic minerals mostly from Hokuriku-Niigata region, Japan, *Science Report of Niigata University, Ser. E* No.11, 43-66 (1996).
6. A.Iida and J. Akai. Crystalline sulfur inclusions in magnetotactic bacteria. *Science Report of Niigata University, Ser.E* No.11, 35-43 (1996).

7. H.A. Lowenstam. Minerals formed by organisms. *Science* 211, 1126-1131
8. S. Mann, N.H.C. Sparks, R.B.Frankel, D.A. Bazylinski and H.W. Jannasch Biomineralization of ferrimagnetic greigite (Fe_3S_4) and iron pyrite (FeS_2) in a magnetotactc bacterium. *Nature* 343, 258-261 (1990).
9. S. Mann, N.H.C. Sparks, V.J. and Wade. Crystallochemical control of iron oxide biomineralization. In: Frankel R. B., Blakemore R. P. ed. *Iron Biominerals*. New York: Plenum Press, 21-49 (1991).
10.T. Matsuda , J.Endo,N. Osakabe, N and A. Tonomura. Morphology and structure of biogenic magnetite particles. *Nature* 302, 411-412.(1983).
11. N. Petersen, T. von Doveneck and H. Vali. Fossil bacterial magnetite in deep sea sediments from the South Atlantic Ocean. *Nature* 320, 611-615 (1986).
12.U. Schwertmann,and R.W. Fitzpatrick. Iron minerals in surface environments. In Skinner,H.C.W. and Fitzpatrick, R.W.(Eds) *Biomineralization processes of iron and manganese*, 7-30 (1992).
13. N. Sparks. Structural and morphological characterization of biogenic magnetite crystal. In: Frankel R B, Blakemore R P(Eds) *Iron Biominerals*. New York: Plenum Press, 167-178 (1991).
14. J. F.Stolz,D. R. Lovley, .and S.E. Haggerty. Biogenic magnetite and the magnetization of sediments. *J. Geophys Res.*. B495, 4355- 4361 (1990).
15. H. Vali, O. Foerster, and N. Petersen. Magnetotactic bacteria and their magnetofossil in sediments, *Earth Planet. Sci. Lett.*. 86, 389-400 (1987).

Proc 30th Int'l. Geol. Congr., Vol. 16, pp. 39-48
Huang Yunhui and Cao Yawen (Eds)
VSP 1997

TEM Study of Microstructure of Omphacite in High-Pressure and Ultrahigh-Pressure Eclogites

MENG DAWEI①② WU XIULING①② HAN YUJING① LIU RONG①

① Department of Physics, Test center and Department of Geology, China University of Geosciences, Wuhan 430074, P. R. China

② Laboratory of Atomic Imaging of Solids, Institute of Metal Research, Academia Sinica, Shenyang 110015, P. R. China

Abstract

The microstructural characteristics of naturally deformed omphacite from the high-pressure (HP) and ultrahigh-pressure (UHP) eclogites in the Qinling-Dabie Mountains ,central China, have been studied using electron diffraction analysis and high resolution image observation techniques. It has been found that the principal structural form of omphacite has an ordered structure P2, cell parameters: a=0.96nm, b=0.88nm, c=0.53nm and β=106°, and there is also a local secondary ordered structure P2/n and disordered structure C2/c. Therefore, omphacite in eclogite from the Qinling-Dabie Mountains formed at low temperature and high or ultrahigh pressure. The structures of the crystal domiains in omphacite have been investigated, and it has been found that 1/2(**a**+**b**) or 1/2<110> antiphase domains have a C2/c structure and (200) microtwin domains in the P2/n structure. The size of the antiphase domains can be regarded as an important indicator of metamophic temperature and time. The ultrastructures of omphacite, including dislocations, faults and distortions etc. have been observed. These heterogeneous textures are possibly related to the physical, chemical conditons in which the mineral crystallized and the metamorphic deformation conditions.

Keywords: eclogite, omphacite, microstructure, ultrastructure, antiphase domain ,twin domain, high resolution electron microscopy

INTRODUCTION

Eclogite is one of the diagnostic rocks of the high-pressure and ultrahigh-pressure metamorphic belt in the Qinling-Dabie Mountains, central China, and its essential rock-forming mineral is omphacite [(Na,Ca)(Mg,Fe,Al)(S_2O_6)], a sodian clinopyroxene mineral formed by solid state reaction in high pressure conditions (P>0.8GPa), which has a composition approximately between jadeite (25-28%) and diopside . Its structure is more complex than the pure end-members because of ordered and disordered arrangements of the chains of silicon-oxygen tetrahedra and bands of cation polyhedron. Clark [8,9], Davidson[10], Matsumoto[14], Champness[7], Carpenter[1,4-6] and others have studied the crystal structure, crystal chemistry and thermodynamics of this mineral and concluded that there are three structural forms, a disordered structure C2/c (high temperature and low pressure), ordered structures P2 and P2/n (low temperature, mesotherm and high pressure). It is evident that the structural forms of omphacite are closely related to its chemical composition and the temperature-pressure conditions. In this paper, we describe the investigation on the microstructural and ultrastructural characteristics of omphacite in high-pressure and ultrahigh-pressure eclogites from the Qinling-Dabie Mountains, China, by means of transmission electron microscopy (TEM) and high resolution transmission electron microscopy (HRTEM). The results of this study reveal the physical and chemical environment when the eclogite minerals crystallized and the displacement of their lattices after crystallization.

SAMPLES AND EXPERIMENTAL METHODS

The samples for this investigation are omphacites chosen from high-pressure and ultrahigh-pressure eclogites in the Qinling-Dabie Mountains metamorphic belt, central China. Glaucophane eclogites are representative of the high-pressure eclogites, their major mineral association is garnet (Pyp 12%~28%)+ omphacite (Jd 32%~48%)+ glaucophane + rutile + phengite + epidote + quartz and the metamorphic conditions are T=590~685°C, P>1.34 ~1.54GPa (sample DX1-1 from Xiongdian district of Henan Province). The coesite and diamond-bearing eclogites are representative of ultrahigh-pressure eclogites, their typical mineral association is garnet (Pyp 30%~40%) + omphacite (Jd 40%~47%) + coesite (quartz) + phengite (mica) + talc, the metamorphic conditions being 650~800°C in temperature and >2.6~2.8GPa in pressure (samples TB23-1 and S090 from Bixiling and Shima districts of Anhui province).

Standard petrographic thin sections were made from selected fresh eclogite samples, in the plane parallel to L and normal to S, from which the specimens for transmission electron microscopy (TEM) were prepared by LBS-1 type ion-beam thinning device working at accelerating voltage V=4KV and ion beam current I<0.8mA.

Particles of omphacite (about 0.5-1.2mm in diameter), were chosen carefully using the binocular microscope. These particles were crushed into fine fragments in an agate mortar and suspended in absolute alcohol, and an ultrasonic vibrator was used to disperse them uniformly. A drop of the suspension was put on a copper grid coated with a holey carbon film with evaporated gold on it and then examined at 200KV in a JEOL-2000EX II electron microscope equipped with a top-entry goniometer stage(±10° tilt) and a ultra-high-resolution pole piece (Cs=0.7mm) with an interpretable point resolution of 0.21nm. Electron diffraction patterns were observed at 200KV with a JEM-200CX analytical electron microscope which has a side-entry, double-tilting sample holder (±45° titlt). All high-resolution images were taken under symmetrical illumination condition for the electron beam.

RESULTS AND DISCUSSION

Electron diffraction analyses and high resolution images observations of omphacite in ultrahigh-pressure eclogite

Selected-area electron diffraction (SAED) patterns were obtained by tilting the sample about b^*. Fig. 1(a-e) shows the five representatives of eight patterns from Bixiling (or Shima) omphacite (TB23-1 or S090) with the electron beam parallel respectively to [100](a), [201](b), [101](c), [102](d) and [103](e). The angles between successive zones are 14° ,11.5° ,29° and 16° . Fig.1(f) is the [010] two-dimension reciprocal lattice plane deduced from the eight SAED patterns including above five (a-e) and their mutual angular relationships, the eight straight lines are the directions perpendicular to b^* in each of the sections in a-e. The rotating angles of the eight electron diffraction patterns in the experiment are identical with the theoretical calculation. The results show that the crystal parameters of omphacite in S090 and TB23-1 are as follows:

a=0.96nm , b=0.88nm, c=0.53nm, β=106°, which are as consistent with the JCPDS card. In order to determine the distribution pattern of the *h0l(h+l=2n+1)* spots on the (010)* plane, the crystal was tilted about h0l spot line (*h+l=2n+1*, such as $00\bar{1}$, $10\bar{2}$, $20\bar{1}$ etc.). It was observed under electron microscope that there is not a disappearance phenomenon for $00\bar{1}$,$10\bar{2}$,$20\bar{1}$ etc. spots in *h0l(h+l=2n+1, n*=0,+1,+2---). According to the diffraction condition of the spots in (010)* reciprocal plane and the distribution in a simple rectangle of the spots in the eight diffraction patterns, it was determined that there is not an n glide plane. Therefore, the omphacite from Bixiling (TB23-1) and Shima (S090) belongs to the monoclinic crystal system, and its space group is P2. A small amount of secondary ordered structure P2/n and disordered structure C2/c were observed once in a while. They indicate the characteristics of the low temperature and high pressure ordered structure of omphacite [2,3,6,11,13,17]. Fig.2 shows a set of selected-area electron diffraction (SAED) patterns and corresponding high resolution electron microscopic (HREM) images of the ordered P2 structure of omphacite (TB23-1 and S009) with different orientations, with the projection of a cell, the interplanar spacing and directions indicated. Fig.3(c) shows the [101] SAED pattern and high resolution structural image of the omphacite P2/n structure, Fig.4 shows the [100] SAED pattern and high resolutions structural image of omphacite C2/c structure.

Twin domain structure of omphacite

Microtwin domains are rather well-developed in omphacite . (200)microtwin domains of P2/n structure have been found (fig.3(a-c)), which is the microtwin domain structure formed by the conversion of lattice deformation from the high symmetry form of omphacite (high temperature variant C2/c) into low symmetry form (low temperature variant P2/n) with the twin plane parallal to the {100} direction. The transition temperature is about 725°C[12], and the cell parameters of the two structures (C2/c and P2/n) are the same. In Fig.3(b), the twin plane (200) consists of silicon-oxygen tetrahedron layers, and the twin plane (200) indicated by an arrow in the middle part of the photograph is a cation polyhedron layer.

Antiphase domain structures of omphacite

Antiphase domain structures in minerals are potential recorders of geological processed and events . A melt of omphacite composition undergoes a phase transformation from order to disorder when the temperature is over about 865°C,in which, when the inner crystal has the lowest energy structure, antiphase domain structures are formed in that crystal.

Fig.5 shows the SAED pattern and high resolution image of antiphase domain structure in disordered omphacite C2/c (S090) corresponding to the zone [001]. The crystal has a displacement vector 1/2(**a**+**b**) or 1/2<110> on the antiphase domain boundary, so the diffraction spots occur dispersed along the **a*** direction. There are a pair of edge dislocations with a sliding vector 1/2<110> in the area "D".

Fig.6(a) and (d) shows a TEM picture of the antiphase domain boundaries of ordered omphacite from two areas, where (a) shows the TEM dark image of sample DX1-1 corresponding to the zone [110], and (d) shows that of sample TB23-1 corresponding to the zone [$\bar{3}$14]. There is a phase transformation from disorder to order for omphacite at about 725° C, so the antiphase domain structures of ordered omphacite are formed in the process of phase trasformation from disordered structure to ordered structure. According to the character and mechanism of formation of crystal domains in mineral [4,7,15,16], the antiphase domain boundaries of omphacite are caused by faults in ordered atom planes of the crystals. The dishevelled, crooked veins reflect a π fault with phase angle $\alpha=\pi$ on antiphase domain boundaries in thin crystal.

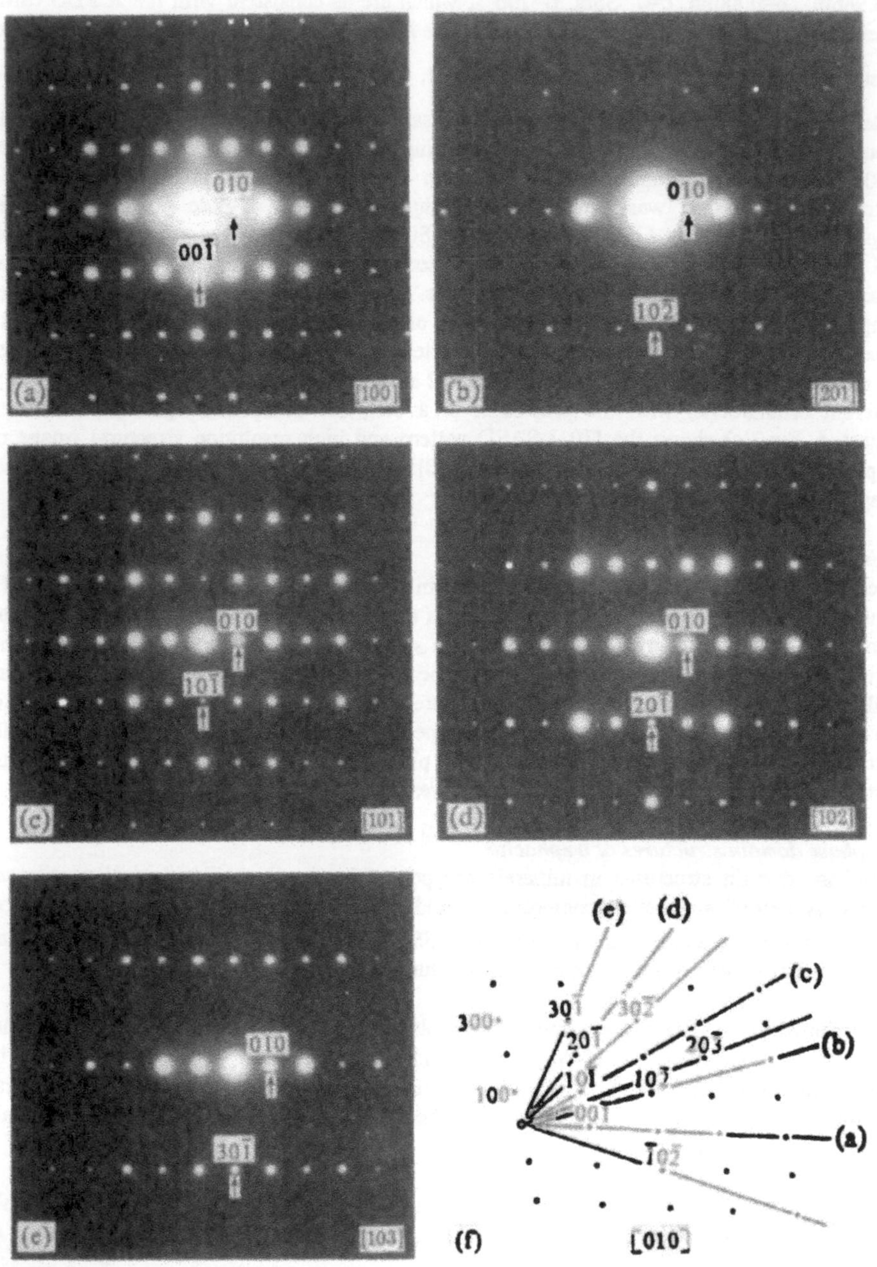

Figure 1. A set of SAED patterns [100](a), [201](b), [101](c), [102](d) and [103](e) of omphacite (samples TB23-1 or S090) obtained by tilting the crystal about **b***. (f) Reciprocal lattice plane deuced from eight SAED patterns including the above five (a-e) and their mutual angular relationships.

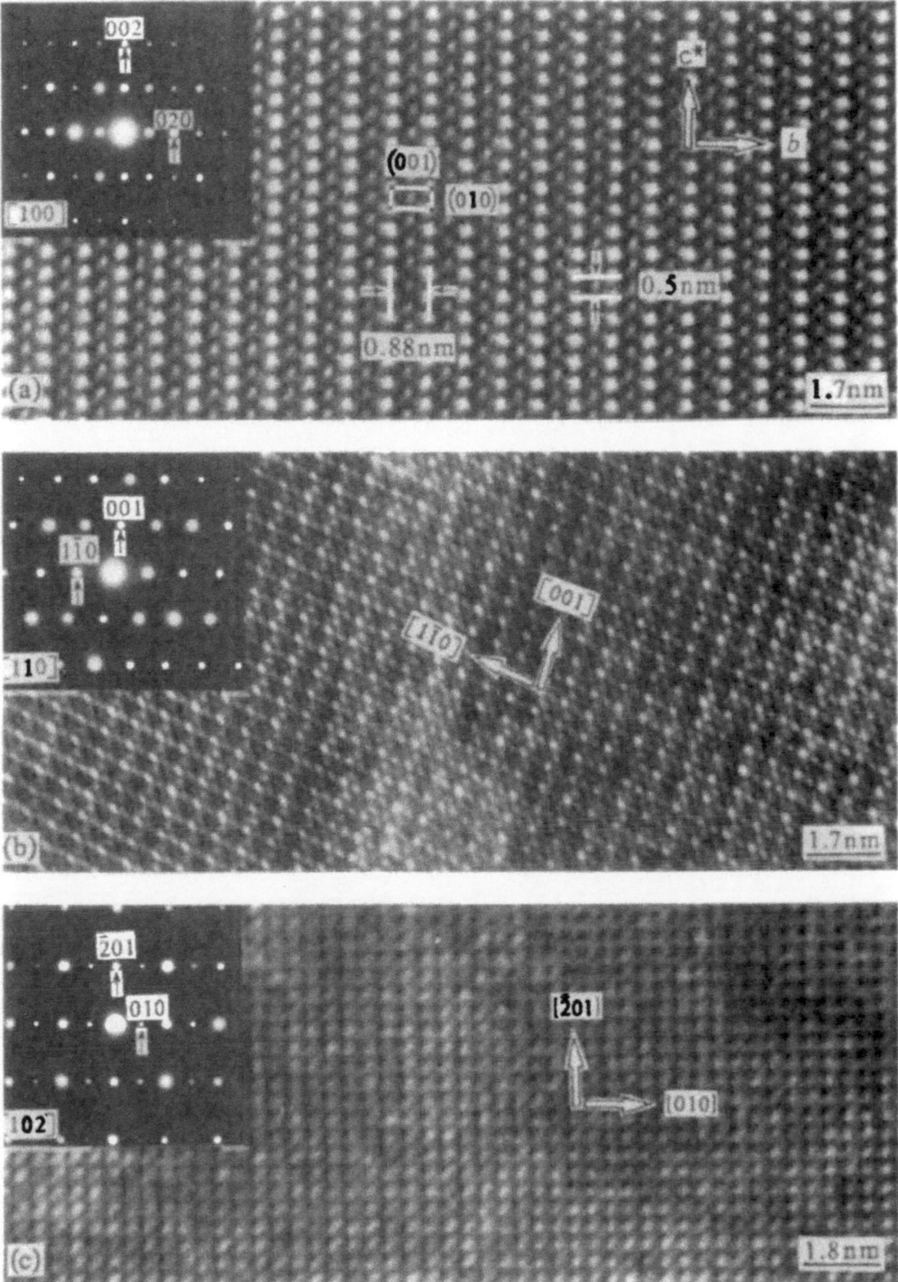

Figure 2. A set of SAED patterns and corresponding high resolution structural images [100](a), [110](b) and [102](c) of P2 omphacites (samples TB23-1, S090 and DX1-1).

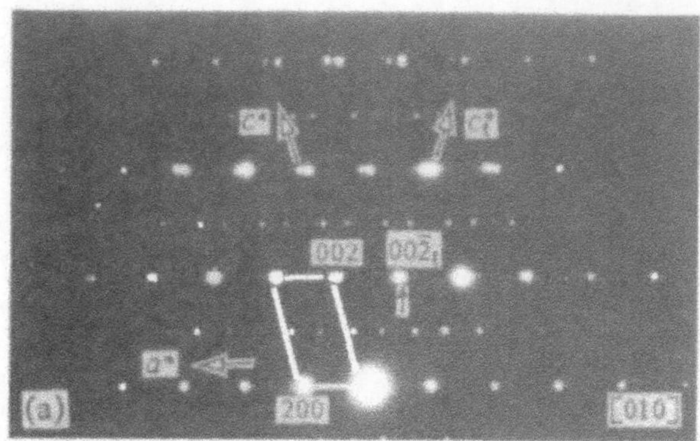

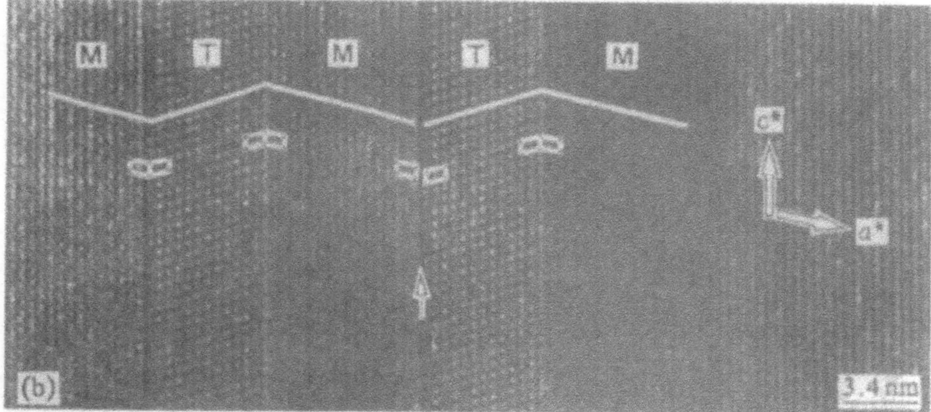

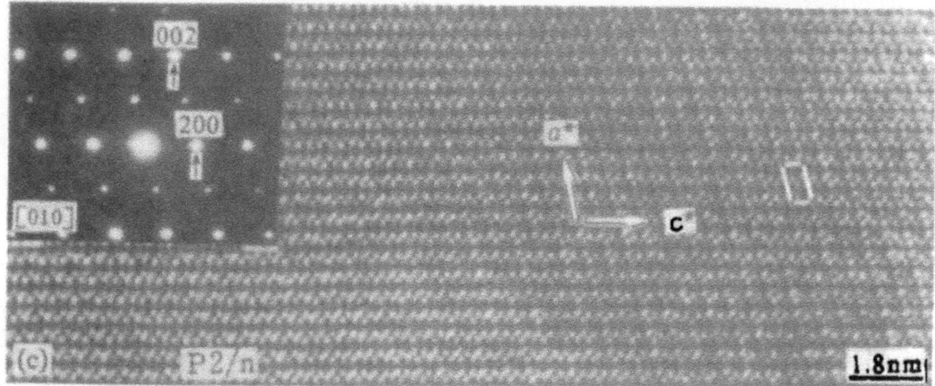

Figure 3. SAED pattern (a) and high resolution structural image (b) of (200) microtwin domain structure in P2/n omphacite corresponding to the zone [010](sample S090), where "M" shows matrix, "T" shows twins. The [010] SAED pattern and high resolution structural image(c) of P2/n omphacite (sample S090).

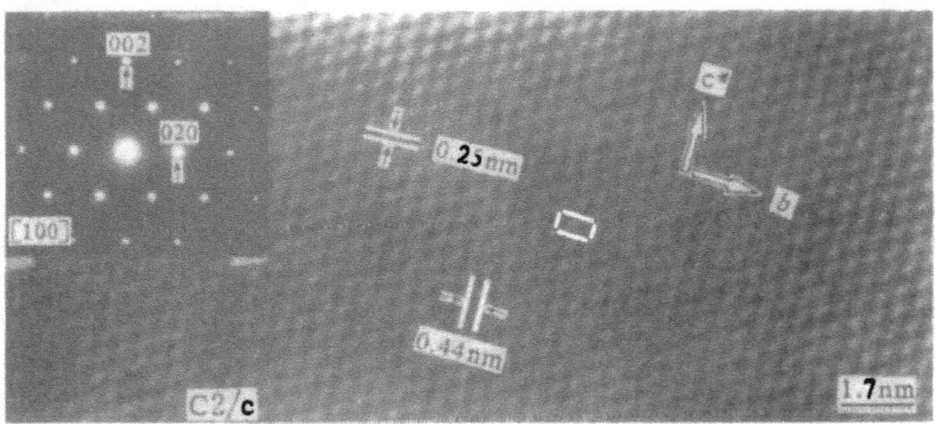

Figure 4. SAED pattern and high resolution structural image of C2/c omphacite corresponding to the zone [100] (sample DX1-1)

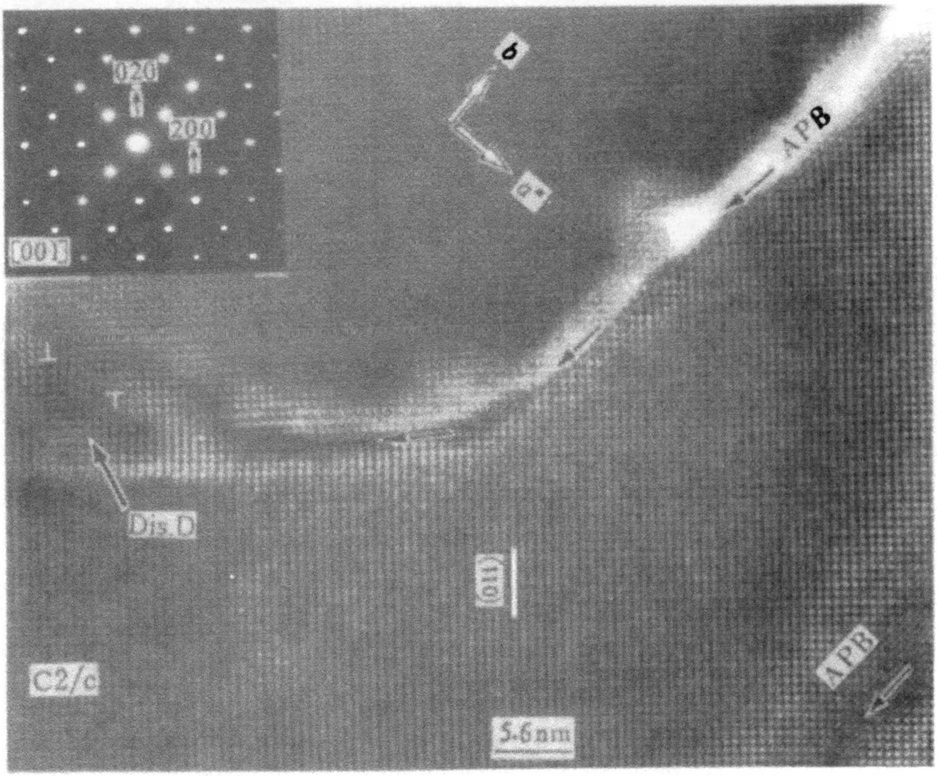

Figure 5. [001]SAED pattern and high resolution structural image of antiphase domain structure in C2/c omphacite, where "APB" shows the antiphase domain boundary (sample S090).

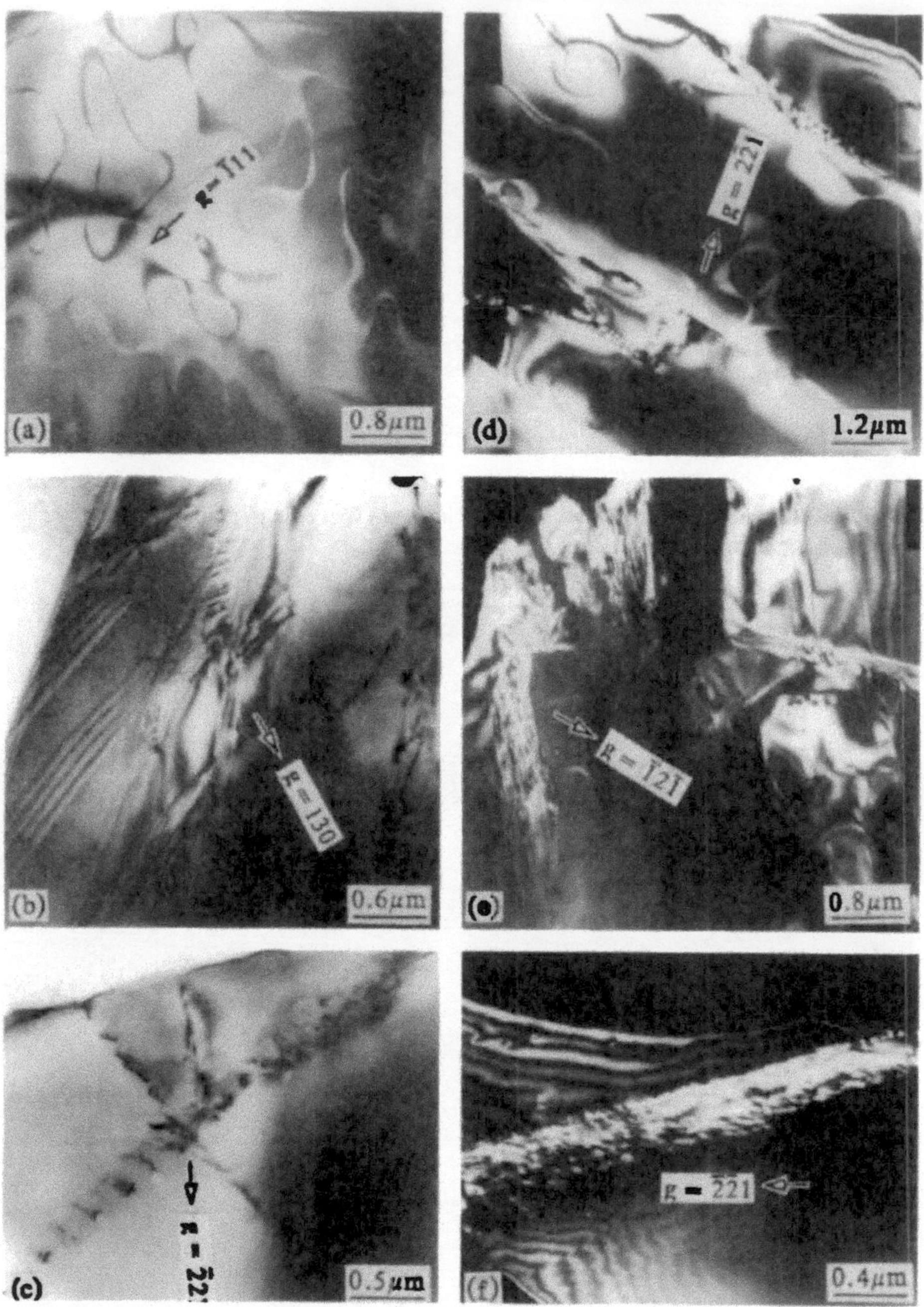

Figure 6. A set of TEM pictures of antiphase domain boundaries and dislocation substructures in ordered omphacite from the three areas. (a) TEM dark field image of zone [232]; (b) and (c) the [$\bar{3}$14] and [110] TEM bright field image (sample DX1-1); (d), (e) and (f) the [232], [210] and [102] TEM dark field image respectively (sample TB23-1).

Dislocation substructures

Ultra-microstructures preserved within omphacite crystals are dislocation substructures: stacking faults (Fig. 6b and 6e), dislocation tiltwalls(Fig.6d and 6e), dislocation networks (Fig.6c,6d and 6f), low angle subgrain boundaries and subgrains (Fig.6c,6d and 6e). Free dislocations are also observed within some subgrains. Most of the individual or free dislocations have a Burgers vector 1/2<110> or [001], corresponding to the activation of slip systems 1/2<110>{110} and [001]{110}. Dislocation densites and subgrain sizes have been calculated (see Table 1).

Table 1. Microstructural characteristics of omphacite in eclogite from the Qinling-Dabie Mountains

Sample No.	Xiongdian (DX1-1)	Shima (S090)	Bixiling (TB23-1)	(JCPDS)
a_0(nm)	0.965	0.960	0.962	0.966
b_0(nm)	0.883	0.881	0.882	0.882
c_0(nm)	0.529	0.525	0.528	0.523
β	107.0°	106.5°	106.0°	106.6°
Space group	P2,P2/n,C2/c	P2,P2/n,C2/c	P2,P2/n,C2/c	C2/c
Rock name	HP Kyanite eclogite	UHP Coesite-bearing eclogite	UHP kyanite eclogite	
Dislocation density(cm^{-2})	$4.5x10^9$	$5.2x10^8$	$1.1x10^{10}$	
Subgrain size(mm)	1.2~4.5	0.8~5.0	2.0~4.0	

CONCLUSIONS

The principal structural form of omphacite from the three districts (Xiondian DX1-1, Bixiling TB23-1 and Shima S090) is the ordered structure P2, accompanied by secondary ordered structure P2/n and disordered structure C2/c. The electron diffraction and high resolution TEM images show that omphacite in eclogite from the Qinling-Dabie Mountains, central China, formed under the conditions of high-pressure and ultrahigh-pressure.

(200) microtwin domains of P2/n structure have been found, which formed by the conversion of lattice deformation from the high symmetry crystals of omphacite (high temperature variant C2/c) into low symmetry crystals (low temperature variant P2/n) with the twin plane parallel to the {100} direction. The conversion temperature is about 725°C.

1/2(a+b) or 1/2<110> antiphase domains have been found with C2/c sturcture. The TEM dark images of the antiphase domain boundaries have been observed for samples DX1-1 and TB23-1. There is a phace transformation from disorder to order at about 725 °C

The dislocation substructures of omphacite have been investigated, including stacking faults, free dislocations, dislocation tiltwalls, dislocation networks,low angle subgrain boundaries and subgrains etc. Dislocation densities and subgrain sizes have been calculated. These heterogeneous textures are possibly related to the phydical, chemical conditions when the minerals crystallized and metamorphic deformation conditions.

Acknowledgements

This work was supported by the Doctorate Special Foundation from the Nation Education Committee of China (Projct No.9349101) and in part by the Laboratory of Atomic Imaging of Solids, Institute of Metal Research, Academia Sinica.

REFERENCES

1. L. P. Aldridge et al. Omphacite studies,II. Mossbauer spectra of C2/c and P2/n omphacites. *Am.Mineral*, **63**, 1107-1115 (1978).
2. J. N. Boland and T.E. Tullis. Deformation behavior of wet and dry clinopyroxenite in the brittle to ductile transition region, In:Mineral and Rock Deformation Lab. Studies, *The patterson Volume. Geoph. Monograph. Ser.*, **36**,35-49 (1986).
3. M. Buatier, H. L. M. Van Roermund, M. R. Drury and J. M. Lardeaux. Deformation and recrystallization mechanism in naturally deformed omphacites from the Sesia-Lanzo zone: Geophysical consequences. *Tectonophysics*, **195**,11- 27 (1991).
4. M. A. Carpenter. Kinetic control of ordering and exsolution in omphacite. *Contrib. Mineral. Petrol.*, **67**,17-24 (1978).
5. M. A. Carpenter. Omphacites from greece, turkey and guatemala: composition limits of cation ordering. *Am. Mineral.*, **64**,102-108 (1979).
6. M. A. Carpenter. Omphacite microstructures as time-temperature indicators of blueschist and eclogite-facies metamorphism. *Contrib. Mineral. Petrol.*, **78**,441-451 (1981).
7. P. E. Champness. Speculation on an order-disorder transformation in omphacite, *Am. Mineral.*, **58**,540-542 (1973).
8. J. R. Clark and J. J. Papike. Eclogitic pyroxenes, ordered with P2 symmetry, *Science*,**154**,1003-1004 (1966).
9. J. R. Clark and J. J. Papike. Crystal-chemical characterization of omphacite, *Am. Mineral.*, **53**,840-868 (1968).
10. P. M. Davidson and B. P. Burton. Order-disorder in omphacitic pyroxenes: A model for coupled substitution in the point approximation, *Am. Mineral.*, **72**, 337-344 (1987).
11. D. J. Ellis and D. H. Green, An experimental study of the effect of Ca upon garnet-clinopyroxene Fe-Mg exchange equilibria, *Contrib, Mineral. Petrol.*, **71**,13-22 (1979).
12. M. E. Fleet et al. Omphacite studies, I. The P2/n-C2/c transformation, *Am. Mineral.*, 63,1100-1106 (1978).
13. T. J. B. Holland. The experimental determination of activities in disordered and short-range ordered jadeitic pyroxenes, *Contrib. Mineral. Petrol.*, **82**,214-220 (1983).
14. T. Matsumoto. The crystal structure of omphacite, *Am.Mineral.*, **60**, 634-641 (1975).
15. P. P. Phakey and S. Ghose. Direct observation of a phase domain structure in omphacite, *Contrib. Mineral. Petrol.*, **39**, 239-245 (1973).
16. S. Sasaki et al. The influence of multiple diffraction on the space group determination of orthopyroxene, spodumene, low omphacite and pigeonite, *Phys. Chem. Mineral.*, **7**, 260-267 (1981).
17. H. L. M. Van Roermund. and J. M. Lardeaux. Modification of antiphase domain sizes in omphacite by dislocation glide and creep mechanisms and its petrological consequences, *Mineralogical Magazine*, **55**, 397-407 (1991).

Proc. 30th Int'l. Geol. Congr., Vol. 16, pp. 49-57
Huang Yunhui and Cao Yawen (Eds)
VSP 1997

Ordered - Disordered Stacking Structure Along the c-Axis in Calcium Rare -Earth Fluorocarbonate Minerals

WU XIULING[1][2] MENG DAWEI[1][2] LIANG JUN [1] PAN ZHAOLU[1]

[1] Department of Physics, Test center and Department of Geology, China University of Geosciences, Wuhan 430074, P. R. China

[2] Laboratory of Atomic Imaging of Solids, Institute of Metal Research, Academia Sinica, Shenyang 110015, P. R. China

Abstract

The ultrastructural characteristics of members of the calcium rare-earth fluorocarbonate mineral series from Sichuan Province, Southwest China, have been studied using selected area electron diffraction (SAED) and high resolution transmission electron microscopy (HRTEM). Microstructural and ultrastructural information of this mineral series has been obtained from derived polycrystal of parisite, such as crystal structural defects in parisite and regular mixed-layer structures B_mS_n(m>n), antiphase domains and disordered domains in the two new polytypes B_2S-2H and 12R, syntactic intergrowths and disordered stacking among the structural unit layers of B_mS_n (m>n) type along the c- axis direction . Five new, regular, mixed-layer structures of the B_mS_n(m>n) type with long periods and different stacking sequences, B_6S_2, $B_{19}S_{10}$, $B_{13}S_8$, B_2S and B_2S_2 were found, their structural symmetry, cell parameters, chemical formulas of crystal and stacking models of structural unit layer were determined . The results of this investigation reveal the complexity of the structure in syntactic polycrystals of such a mineral series. The new, regular, mixed-layer structures and the new polytypes are composed of structural unit layers of the two end elements of this mineral series -- bastnaesite(B) and synchisite(S) with ordered stacking in varying proportions along c axis. The different, regular, mixed-layer structures and polytypes have various compositions and structural stacking models.

Key words: mixed-layer structure, syntactic intergrowth, antiphase domain, high resolution electron microscopy

INTRODUCTION

The members of the calcium rare-earth fluorocarbonate mineral series [bastnaesite(B), parisite(BS), roentgenite(BS_2) and synchisite(S)] are layered rare-earth minerals with complex syntactic intergrowth structures, which were studied by the earlier workers [2-4,5-8, 10-12], using x-ray diffraction and TEM. Until now no fine structure parameters have been found in the members of this mineral series except bastnaesite[12]. In this paper, we describe a HRTEM study of the microstructural and ultrastructural characteristics of calcium rare-earth fluorocarbonate minerals and their derived polycrystals from Sichuan Province, Southwest China, by means of HRTEM, in which parisite and roentgenite are determined as well as bastnaesite. Five new, regular, mixed-layer structures of B_mS_n(m>n) type and two new polytypes[1-3,7,9] have been found. Some microstructural and ultrastructural characteristics are also observed, which are syntactic structures among the different polytypes, disordered stacking and syntactic intergrowth structures among different minerals and structural unit layers of different B_mS_n(m>n) structure along the c- axis direction, and sundry defects, such as dislocations, faults, antiphase domains, disordered domains etc..

SAMPLES AND EXPERIMENTAL METHODS

The samples for this investigation are the polycrystal grains of parisite (about 0.08-0.2mm in diameter) that have been collected from a rare-earth mineral deposit within an aegirine alkali granite massif in Sichuan Province, Southwest China. They were chosen carefully under the binocular microscope, crushed into fine fragments in an agate mortar and suspended in absolute alcohol, and an ultrasonic vibrator was used to disperse them uniformly. The suspension was dropped on a copper grid coated with a holey carbon film with evaporated gold on it and then examined at 200kV in a JEOL-2000EXII electron microscope equipped with a top-entry goniometer stage (±10° tilt) and a ultra-high-resolution pole piece (Cs=0.7mm) with an interpretable point resolution of 0.21nm. Electron diffraction patterns were observed at 120kV with a Philips-CM12 transmission electron microscope which has a side-entry, double-tilting sample holder (±45° tilt). Crystals with <0001> parallel to the support film were chosen so that the images would always contain the c-axis. All high-resolution images were taken under symmetrical illumination conditions for the electron beam.

OBSERVATION RESULTS AND DISCUSSION

All members of calcium rare earth fluorocarbonate mineral series have the same cell parameter a =0.710nm and the smallest subcell height c''' =0.471nm, but they have different heights for secondry substructure c'', substructure c' and superstructure c. Observed under TEM, the SAED patterns of this mineral series have following characteristics: the position of diffraction spots and their distribution density between 0000 and 0001''' are different for every mineral member along c^* and the hohl$_i'''$ (i =1,2,......, h=|$\bar{h}$| =3n, n=0, ±1, ±2,......) spot line, which reflect the distinct crystal structures of the different minerals.

Parisite and its crystal structural defects

As an an important member of the calcium rare-earth fluorocarbonate mineral group , parisite, $2CeFCO_3.CaCO_3$, has space group R3, a=0.712nm,c=8.410nm,z=18 [2]. Its chemical composition can be considered as an intermediate compound (BS) between bastnaesite($CeCO_3F$) and synchisite [$CeCa(CO_3)_2F$], the two end-elememts among this mineral group in a 1:1 ratio . By means of SAED and HRTEM, we have determined the cell and subcell parameters, point group, crystal system, etc. of parisite. which were quite concordant with the published result (see Table.1). Fig.1(a) shows the electron diffraction pattern and corresponding high-resolution strcture image from parisite with the electron beam parallel to the [$\bar{1}2\bar{1}0$] direction. Fig.1(a) was indicated that the projection of a cell along the [$\bar{1}2\bar{1}0$] direction ,the spacing of structural unit layer (1.413nm), and the height of a cell (c=1.413nm x 6 = 8.478nm), and the high-resolution structure image shows that the CO_3 ionic group between Ce-F ionic layer and Ca ionic layer differ from the CO_3 ionic groups between two Ce-F ionic layers in space orientation.

Fig.1(b)-(c) shows [$\bar{1}2\bar{1}0$] SAED patterns and HRTEM images of the disordered structure in the derived polycrystal of parisite and regular mixed-layer structures B_2S_2 and B_4S_4. B_2S_2 is one of a series of regular mixed-layer structures of the BmSn type, which were found by Wu Xiuling et al. in the calcium rare-earth fluorocarbonate minerals, and its spacing of the structural layer, 2.826nm, is composed of two Ca-Ce layers and two Ce-Ce layers. Fig. (b) shows the compound SAED pattern and the corresponding HRTEM image formed by two lattices, the regularly mixed-layer structure B_2S_2 and parisite. In the SAED pattern of the [$\bar{1}2\bar{1}0$] zone , a set of stronger diffraction spots is contributed by both B_2S_2 and BS, and a set of weaker diffraction spots is contributed by the B_2S_2

structure ,whose characteristic is that there are five diffraction spots between 0000 and 0001_1^{m} (c^{m} *), two of them overlapping the diffraction spots of BS. The diffraction spots of B_2S_2 have twice the distributive density as BS along the 0001 and $h0\bar{h}l$ (h=|$\bar{h}$|=3n) spot lines. The weaker dispersion occurring at some diffraction spots along the c* direction is due to disordered interlayers of B_3S_2 and B_4S_4 inserted into B_2S_2 or BS structures and the stacking faults. The lattice image (Fig.1(b)) shows that in about a quarter of Ce-F ionic layers with BS structure, a relative translation occurs along the c direction (the translational distance equals *c*/18 of BS or 0.471nm). Two different structures, B_2S_2 and BS, are respectively formed above and below the distorted area "D" because of this translation. Fig. 1(c) shows a compound SAED pattern and the corresponding HRTEM image formed by three lattices, the regular mixed-layer structures B_2S_2, B_4S_4 and parisite (BS). In the SAED pattern of the [$\bar{1}2\ \bar{1}0$] zone, a set of stronger diffraction spots is contributed by the B_2S_2, B_4S_4 and BS structures, a set of weaker diffraction spots is contributed by the structures of B_2S_2 and B_4S_4, and the HRTEM image shows a fault defect among the three different structures, where B_4S_4(BSBSSBBS, spacing 5.652nm) is the transition phase.

Fig.2(a-d) shows compound SAED patterns and corresponding HRTEM images of syntactic intergrowth structure formed by disordered stacking of parisite (BS) and unit layers of different B_mS_n($m>n$) in calcium rare-earth fluorocarbonate minerals along the c direction.

Table 1 . Crystal structure of the calcium -rare-earth fluorocarbonate minerals

<table>
<tr><th>name and polytype</th><th>cell(nm) a</th><th>c</th><th>height of unit layer (nm)</th><th>stacking mode unit of layer</th><th>chemical formula</th><th>CaCO3 (%)</th></tr>
<tr><td>synchisite</td><td>0.710</td><td>5.472</td><td>0.912</td><td>|S|......</td><td>$CaCe(CO_3)_2F$</td><td>50.00</td></tr>
<tr><td>roentgenite</td><td>0.713</td><td>7.065</td><td>2.355</td><td>|BSS|......</td><td>$Ca_2Ce_3(CO_3)_5F_3$</td><td>40.00</td></tr>
<tr><td>parisite</td><td>0.717</td><td>8.478</td><td>1.413</td><td>|BS|......</td><td>$CaCe_2(CO_3)_3F_2$</td><td>33.33</td></tr>
<tr><td>B_2S_2</td><td>0.714</td><td>8.748</td><td>2.826</td><td>|BBSS|.......</td><td>$Ca_2Ce_4(CO_3)_6F_4$</td><td>33.33</td></tr>
<tr><td>B_4S_4</td><td>0.715</td><td>16.956</td><td>5.652</td><td>|BSBSSBBS|......</td><td>$Ca_4Ce_8(CO_3)_{12}F_8$</td><td>33.33</td></tr>
<tr><td>B_5S_4</td><td>0.712</td><td>6.123</td><td>6.123</td><td>|BBSBSBSBS|......</td><td>$Ca_4Ce_9(CO_3)_{13}F_9$</td><td>30.76</td></tr>
<tr><td>B_3S_2(3R)</td><td>0.710</td><td>9.891</td><td>3.297</td><td>|BBSBS|......</td><td>$Ca_2Ce_5(CO_3)_7F_5$</td><td>28.57</td></tr>
<tr><td>B_3S_2(2H)</td><td>0.713</td><td>6.594</td><td>3.297</td><td>|BBSBS|......</td><td>$Ca_2Ce_5(CO_3)_7F_5$</td><td>28.57</td></tr>
<tr><td>$B_{13}S_8$</td><td>0.706</td><td>40.977</td><td>13.659</td><td>|BBSBBSBBSBBS BSBSBBSBS|......</td><td>$Ca_8Ce_{21}(CO_3)_{29}F_{21}$</td><td>27.59</td></tr>
<tr><td>$B_{19}S_{10}$</td><td>0.719</td><td>55.107</td><td>18.369</td><td>|BSBBSBBSBBSB BSBBSBBSBBSBB SBBS|......</td><td>$Ca_{10}Ce_{29}(CO_3)_{39}F_{29}$</td><td>25.64</td></tr>
<tr><td>B_2S(2H)</td><td>0.706</td><td>3.768</td><td>1.884</td><td>|BBS|......</td><td>$CaCe_3(CO_3)_4F_3$</td><td>25.00</td></tr>
<tr><td>B_2S(12R)</td><td>0.711</td><td>22.608</td><td>1.884</td><td>|BBS|......</td><td>$CaCe_3(CO_3)_4F_3$</td><td>25.00</td></tr>
<tr><td>B_2S(4H)</td><td>0.705</td><td>7.536</td><td>1.884</td><td>|BBS|......</td><td>$CaCe_3(CO_3)_4F_3$</td><td>25.00</td></tr>
<tr><td>B_6S_2</td><td>0.701</td><td>4.710</td><td>4.710</td><td>|BBBBSBBS|......</td><td>$Ca_2Ce_8(CO_3)_{10}F_8$</td><td>20.00</td></tr>
<tr><td>bastnaesite</td><td>0.710</td><td>0.978</td><td>0.489</td><td>|B|......</td><td>$Ce(CO_3)F$</td><td>0.00</td></tr>
</table>

* Donnay, G., Donnay, J.D.H.(1953), R=rhombohedral, H=hexagonal

In Fig.2(a-b), the [$\bar{1}2\ \bar{1}0$] SAED patterns are compound electron diffraction patterns taking BS as the principal diffraction pattern and superimposing B_3S_2 (BBSBS) and bastnaesite (B, Fig.2(b)) diffractions, B_3S_2 and B are respectively a structural interlayer with local order and overall disorder (see the HRTEM images in Fig.2(a-b)), the weaker dispersion occurs around the diffraction spots in the pattern because of their smaller areas. The crystal structure of parisite (BS) and B_3S_2 is composed of a ordered stacking of bastnaesite(B) layers in different proportions with synchisite (S) layers along the c direction, Therefore, the diffraction spots of BS and B_3S_2 partially overlap each other on the 0001 or hohl (h=|$\bar{h}$|=3n) spot line. Fig.2(c-d) shows the disordered stacking structures with regular mixed-layer structures B_5S_4 (3BS + B_2S) as the principal pattern and superimposed BS and B_2S, in which the B_5S_4 is formed by ordered stacking of the structural unit layers of parisite and B_2S along the c direction. In the lattice images, the most narrow fringes have a spacing 1.413nm, which equals the spacing of parisite in the c direction, another wider fringes have a spacing 1.884nm, equaling the spacing of regular mixed -layer structure B_2S in the c direction.

New regular mixed-layer structures --$B_{13}S_8$, $B_{19}S_{10}$ *and* B_6S_2 *in calcium rare-earth fluorocarbonate minerals*

Fig.3(a-b) show selected-area electron diffraction (SAED) patterns and high resolution electron microscopy (HRTEM) images taken of $B_{13}S_8$ and $B_{19}S_{10}$ with the electron beams parallel to $[\bar{1}2\bar{1}0]$ directions. The diffraction intensity of spots in SAED patterns displays a significant periodic fluctuation, the strongest diffraction spots in the 0001 spot line are contributed mainly by the substructure of the crystal (c'''=0.471nm), and the stronger diffraction spots nearest to the transmission spot are contributed by the spacing between superstructural crystals of 13.659nm (c' =d_{0003}, $B_{13}S_8$) and 18.369nm (c' =d_{0003}, $B_{19}S_{10}$). Because there are twenty-eight ($B_{13}S_8$) and thirty-eight ($B_{19}S_{10}$) weak diffractions respectively between the strongest diffraction spots (positions shown in Fig.3(a-b)) and the transmission spots, the strongest diffraction spots were indexed as 00087($B_{13}S_8$) and 000117($B_{19}S_{10}$) with the composition of geometrical pictures. Fig.3(a-b) show the high-resolution images taken in the $[\bar{1}2\bar{1}0]$ orientation and that $B_{13}S_8$ and $B_{19}S_{10}$ possess an ordered stacking structure and the thickness of a structure unit layers respectively are 13.659nm and 18.369nm (corresponding to the least reciprocal spacing d_{0003}=13.659nm ($B_{13}S_8$) and 18.369nm($B_{19}S_{10}$) between the spots in c spot line), where a thick black fringe corresponds to three Ce-Ce basic layers of bastnaesite, a thick white fringe corresponds to one Ca-Ce basic layer of synchisite, and a thin black fringe corresponds to one Ce-Ce basic layer of bastnaesite . Therefore, a structure unit layer of $B_{13}S_8$ is composed of thirteen Ce-Ce basic layers of bastnaesite (B) and eight Ca-Ce basic layers of synchisite(S), and the stacking order is |BBSBBSBBSBBSBSBSBBSBS|...... (see Tab.1 and Fig.3), and a structure unit layer of $B_{19}S_{10}$ is composed of nineteen Ce-Ce basic layers of bastnaesite and ten Ca-Ce basic layers of synchisite and the stacking order is |BSBBSBBSBBSBBSBBSBBSBBSBBSBBS|.......

The composition and the block sequence of structural unit layers decide the types and the repeated periods of regular mixed-layer structures , hence for $B_{13}S_8$ and $B_{19}S_{10}$, it was determined from the analyses of the SAED and HREM images that the chemical formula are $Ca_8Ce_{21}(CO_3)_{29}Fe_{21}$($B_{13}S_8$) and $Ca_{10}Ce_{29}(CO_3)_{39}F_{29}$($B_{19}S_{10}$) respectively, the types of structural basic layer and the proportions are 8[Ca-Ce]:13[Ce-Ce]($B_{13}S_8$) and 10[Ca-Ce]:19[Ce-Ce]($B_{19}S_{10}$) , and the crystals possess rhombohedron symmetry. The diffraction spots in SAED patterns in different orientations are in accordance with the reflection conditions: hkil, h-k+l=3n; $hh2\bar{h}l$, l = 3n; $h\bar{h}0l$, 2h+l=3n; 000l, l=3n. There are two groups of fringe spacings in the lattice image (Fig. 3(b)), where the thinner one (spacing 1.413nm) is the height of a unit layer in parisite(BS), which is equal to the height of one Ce-Ce basic layer plus the height of one Ca-Ce layer, while the thicker one (spacing 1.884nm) is BBS equal to the height of two Ce-Ce layers plus the height of one Ca-Ce basic layer. The structural unit layer or repeated period in the c axis direction of $B_{19}S_{10}$ is composed of one thinner fringe (BS x 1) plus nine thicker fringes (BBS x 9) (1.413nm x 1 + 1.884nm x 9 = 18.369nm =c'), as it is indicated in the lattice image.

Fig.3(c) shows the SAED pattern and corresponding lattice image of B_6S_2 in $[\bar{1}100]$ orientation, in which the distribution of diffraction spots has a hexagonal symmetry, and is in accordance with the space groups: P6/m, P6, $P\bar{6}$. In lattice images taken from different crystal fragments, all these fringes are arranged in a regular-order form. The repeated period along the c-direction of the structural unit layer in B_6S_2 (spacing 4.71nm) is composed of the unit layer of one BBBBS (spacing 2.826nm) plus the unit layer of one BBS (spacing 1.884nm). The height of cell is c =2.826nm + 1.884nm =4.71nm, and the chemical composition is $Ca_2Ce_8(CO_3)_{10}F_8$ (see Table1 and Fig.3(c)).

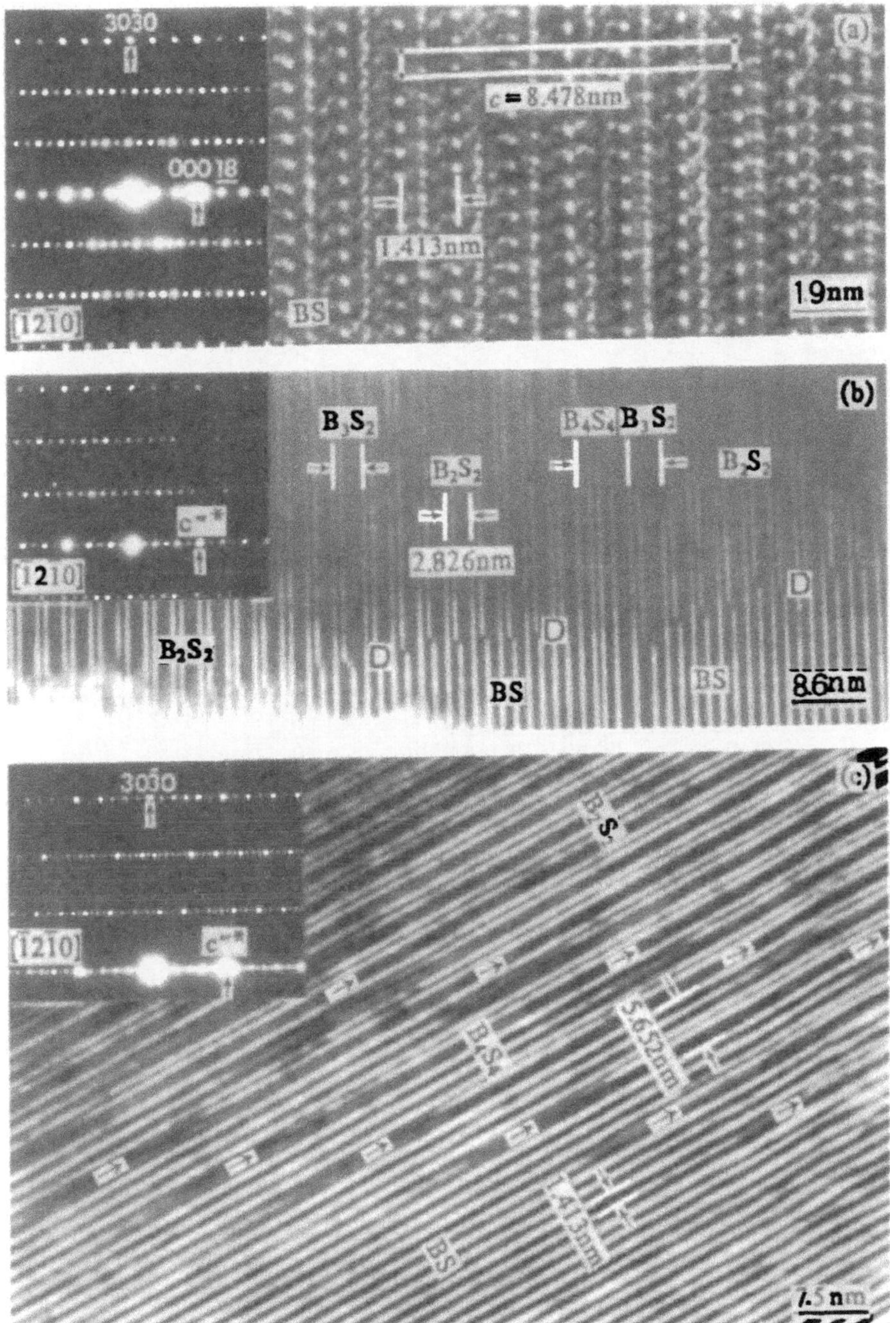

Fig.1. $[\bar{1}2\bar{1}0]$ SAED patterns and high resolution electron microscopic (HRTEM) images of parisite (BS-6R) and its crystal structural defects.

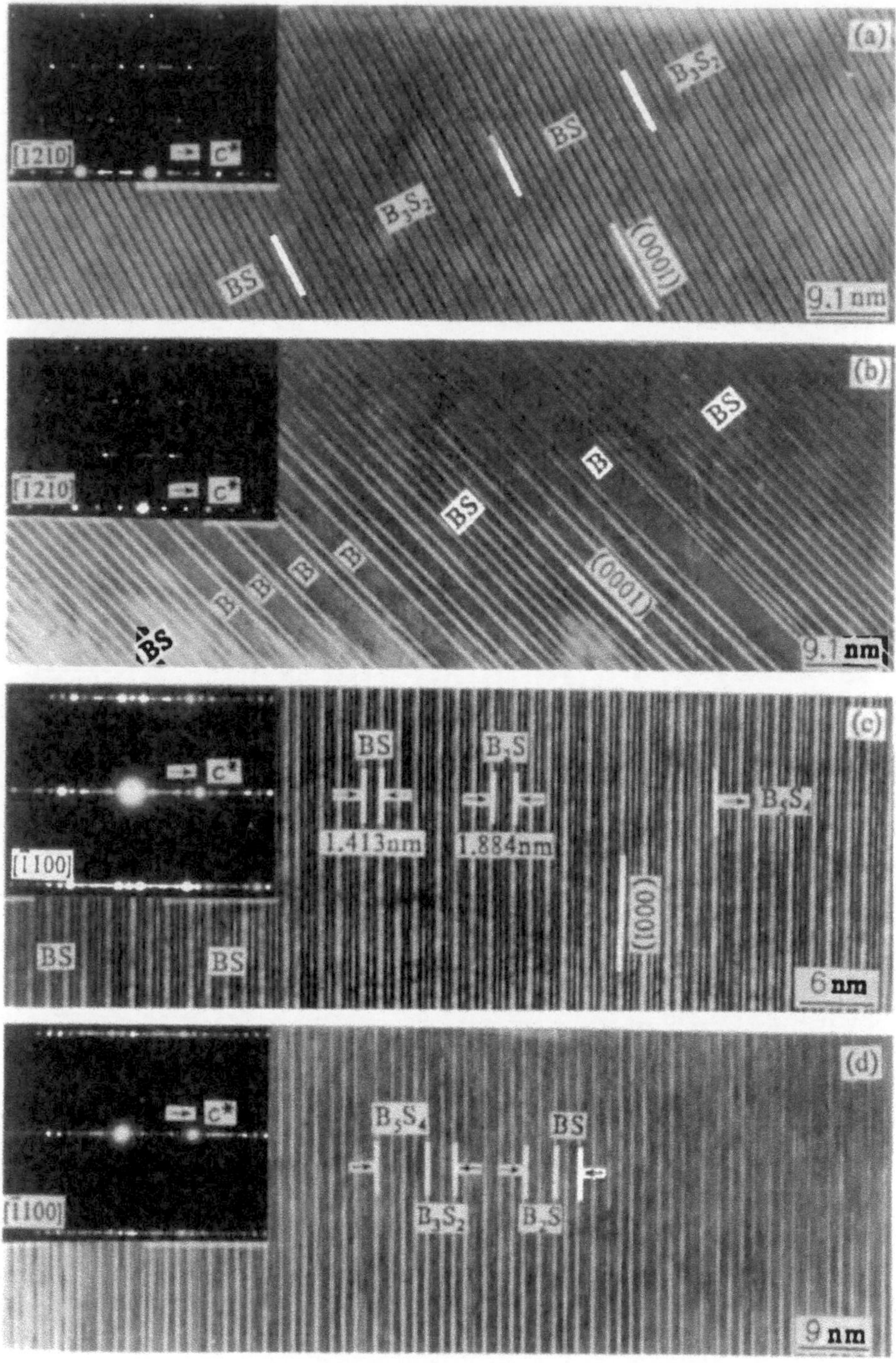

Fig. 2. High resolution images of disordered stacking structures of the different mineral unit layers in calcium rare-earth fluorocarbonate along c axis direction .

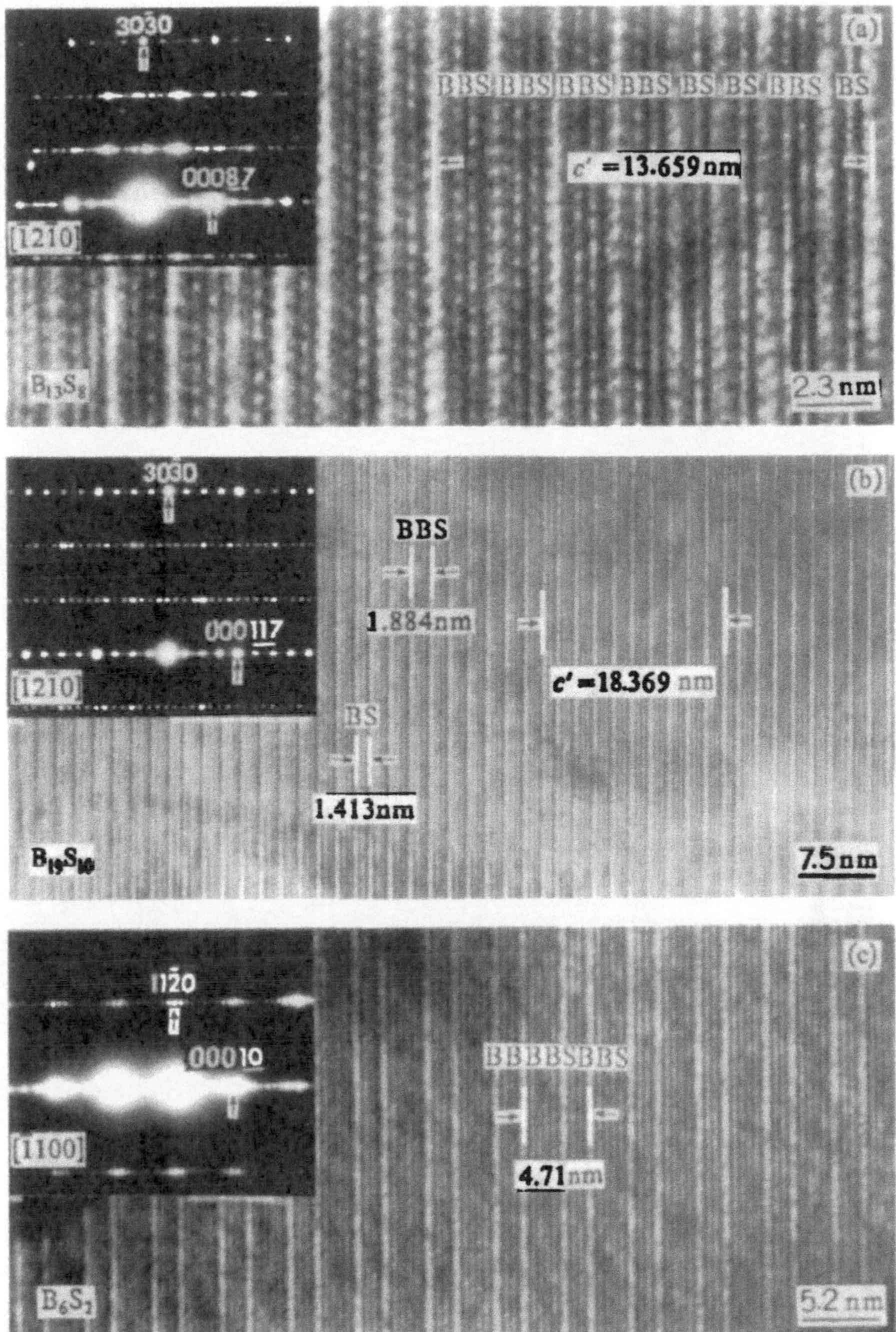

Fig.3. The three new, regular, mixed -layer structures of B_mS_n(m>n) type in calcium rare-earth fluorocarbonate minerals. (a)-$B_{13}S_8$ (rhombohedral, $[\bar{1}2\bar{1}0]$); (b)- $B_{19}S_{10}$ (rhombohedral,$[\bar{1}2\bar{1}0]$); (c) B_6S_2 (hexagonal,$[\bar{1}100]$)

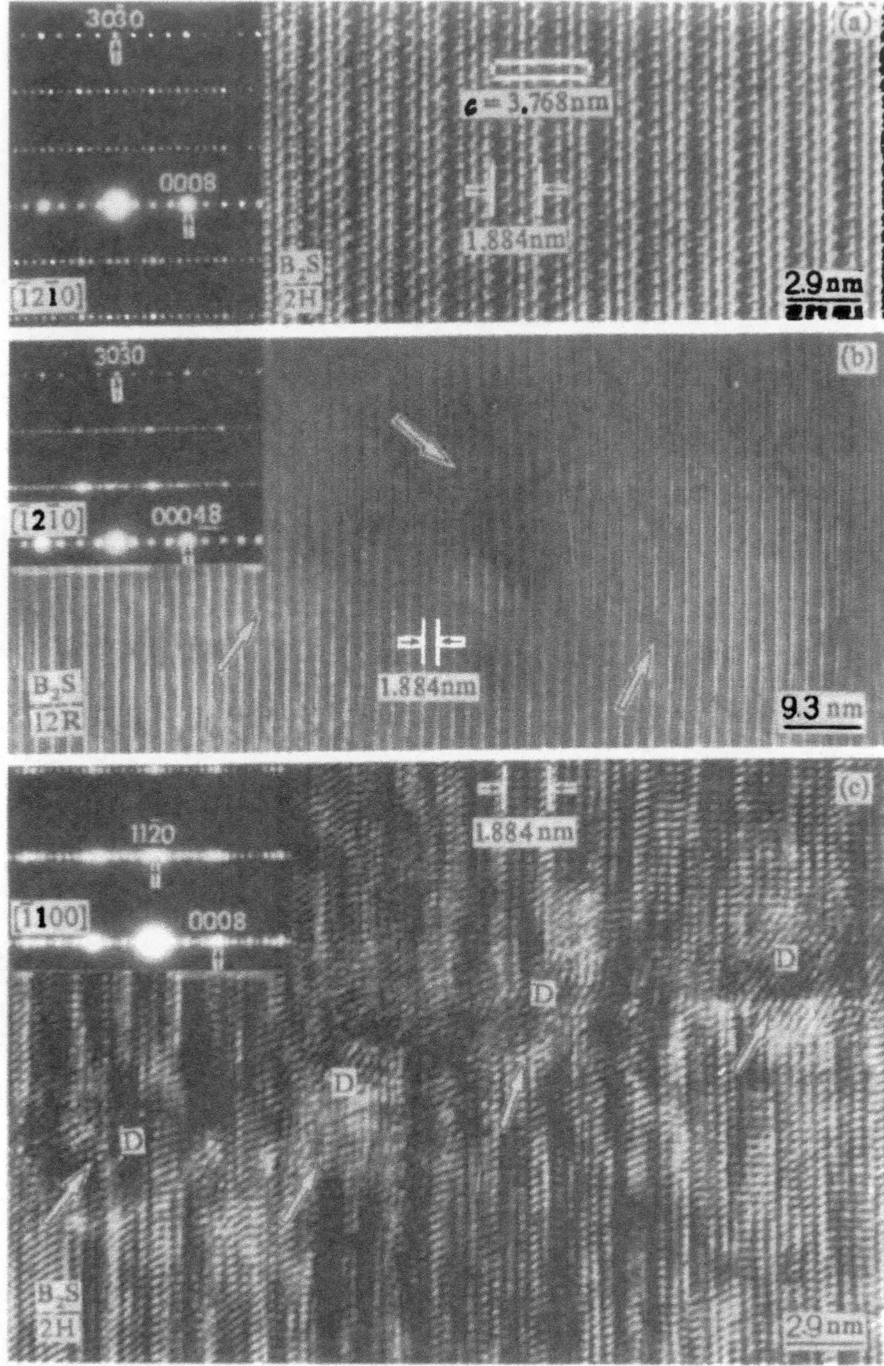

Fig.4. SAED patterns and HREM images of the new, regular, mixed-layer structure and polytypes of B_2S-2H (a) and B_2S-12R(b) corresponding to the zone [$\bar{1}2\bar{1}0$], (b,c) antiphase domains and disordered crystal domains structures of B_2S-12R and B_2S-2H.

The two new polytypes B_2S-2H and 12R and their crystal domain structures

Many selected-area electron diffraction patterns were obtained by changing the incident-beam direction. Fig.4(a)-4(c) show the SEAD patterns taken from the two new polytypes B_2S-2H and B_2S-12R with the electron beam respectively parallel to $[\bar{1}2\bar{1}0]$ (Fig.4(a-b)) and [1100] (Fig.4(c)) direction and their corresponding high-resolution images. It can be determined that B_2S-2H belongs to the hexagomal system (Fig. 4(a),4(c)), the possible space groups are $P6_3/mmc$, $P6_3mc$ or $P\bar{6}2c$, with the cell parameter c=3.768nm (see Table1). The high-resolution structural image (Fig.4(a)) with $[\bar{1}2\bar{1}0]$ orientation shows that a repeated period is composed of the two structural unit layers (1.884nm), that is the height of a cell c=1.884nm x 2 = 3.768nm. Thus this new polytype is named B_2S-2H. Fig.4(c) shows the antiphase domains and disordered crystal domains (in area "D") with the B_2S-2H structure. The spots of l=2n(n=0,1,2,.....) on the 0001 spot line in Fig.4(c) are caused by multiple diffraction, and should be extinction points, and the thicker white fringes in Fig.4(c) are the diffraction effect formed by the 0002 spot.

Fig.4(b) shows that the new polytype B_2S-12R belongs to the trigonal system with rhombohedral symmetry, the possible space group is R3, $R\bar{3}$ R32, R3m or $R\bar{3}m$, the height of a cell c=1.884nm x 12 = 22.608nm is composed of twelve structural unit layers (1.884nm), a subcell c'=1.884 x 4 = 7.536nm is composed of four structural unit layers.

Acknowledgements

The project was supported by the National Natural Science Foundation of China (No.49102019) and in part by the laboratory of Atomic Imaging of Solids, Institude of Metal Research, Academia, Sinica.

REFERENCES

1. R.J. Angel, Polytypes and polytypism, *Zeir. Krist.*, **176**, 193-204(1986).
2. G.Donnay and J.D.H. Donnay. The crystallography of bastnaesite, parisite, roengtenite and synchisite. *Am. Mineral.*, **38**, 932-963(1953).
3. G.Donnay,Roengtenite,a new mineral from Greenland, *Am. Mineral.*,**38**,868-870(1953).
4. G.Donnay, The polycrystal, a product of syntaxic intergrowth: *A.C.A. meeting, Ann Arbor Mich*, June 22-26(1953).
5. H. J. Fan and F.H. Li, Intergrowth and disorder stacking in the crystal structure of alkaline-earth-cerium fluorocarbonate minerals. *Acta physica Sinica*, **31**, 680-684(1982).
6. J. J. Glass and R.G. Smalley, Bastnaesite. *Am. Mineral.*, **30**, 601-615(1945).
7. J.V. Landuyt and S. Amelinckx, Multiple beam direct lattice imaging of new mixed-layer compound of the bastnaesite-synchisite series. *Am, Mineral.*, **60**, 351-358 (1975).
8. A.A. Levinson, A system of nomenclature for rare-earth minerals. *Am. Mineral.*, **51**, 152-157(1966).
9. F. H. Li and H.J. Fan , Electron diffraction and lattice image study on cebaite. *Acta Physica Sinica*, **31**, 206-1214 (1982).
10. F. Michael, Relative proportions of the lanthanides in minerals of the bastnaesite group. *Candian Mineralogist*, **16**, 361-363(1978).
11. Y.X. Ni, J. M. Hughes and A.N. Mariano, The atomic arrangement of bastnaesite-(Ce), $Ce(CO_3)F$, and structural elements of synchisite-(Ce), roentgenite-(Ce) and parisite-(Ce). *AmMineral.* **78**,415-418(1993).
12. I. Ofdetal, Zur Kristallstruktur von bastnaesite . *Zeits, Krist.* **78**, 462-469(1931a)

Proc. 30th Int'l. Geol. Congr., Vol. 16, pp. 59-65
Huang Yunhui and Cao Yawen (Eds)

Thermodynamic Properties of Minerals: the Application of Kieffer's Model

JI-AN XU
Institute of Earth Sciences, Academia Sinica, Nankang PO Box 1-55, Taipei City, Taiwan 11529

Abstract

Information on thermodynamic properties of minerals at various temperature, pressure conditions, are important in the study of Earth's sciences. Kieffer's model is, in one hand, a good and powerful tool for determination on these properties of various minerals. In the other hand, the studies using various optical spectroscopic methods grow rapidly in Earth sciences, the usage of increasing data is also questionable. The Kieffer's model obviously provides an area where these data are needed badly. Based on this understanding, we performed the Raman measurements on corundum and grossular up to 24 and 42 GPa pressures, respectively, and found that Kieffer's model can be used to determine various thermodynamic properties not only in ambient conditions, but also in high pressure and high temperature conditions. Therefore, the use of Kieffer's model will possibly shed a light to the understanding on the structure and movement in the deep interior of the Earth.

Keywords: Mineral, Thermodynamics, Kieffer's model, Corundum, Grossular

INTRODUCTION

Thermal energy supplied by a heat source is absorbed by various vibrational modes in a solid. Although Raman (or infrared) spectroscopic studies provide information only on the Raman-active (or infrared-active) lattice vibrations, understanding of the thermodynamic properties of the solid can be obtained by using a suitable model on these modes. However, the classic models such as the famous Debye or Einstein model are confirmed to be too simple to deal such a problem. Thus, Kieffer's model was proposed [1-5]. Recently, we have found an useful relationship to process different modes in the optic continuum in this model and used such a model on minerals at high-pressure and high-temperature environments successfully [6, 7].

Since last decade, using the diamond anvil cell Raman and infrared spectroscopic studies on various minerals at high pressures have grown up rapidly [6 - 9], and thus the increasing spectroscopic data were produced. Kieffer's model provides an suitable way to use such data. More information on the Earth's materials in the environment of the deep interior of the Earth could shed a light to the understanding on the status and dynamical processes of the Earth's interior. In the present work, a brief review on the model and its application concerning corundum and grossular are given.

KIEFFER'S MODEL

Kieffer's model has been used successfully in the calculation of the thermodynamic parameters of various minerals since it was first proposed more than ten years ago [1-5]. The fundamental basis of this model is the partitioning of the contribution of lattice vibrations by using three frequency distributions for three mode groups: the acoustic, the optic continuum, and the Einstein modes. The acoustic modes are determined from three acoustic (a compressional and two shear) vibrations in the solid by either of acoustic or optical measurement. The other two kinds of modes are measured by mainly optical methods (infrared absorption, Raman scattering ......). The frequencies of optic continuum normally distribute in the frequency range lower than 1000 cm^{-1}. The Einstein modes are defined as those vibrations that can be considered as the internal modes, such as molecular-like vibrations. and their frequencies are much higher than those observed in the Debye frequency range, which is related to the Si-O and O-H stretching vibrations in silicate and hydrous minerals, respectively [4-6].

According to Kieffer's model, all the thermodynamic parameters are expressed as the summation of the contribution from these three groups of modes. As an example, if we express the frequencies of various groups of modes ω_i (the frequency of the ith acoustic mode), ω_l and ω_u (the lower and upper limits of optic continuum) and $\omega_{e,i}$ (the frequency of the ith Einstein mode) as following variables: $x_i = h\omega_i/kT$; $x_l = h\omega_l/kT$ and $x_u = h\omega_u/kT$; and $x_{e,i} = h\omega_{e,i}/kT$, h is Plank's constant, and T is (absolute) temperature, the heat capacity at constant volume C_v and entropy S of a solid are calculated by

$$C_v = (3N_Ak/s)\ \Sigma\ S_1(x_i) + 3N_Ak\ [1-(1/s)-q]\ K_1(x_l, x_u) + 3N_Akq\ \Sigma\ E_1(x_{e,i}) \quad (1)$$

$$S = (3N_Ak/s)\ \Sigma\ S_2(x_i) + 3N_Ak\ [1-(1/s)-q]\ K_2(x_l, x_u) + 3N_Akq\ \Sigma\ E_2(x_{e,i}) \quad (2)$$

where N_A is Avogadro's number, k is Boltzmann's constant, q is the proportion of Einstein vibrational modes, s is the total number of atoms in a Bravais unit cell, and the integrals S_1, S_2, K_1, K_2, E_1 and E_2 are Kieffer's integrals, their definitions and corrected forms can be found elsewhere [3, 6].

From the equations listed above, it is evident that in this model the contributions from the acoustic and Einstein modes are taken into account individually, whereas the contributions from the optic modes are considered as an uniform packet, the optic continuum [3, 6]. The thermodynamic properties that can be derived from these vibrational modes depend only on the lower and upper frequencies (ω_l, ω_u, thus x_l, x_u). Although this model is used successfully in ambient conditions for a large numbers of minerals, there are several thermodynamic parameters, such as the Gruneisen parameter γ, that can be individually related to each mode, it is thus important to understand the contribution of each mode in the optic continuum. For such a purpose, we have shown that if the whole frequency range in optic modes is divided into several subintervals, Kieffer's integral, K_j, equals the weight sum of the same integral over the subintervals [6] as

$$K_j = (\Sigma\, w_i\, K_{j,i})/w \tag{3}$$

where the summation is taken over all subintervals (i = 1 to n), and $w = x_u - x_l$, $w_i = x_{i+1} - x_i$. The equation (3) is more important when the Kieffer's model is used in the extreme conditions, such as high-temperature and high-pressure environments because the contribution of the individual mode with temperature or pressure differs. In previous work [6], we have used the Gruneisen parameter γ at ambient conditions as a constrain to select a suitable combination of this optic continuum, thus more confidence can be obtained in the calculation of various thermodynamic properties at such environments as in the deep interior of the Earth.

THERMODYNAMIC PROPERTIES OF CORUNDUM

Using Kieffer's model as described above and experimental Raman scattering data up to 24 GPa, thermodynamic properties, such as heat capacity at constant volume C_v, or the heat capacity at constant pressure $C_p = C_v\,(1+\alpha\gamma T)$, entropy S, the coefficient of thermal expansion α and the Gruneisen parameter γ of various temperature-pressure conditions can be determined [6]. The experimental value (1.22 - 1.33) of Gruneisen parameter γ at ambient conditions [10, 11], was used as a constrain to select a suitable frequency distribution for the optic continuum. In the case of corundum, the value was thus calculated from the most feverable distribution (using the total of the optical frequencies measured from Raman spectroscopic method [6] in Equation 3) as 1.12. It is much better than the value of 0.98 obtained from the simple frequency distribution (it is defined as such a distribution: only the lower and upper frequencies are considered).

Using Kieffer's model, thermodynamic properties at various conditions can be calculated. However, only experimental thermodynamic data at atmospheric pressure are available, so in Figure 1 (A) and (B), we have plotted the calculated C_p, S, α and γ_0 at various temperatures, a good agreement between the calculated curve and experiment data is shown.

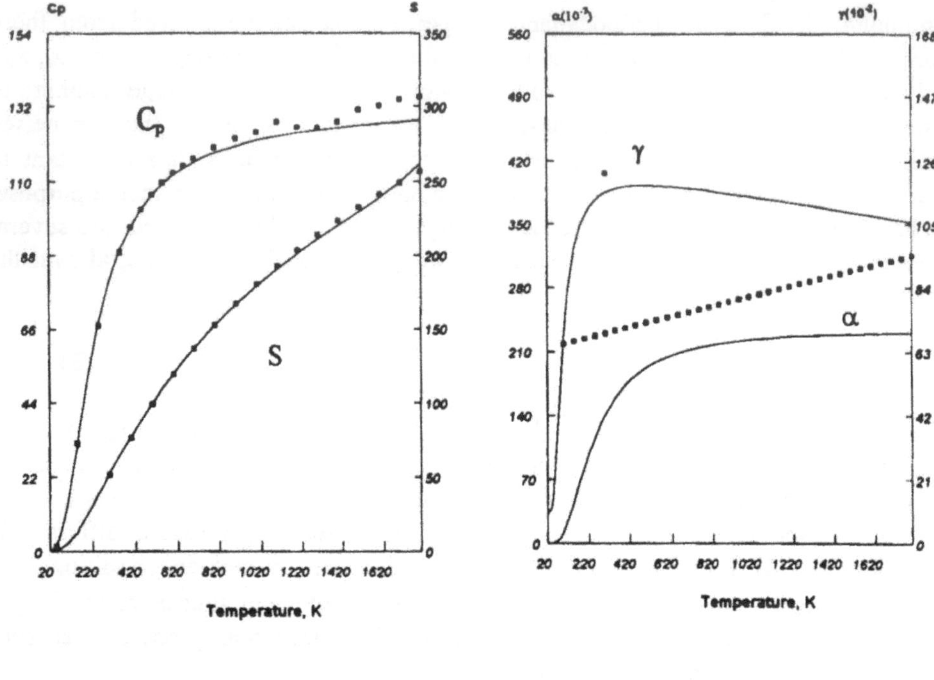

Figure 1 (A)

Figure 1 (B)

Figure 1 (A):Comparison of the calculated (solid lines) and experimental (open squares) heat capacities (Cp, Unit on the left: J/Mol•K) and entropies (S, Unit on the right: J/Mol•K) of corundum at atmospheric pressure and temperatures up to 1800 K.

Figure 1 (B):Comparison of the calculated (solid lines) and experimental (open squares) coefficient of thermal expansion (α, Unit on the left: 10^{-7}/K) and Gruneisen parameter (γ, Unit on the right: 10^{-2}) for atmospheric pressure and temperatures up to 1800 K. Thermal expansion approaches zero when absolute temperature is close to zero.

THERMODYNAMIC PROPERTIES OF GROSSULAR

As an other example, we have performed similar process on grossular using the high pressure experimental (X-ray diffraction and Raman spectroscopic) data up to 42 GPa [7]. Different with the case of corundum, in grossular, we found that the most feverable distribution of the optic continuum is the simple frequency distribution. As in the case of corundum, only the experimental data at ambient conditions are available, the comparison between the calculated curve and experimental data is shown in Figure 2 (A) and (B).

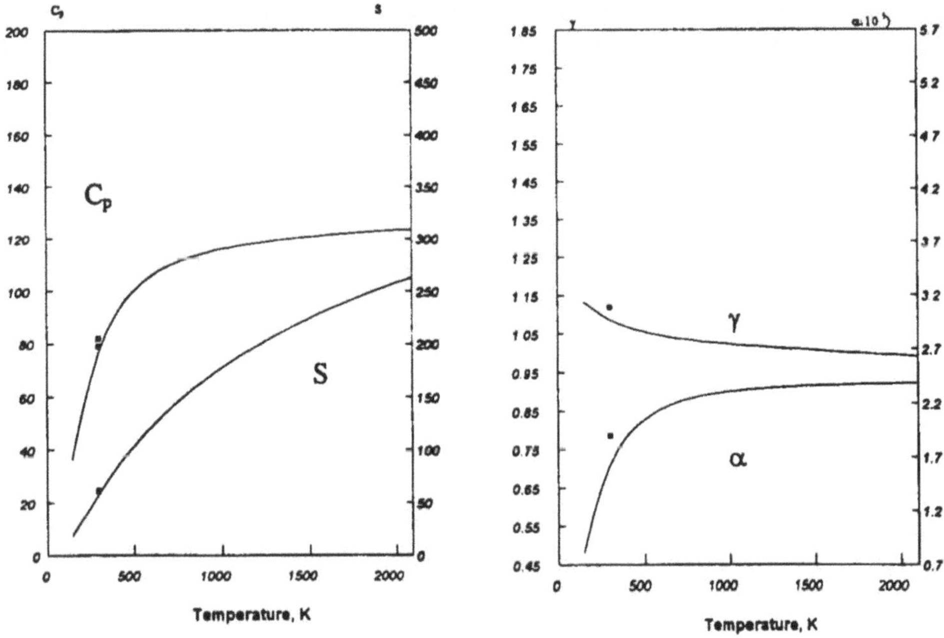

Figure 2 (A):

Figure 2 (B)

Figure 2 (A):Comparison of the calculated (solid lines) and experimental (open squares) heat capacities (Cp, Unit on the left: J/Mol•K) and entropies (S, Unit on the right: J/Mol•K) of grossular at atmospheric pressure and temperatures up to 1800 K.

Figure 2 (B):Comparison of the calculated (solid lines) and experimental (open squares) coefficient of thermal expansion (α, Unit on the right: 1/K) and Gruneisen parameter (γ,Unit on the left) for grossular at atmospheric pressure and temperatures up to 1800 K.

DISCUSSION

As described, Kieffer's model is so effective to determine thermodynamic properties at various pressure/temperature conditions even using a part of vibrational modes which were obtained by any optical spectroscopic method. Using such data, thermodynamic properties of minerals in various environments can be predicted.

The thermal properties in the interior of the Earth are directly related to the heat capacities of its constituent minerals at various high-temperature and high-pressure environments. Although for a long time, the classic point of view has been used to deal with the problem, one question has always remained: to what extent does high pressure affect the thermal properties of minerals? [10]

A quite good estimation on the values of temperature and pressure in the interior of the Earth are given as shown in Figure 3 [12]. In the figure, there are also shown the calculated heat capacity C_v of corundum at such environments according to the Kieffer's model. It is clearly showing that the classic Dulong-Petit limit is a good approach for the heat capacity of corundum beneath a depth of 50 km. Such an approach will be more feverable for other materials which Debye temperature are lower than corundum as most Earth's materials does.

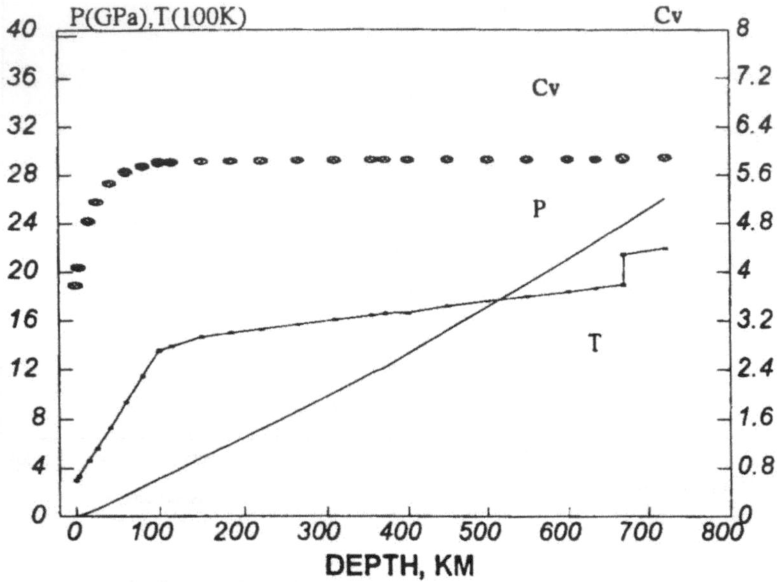

Figure 3: Heat capacity Cv (Unit on the right: Cal Mol•K) of corundum approaches its classic (Dulong-Petit) value in the interior of the Earth at depths of 50 km or more (Unit on left of P: GPa; T: 100K).

Author gratefully acknowledges the editorial help from Mr. Xu, Żuming from the Institute of Geochemistry, Chinese Academy of Sciences, Guiyang, and the support from the National Natural Scientific Foundation, Beijing and the National Science Council, Taipei, through NSC-84-0202-001-015.

REFERENCE:

1. S. W. Kieffer, Thermodynamics and lattice vibrations of minerals: I. Mineral heat capacities and their relationships to simple lattice vibrational modes, *Rev. Geopys. Space Phys.*, 17, 1-19 (1979).

2. S. W. Kieffer, Thermodynamics and lattice vibrations of minerals: II. Vibrational characteristics of silicates, *Rev. Geophys. Space Phys.*, 17, 20-34 (1979).

3. S. W. Kieffer, Thermodynamics and lattice vibrations of minerals: III. Lattice dynamics and an approximation for minerals with application to simple substances and framework silicates, *Rev. Geophys. Space Phys.*, 17, 35-59 (1979).

4. S. W. Kieffer, Thermodynamics and lattice vibrations of minerals: IV. Application to chain and sheet silicates and orthosilicates, *Rev. Geophys. Space Phys.*, **18**, 862-886 (1980).

5. S. W. Kieffer, Thermodynamics and lattice vibrations of minerals: V. Applications to phase equilibra, isotopic fractionation, and high pressure thermodynamic properties, *Rev. Geophys. Space Phys.*, **20**, 827-849 (1982).

6. J. Xu, E. Huang, J. Lin. and L. Xu, Raman study at high pressure and thermodynamic property of corundum: application of Kieffer's model, *Amer. Mineal.*, **80**, 1159-1167 (1995). .

7. J. Xu, E. Huang, R. Chen, R. M. Hazen, H. K. Mao and P. M. Bell, X-ray diffraction and Raman spectroscopic studies on grossular at high pressure - Application of Kieffer's model, *in preparation* (1996).

8. J. Xu, T. C. Hoering, H. K. Mao, P. Wong and P. M. Bell, Fourier transform infrared measurements of methane at high pressure, *Carnegie Institution of Washington (CIW) Yearbook*, 81, 389-390 (1982).

9. J. Xu, H. K. Mao, K. Weng and P. M. Bell, High pressure, Fourier-transform infrared spectra of forsterite and fayalite, *CIW Yearbook*, 82, 350-352 (1983).

10. D. L. Anderson, in *Theory of the Earth*, Blackwell Sci. Pub., Boston, p. 85 and p. 112 (1989).

11. E. Schreiber and O. L. Anderson, Pressure derivatives of the sound velocities of polycrystalline alumina, *J. Am. Ceram. Soc.*, 49, 184-190 (1966).

12. J. -P. Poirier, in *Introduction to the Physics of the Earth's Interior*, Cambridge University Press, Cambridge, p. 211-214 and p. 236-237 (1991).

Proc. 30th Int'l. Geol. Congr., Vol. 16, pp. 67-74
Huang Yunhui and Cao Yawen (Eds)
VSP 1997

The Influence of Water on the Partitioning of Fe and Mg between Olivine and Orthopyroxene *

XU YOUSHENG, XIE HONGSEN, SONG MAOSHUANG, ZHANG YUEMING AND GUO JIE
Institute of Geochemistry, Chinese Academy of Sciences, Guiyang 550002

Abstract

A considerable number of experiments have been done on the partitioning of Fe and Mg between coexisting silicates and many important geothermometers have thus been established. Nevertheless, little has been reported on the partitioning of Fe and Mn between silicate phases. This paper studies the influence of water on the partitioning of elements at 2.0, 4.0 GPa and 1200-1400°C. The research results show that the composition can cause the variation of $\ln K_D$, the addition of water in small amounts into the experimental system would also lead to the obvious increase of $\ln K_D$. But when the content of water in the system is increased, $\ln K_D$ will decrease correspondingly. This indicates that the influence of water on the partitioning of Fe and Mg between olivine and orthopyroxene does not show a simple positive correlation. Experiments demonstrated that when a certain amount of water is present in the system, $\ln K_D$ will increase with increasing pressure; $\ln K_D$ still shows a linear correlation with 1/T. Therefore, a geothermometer can be established on the partitioning coefficients of Fe and Mg for olivine and orthopyroxene.

Key words: high temperature; high pressure; element partition; water

The exchange of Fe and Mg between the coexisting silicates provides the basis for the establishment of a series of important geothermometers and geobarometers [1, 2]. Recent studies indicate that in many cases reequilibrium can be established between Fe and Mg after they have experienced exchange [3, 4]. In order to address this problem in a comprehensive way, more experimental data are needed on the exchange of Fe and Mg between the coexisting silicates. Although research work on the element partitioning between minerals has never been interrupted, little is known about the influence of aqueous fluid on the partitioning of elements between the coexisting silicates. Whether the fluid phase is present or not would have a reat influence on the reaction rate during the experiment as well as on the reaction mechanism [5]. Because of dissolution and participation, even the presence of a small

* This project was financially supported by the National Natural Science Foundation of China.

amount of water can accelerate the reaction rate up to 8-10 orders of magnitudes. According to Rubie, Luttge and Metz [6, 7] in the presence of aqueous fluid the reaction will be effected through dissolution and precipitation. The above authors only considered the influence of fluid on the reaction rate as well as the reaction mechanism, but few of them have paid sufficient attention to the influence of the fluid on the partitioning of elements between minerals.

Recently, we have done research work on the influence of water on the exchange of Fe and Mg between olivine and orthopyroxene. The partitioning of Fe and Mg between the coexisting minerals --- olivine and orthopyroxene can be expressed as following [8-10].

$$\frac{1}{2}Mg_2SiO_4 + \frac{1}{2}Fe_2S_{i2}O_6 = \frac{1}{2}Fe_2SiO_4 + \frac{1}{2}Mg_2S_{i2}O_6 \tag{1}$$

olivine orthopyroxene olivine orthopyroxene

the equilibrium constant for the above expression $K_{(T,P)}$ is:

$$K_{(T,P)} = \frac{(a^{ol}{}_{Fe2SiO4})^{1/2}(a^{opx}{}_{Mg2Si2O6})^{1/2}}{(a^{ol}{}_{Mg2SiO4})^{1/2}(a^{opx}{}_{Fe2Si2O6})^{1/2}} \tag{2}$$

where $a^{\phi}{}_i$ refers to the activity of the constituent i in the phase ϕ, which is related with the mole fraction $X^{\phi}{}_i$:

$$a^{\phi}{}_i = (X^{\phi}{}_i \cdot \gamma^{\phi}{}_i)^v \tag{3}$$

By dividing $K_{(T,P)}$ into two parts, K_D and K_γ; we have:

$$K_{(T,P)} = K_D \cdot K_\gamma \tag{4}$$

where

$$K_D = \frac{X^{ol}{}_{Fe} \cdot X^{opx}{}_{Mg}}{X^{ol}{}_{Mg} \cdot X^{opx}{}_{Fe}} \tag{5}$$

where

$$K_\gamma = \frac{\gamma^{ol}{}_{Fe} \cdot \gamma^{opx}{}_{Mg}}{\gamma^{ol}{}_{Mg} \cdot \gamma^{opx}{}_{Fe}} \tag{6}$$

where $X^{\phi}{}_{Mg} = \left[\frac{n_{Mg}}{n_{Mg}+n_{Fe}}\right]_{\phi}$, $X^{\phi}{}_{Fe} = \left[\frac{n_{Fe}}{n_{Mg}+n_{Fe}}\right]_{\phi}$, where ϕ refers to olivine or orthopyroxene and γ indicates the activity coefficient.

I. EXPERIMENTAL METHOD

The experiments were done on a YJ-3000 press fitted with a wedge type cubic anvil[11]. Pyrophyllite was adopted as the pressure-transfer medium. In the experiment a graphite heater was used with a Pt-$Pt_{90}Rh_{10}$ thermocouple. No correlation was made for the effect of pressure on the thermocouple emf. Iherzolite samples were taken from Hannoba, Hebei Province, which are composed mainly of olivine (70%, Fo_{90}), clinopyroxene (15%) and orthopyroxene (10%). The samples were ground as fine as 200 mesh, followed by the addition of water in different proportions. Then the samples were put into the Pt tubes (5mm in diameter, 3mm in length). As a result of the presence of water, the reaction rate can be greatly accelerated. The samples were kept at equilibrium for 1-5 hours under different temperature and pressure conditions, followed by quenching in high pressure. Within one minute the temperature of the samples can go down to 100°C or below.

The experimental products are composed dominantly of euhedral and semi-euhedral olivine and orthopyroxene. Some clinopyroxene was also observed in the experimental products of lherzolite samples. The minerals grains are almost isogranular in size. Olivine crystals are measured at 10-50 μm in diameter and orthopyroxene crystals at 10-60 μm in diameter. The experimental products were divided into two portions, with one portion for microscopic observation and electron microprobe analysis and the other portion for X-ray diffraction analysis. The results show that most of the experimental resultants contain a small amount of clinopyroxene only, but are composed largely of olivine and orthopyroxene. The electron microprobe analysis was carried out under the condition: voltage 20kV and current 220mA. The experimental resultants of the samples synthesized from chemical reagents should at least be analyzed on 7-9 points where olivine and orthopyroxene are present; those of lherzolite as the starting material on 3-5 points where olivine and orthopyroxene are present.

II. EXPERIMENTAL RESULTS

The partitioning coefficients of Fe and Mg between olivine andorthopyroxene obtained in the experiments at 2.0-4.0 GPa and 1200-1400°C are presented in Table 1.

1. The influence of water content

Shown in Fig.1 is the plot of $\ln K_D$ vs. X_{Fe}^{ol} under the same temperature and pressure conditions. The experimental results show that $\ln K_D$ is strongly controlled by the content of water in the system (Figs.1 and 2). When only a small amount of water is present in the system, $\ln K_D$ is obviously greater than that in the dry system, and then it will decrease with increasing content of water. The addition of water in the system will lead to the variation in composition of the olivine, but the orthopyroxene is less variable in composition (Fig.3). This implicates that the influence of water on the partitioning of Fe and Mg between olivine and orthopyroxene is by no means a positive or a negative correlation.

Table 1. Experimental conditions and electron microprobe analyses

Run No.	Water (%)	P (GPa)	T (℃)	Equil.time (h)	X^{ol}_{Fe}	X^{opx}_{Fe}	K_D	$\ln K_D$
Protolith					0.100	0.098	1.0294	0.0290
52/2		3.0	1200	5	0.097	0.096	1.0112	0.0111
54	1.0	3.0	1200	2	0.100	0.094	1.0701	0.0678
55	3.0	3.0	1200	2	0.101	0.096	1,0551	0.0536
56	5.0	3.0	1200	2	0.105	0.105	1.0022	0.0022
57	3.0	3.0	1300	1.5	0.088	0.087	1.0130	0.0129
60	3.0	3.0	1400	1	0.086	0.090	0.9547	-0.0464
59	3.0	2.0	1200	1.5	0.090	0.088	1.0436	0.0427
61/2	3.0	4.0	1200	1.5	0.107	0.099	1.0986	0.0940

Note: Electron microprobe analysis by Wang Mingzai of the Institute of Geochemistry, Chinese Academy of Sciences

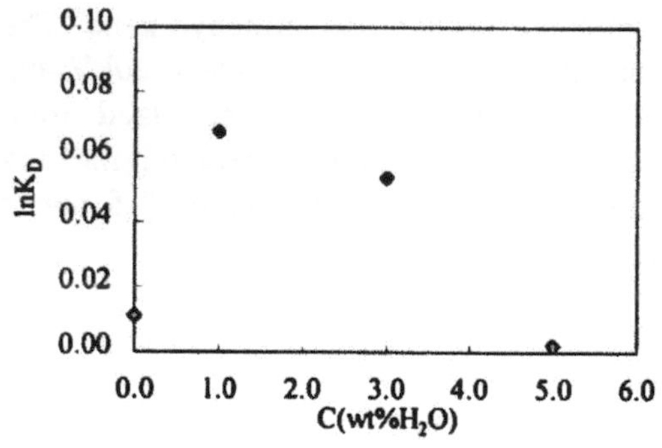

Fig.1. Relationship between $\ln K_D$ and water content at 3.0 GPa and 1200°C.

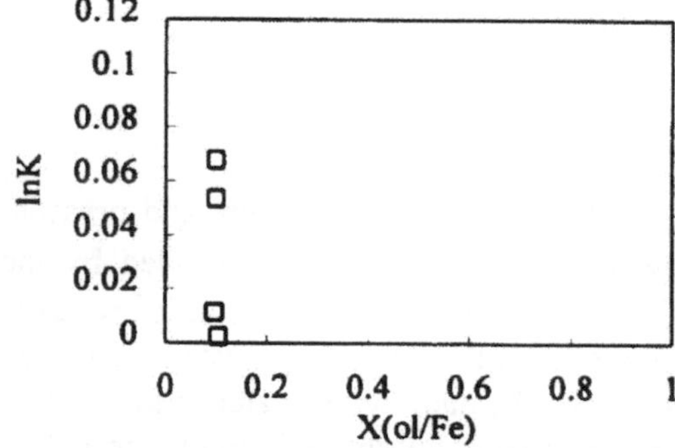

Fig.2. Relationship between $\ln K_D$ and olivine composition at 3.0 GPa and 1200°C.

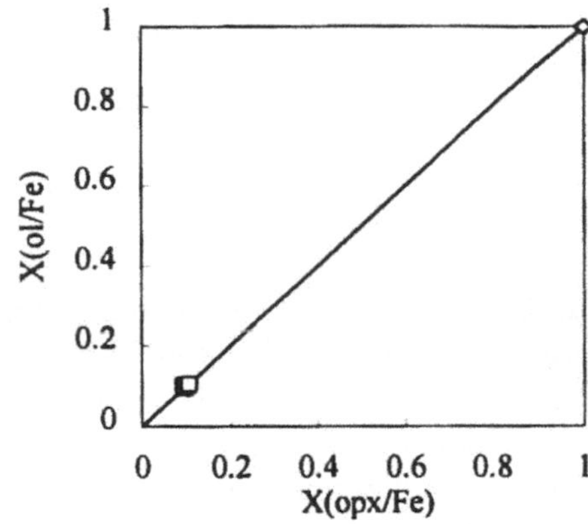

Fig.3. Relationship between the composition of olivine and that of orthopyroxene at 3.0 GPa and 1200°C.

2. Influence of temperature

A linear relationship between $\ln K_D$ vs. 1/T had been confimed theoretically and experimentally by O'Neil and Wood [1]. Same relationships can be found in our experiments in both of a dry system and a system which water accounts for 3%. Such a relationship for the latter case is shown in Figure 4, the $\ln K_D$ decreases linearly with increasing temperature (the 1/T decreases)..

$$\ln K_D = \frac{1225.9}{T} - 0.7747 \quad \text{(Unit: K for T)} \tag{7}$$

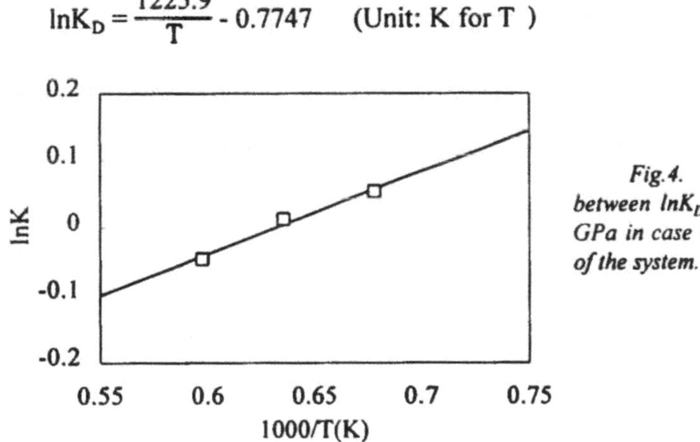

Fig.4. Reciprocal relationship between $\ln K_D$ and temperature at 3.0 GPa in case water accounts for 3wt% of the system.

It can be seen that when water accounts for 3%, the partitioning of Fe and Mg between olivine and orthopyroxene is still fit to the relationship in the case of no water present. When $\ln K_D$ is obtained, the equilibrium temperature between olivine and orthopyroxene at 3.0 GPa can be worked out in accordance with the following formula:

$$T\,(K) = \frac{1225.9}{\ln K_D + 0.7749} \tag{8}$$

3. The influence of pressure

When the pressure dependence of reaction equilibrium is estimated on the basis of the thermodynamic principles, the influence of pressure on the reaction could be easily neglected in case the pressure is as high as up to several kilobars. The variation of reaction volume (ΔV_S) can also be calculated in terms of the unit cell or density measurements [12]. Since they are small and are in equilibrium with each other, the thermal expansion and compression coefficients can be neglected. Then we have [5]:

$$\int_{P_1}^{P_2} \ln K_{(T,P)} = -\frac{\Delta V_S}{RT} \int_{P_1}^{P_2} PdP \tag{9}$$

where P_1 =1 bar, and P_2 refers to the experimental pressure. From the above expression, we can seen that with increasing pressure $K_{(T,P)}$ tends to extremely slightly decrease, and K_D will increase with increasing pressure (Fig.5).

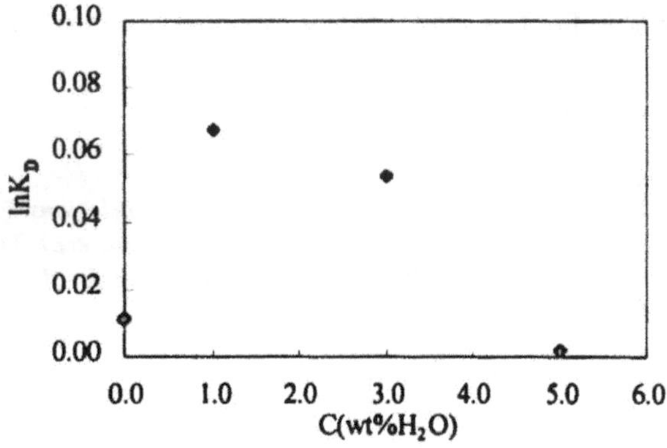

Fig.5. Relationship between $\ln K_D$ and pressure at 1200°C in case water accounts for 3wt% of the system.

III. CONCLUSIONS

From experimental research on the influence of water on the partitioning of Fe and Mg between olivine and orthopyroxene at 2.0-3.0 GPa and 1200-1400°C, the following conclusions can be drawn:

(1) The influence of water on the partitioning of elements does not show a simple positive correlation. The addition of a small amount of water in the system will lead to an increase in $\ln K_D$. When the water content increases in the system, $\ln K_D$ will show an opposite tendency of decreasing.

(2) In case the water content is the same in the systems, $\ln K_D$ will increase with increasing pressure.

(3) In case the water content is the same in the systems, $\ln K_D$ is linearly correlated with 1/T. At 3.0GPa, when water accounts for 3.0wt% of the system, the forming temperature of the coexisting minerals olivine and orthopyroxene can be calculated in accordance with the following expression:

$$T\,(K) = \frac{1225.9}{\ln K_D + 0.7749} \qquad (10)$$

Therefore, in case the system contains a certain amount of water, a thermometer can be established on the basis of the partitioning coefficients of Fe and Mg for olivine and orthopyroxene.

Acknowledgements: The authors wish to thank Prof. Hou Wei, Senior Engineer Xu Huigang and Senior Engineer Wang Sanxue for their help in this work.

REFERENCES:

1. H. St. C. O'Neill and B. J. Wood, An experimental study of Fe-Mg partitioning between garnet and olivine and its calibration as a geothermometer, *Contrib. Mineral. Petrol.*, **70**, 59-70 (1979).

2. S. R. Bohlen and A. L. Boettcher, Experimental investigations and geological applications of orthopyroxene geobarometry, *Am. Mineral.*, **66**, 951-964 (1981).

3. R. H. Nafziger and A. Muan, Equilibrium phase compositions and thermodynamic properties of olives and orthopyroxenes in the system MgO - "FeO"- SiO_2, *Am. Mineral.*, **52**, 1364-1385 (1967).

4. D. R. M. Parttison, Are reversed Fe-Mg exchange and solid solution experiments really reversed?, *Am. Mineral.*, **79**, 936-950 (1994).

5. D. R. M. Parttison and R. C. Newton, Reversed experimental calibration of the garnet-clinopyroxene Fe-Mg exchange thermometer, *Contrib. Mineral. Petrol.*, **101**, 87-103 (1989).

6. D. C. Rubie, The catalysis of mineral reactions by water and restrictions on the presence of aqueous fluid during metamorphism, *Mineral. Mag.*, **50**, 399-415 (1986).

7. A. Luttge and D. Metz, Mechanism and kinetics of the reaction: 1 dolomite + 2

quartz = 1 diopside + 2 CO_2: A comparison of rocksample and powder experiments, *Contrib. Mineral. Petrol.*, **115**, 155-164 (1993).

8. R. Y. Zhang et al., in *Mineralogical Geothermometer and Geobarometry*, Science Press, Beijing(1980).

9. V. V. Seckendorff and H. St. C. O'Neill, An experimental study of Fe-Mg partitioning between olivine and orthopyroxene at 1173, 1273 and 1423K and 1.6GPa, *Contrib. Mineral. Petrol.*, **113**, 196-207 (1993).

10. M. Koch-Muller, L. Cemic and K. Langer, Experimental and thermodynamic study of Fe-Mg exchange olivine and orthopyroxene in the system MgO-FeO-SiO_2, *Eur. J. Mineral.*, **4**, 115-135 (1992).

11. H. S. Xie, W. S. Peng, C. L. Xue, Y. M. Zhang, H. G. Xu and G. K. Liu, Synthesis, infrared spectra and X-ray diffraction of Mg-Fe olivine, *Acta Minerologica Sinica*, **6**, 103-108 (1986).

12. R. A. Robie, in *Handbook of Physical Constants* (ed. S. P. Clark), Geol. Soc. Amer., p437-458 (1966).

Proc. 30th Int'l. Geol. Congr., Vol. 16, pp. 75-83
Huang Yunhui and Cao Yawen (Eds)

Mineralogical Mapping in Jiaodong Gold Province, China

CHEN GUANGYUAN YANG ZHUSEN WANG ZHIJING
China University of Geosciences, Beijing, 100083, P.R. China

Abstract

Mineralogical mapping is much less popular than geographical and geological mappings It is considered to be one of the most important methodological achievements in modern mineralogy In mineralogical mapping the mineralogical phenomena and their realities are qualified, quantified, located and orientated on the map by means of various mapping parameters Its theoretical foundation is Genetic Mineralogy, especially the Theory of Mineral Typomorphism and Its Six Principles It is helpful to the study of fundamental problems of geosciences like petrogenesis and metallogenesis and useful to the solution of practical problems like prospecting and exploitation of minerals History of mineralogical mapping in China is reviewed in short Functions of mineralogical mapping are summarized in detail from more than one hundred large-scale local mineralogical mappings made in 1984-1994 over more than twenty gold deposits and several small-scale regional mineralogical mappings made in 1995-1996 covering an area of 3396 km^2 with 26850 measurements of mineralogical parameters over the three major mineralized belts Three new mineralogical plane mappings at various levels of Jiaojia Mine representative of the wall-rock alteration type of deposits in northwestern Jiaodong Gold Ore Province and one new regional mineralogical mapping over the northern Jiaodong gold mineralized zone with high to superhigh grade of ores up to 20-30wt% are cited as examples In the four cited examples appearing frequency of P conduction type of thermoelectricity of pyrites is taken as mapping parameter with 5343 measurements

Keywords: mineralogical mapping, Jiaodong Gold Ore Province, historical review in China, frequency of P-type pyrites

INTRODUCTION

Most people know geographical and/or even geological map, but few people know mineralogical map fulfilled by means of mineralogical mapping.

Mineralogical mapping is considered to be one of the most important methodological achievements of modern mineralogy [3]. It is not only the continuation of and supplement to geological, lithological, tectonic, geophysical, geochemical, aerophotographic and remote-sensing as well as lithofacies-palaeogeographical mapping etc., but also the deepening, broadening and further developing of the above-mentioned mappings, representing one of the inexorable progressive trend of geological techniques not yet well-known enough.

In mineralogical mapping the mineralogical phenomena are qualified, quantified, located and orientated on the map by various mineralogical parameters. Its theoretical foundation is Genetic Mineralogy [6, 8, 12, 18], especially the Theory of Mineral Typomorphism, the key topic of Genetic Mineralogy [1, 2, 5, 6, 16, 17], and its six basic principles [6, 8-10, 16, 18].

It is not only helpful to the theoretical study of fundamental problems of geosciences like petrogenesis and metallogenesis, but also useful to the solution of practical problems such as prospecting, exploration and exploitation of mineral resourses.

In ancient China the primitive mineralogical mapping can be traced back to more than 1000 B.C. ago. According to records of 'Zhouli' (Rites of the Zhou Dynasy), formerly called 'Zhouguan' (Officials of the Zhou Dynasty), one of the famous ancient Chinese Classics dated 1055-1074 B.C., officers in charge of the mineral resources marked the localities and boundaries of gold, nephrite, tin(cassiterite), stone materials, cinnabar and malachite on maps, gave them to miners to use in mining, and patrolled the mines to inspect how the government orders of forbidding private exploitation were put into execution.

In modern China it was first started in the early 1920s by Li Siguang (J. S. Lee), the outstanding Chinese geologist, with mineral assemblage as mapping parameter on the zonation of the mineralogical halo of the outer contact zone around the stock of the so-called Fangshan Granite in the Western Mountains on the outskirts of Beijing.

In the early 1960s, under supervision of Chen Guangyuan (K. Y. Chen) mineralogical profiles were made by means of stress effects on minerals of the Shachang BIF deposit of Miyun County on the outshirts of Beijing, disclosed its structure being a southward plunging isoclinal synclinorium instead of the then current speculation of monocline, and thus led to a considerable enlargement of its prospect and reserve [6, 18].

In the middle of the 1970s the application of the genetic-structural model of the Shachang BIF deposit to the study of Qianan-Luanxian BIF orefield in Jidong (eastern Hebei Province) resulted in the evaluation of the latter to be the second largest in China, next only to the Anshan iron ore base in Liaodong (eastern Liaoning Province), which was proved to be true by later exploration.

In the end of the 1970s a number of phylogenic mineralogical columnar sections of almandites, biotites,chlorites and amphiboles based on mineral typomorphism were made by the first author and his collaboraters in Gongchangling Iron Mine, Liaoyang County, eastern Liaoning Province, revealing the stratified iron ores, whether low- or high-graded, magnetite- or hematite-banded, or silica- or carbonate-banded, to be all of them the evolutional products of different evolutional stages of the Neo-Archaean greenstone belt of the Anshan Group [4].

Owing to its remarkable effectiveness in solution of the theoretical and practical problems, mineralogical mapping was systematically and purposefully concentrated on gold deposits in Jiaodong Gold Ore Province, the topmost gold-producing region in both ancient and modern China.

GENERAL REVIEW

Since the eastern half of Shandong Province in North China is occupied by the Shandong or Jiaodong Peninsula, it is called Jiaodong in brief. Its gold-mining history started about 1400 years ago in 598 A.D. during the early Sui Dynasty. Its gold production reached 89.5% of the whole country in 1078 A.D. in the middle of Song Dynasty. In modern China, 40% of the country's gold production came from Jiaodong in 1995 while in 1994 one fourth of gold production of the whole country came from Yantai Municipality, the most productive municipality of gold in Jiaodong.

More than twenty individual gold deposits in Jiaodong Gold Ore Province were in the late ten years (1984-1994) mapped mineralogically mostly on the scale of 1:1000-1:5000 under the first author's supervision [8, 15, 17, 19]. They include Linglong-Dongshan , Linglong-Xishan , Jiehe , Fujia, Yuantang, Shilipu, Luofeng, Xiadian and Lingshangou of Zhaoyuan Municipality, Jiaojia, Xincheng and Sanshandao of Laizhou Municipality , Denggezhang , Jinniushan and Xiayucun of

Muping County, Jinqingding, Lujia, Chujiagou and Tongling of Rushan County.

Besides several regional mineralogical mappings on the scale of 1:200,000-1:300,000 in the northwestern, northern and northeastern mineralized belts of Jiaodong were proceeded in the recent two years (1994-1995). Among them the three regional mineralogical maps with thermoelectric P-type % of pyrites as mapping parameter covering an area of 1350 km^2, 246 km^2 and 1800 km^2 and comprising 63, 28 and 59 deposits and occurrences respectively were based on 26850 thermoelectric measurements of pyrite grains in total [17].

Units or parameters used for mineralogical mapping include mineral assemblages, mineral species, mineralization stages, single forms and combinate forms, crystallomorphological ballotage scores (with distinction between F-faced and S-faced cubes as well as S-faced and F-faced pentagonal dodecahedrons), zonal structure, major elements and their ratios, minor components, isotopic ratios, cell parameters, infrared spectral peak values, thermoelectric coefficients, frequency of thermoelectric conduction types, thermoelectric ballotage scores, integral intensity of natural thermoluminescence, susceptibility, resistivity, induced polarization potential, decrepitation activity of fluid inclusions, decrepitation temperature of fluid inclusions, homogenization temperature of fluid inclusions, pressure of fluid inclusions, average diameter of fluid inclusions, distribution density of fluid inclusions, D_{H2O} of fluid inclusions, D_{CO2} of fluid inclusions, $D_{CO2/H2O}$ of fluid inclusions.

From mineralogical mapping by means of the above-mentioned mapping units or parameters a lot of information can be extracted like the following, namely genetic type of the deposit, delimitation of the orebody boundaries of the wall-rock alteration type of deposits, visible zonation of wall-rock alteration, invisible zonation of auriferous quartz veins, size or extent of the deposit, mineralization intensity or grades of the deposit, visible and invisible postore faults and their natures, degree of denudation of the deposit, ore-forming fissure systems, channelway of hydrothermal fluids, extension tendency and plunging direction of orebodies, positioning of ore shoots and ore nuggets, location of deep-seated and outlying orebodies in surroundings of individual deposits as well as prognostic occurrence of new deposits in orefields and ore belts.

It is thus a necessary profitable means for prospecting, exploration and exploitation of mineral resources. In exploited mines it is possible to increase the output by lowering the percentage of ore loss and ore dilution and to lengthen the mine life by enlarging the ore reserve and delivering the mines out of the crisis of their ore reserves.

Through practice of more than one hundred mineralogical mappings over more than twenty gold deposits and several regional mineralogical mappings in the three mineralized belts of Jiaodong Gold Ore Province in the past twelve years, it is found that polyparametric mineralogical mappings show very well correspondence of mineral typomorphism and provide very well the necessary mutual checkup of the precision of the measurements of the mapping parameters and the preciseness of the extracted information.

Four new mineralogical maps compiled in 1996 after the 30th International Geological Congress are presented in this article as examples. All of them are mapped with the frequency of P-type pyrites as mapping parameter which is one of the effective invisible typomorphic characteristic features of the most common and most important through minerals in gold deposits. All measurements were made by means of the portable RD_2 probe-type thermoelectrometer (China National Patent No. 882156802) designed and manufactured by Shao Wei in 1986.

LOCAL MINERALOGICAL MAPPING

The first three mineralogical maps (Fig. 1: a, b, c) were made at three successive levels in available galleries of the Jiaojia Gold Mine. They display not only the respective horizontal zonations individually but also the vertical variation of the deposit totally so as to give a stereo perspective view of the prospect of the deposit as a whole. Further information can be extracted in detail from the maps as the following.

1.The Jiaojia Fault dips to the northwest with low dip angles. This means that low-angle thrust is favourable for the production of fissure systems in wall rocks and the formation of wall-rock alteration type of gold deposit.

2. Both the principal and the subordinate mineralized zones represented by pyritization are roughly in parallel with the Jiaojia Fault. This means that the Jiaojia Fault is the ore-controlling fault.

3. However the isoline curves of frequency of P-type pyrites in the principal mineralized zone are not closed northwestward but cut through by the Jiaojia Fault. This means that the Jiaojia Fault is a long-lived fault, being both pre-ore and post-ore. The post-ore gouge is found to be 2.0-3.5 cm in thickness and grey to greyish black in colour depending on the grade of ore cut through.

4. The distribution of the various isolines of frequency of P-type pyrites displays the invisible zonation of mineralization of the wall-rock alteration type of deposit. Low, medium and high frequency zones are expected to yield low-, medium- and high-grade ores. Locations with 100% P-type pyrites are expected to yield superhigh-grade ores.

5. On the southeastern side of the Jiaojia Fault the total mineralized zones are enlarged downwords from -110 m level through -150 m to -190 m level with figures of 160,000 m^2 through 200,000 m^2 to 290,000 m^2. It gives an increment of mineralized area of 130,000 m^2 within a vertical drop of 80 m, being 1625 m^2 of increment per 10 meters vertical drop. This indicates that there is very good prospect for deep-seated ore reserves below -190 m level underground on both sides of the median base line.

6. The enlargement of mineralized areas with both $f \geq 80$ and $f \geq 50$ appears downward southwesterly for both the principal and the subordinate mineralized zones. This points out that the plunging direction of the orebodies is toward the southwest, and the deep-seated ore-finding below -190 m should be toward the southwest, but not toward the northeast.

7. The isolines of frequency of P-type pyrites of the principal mineralized zone are open on both terminals from -110 m to -190 m. Therefore there are horizontal extensions of orebodies both southwesterly and northeasterly on both terminals of the principal mineralized zone. However the northeastern terminal is more favourable than the southwestern terminal, since the plunging direction of the orebodies is toward the southwest and the uprising channelway of hydrothermal fluids is toward the northeast.

8. Laterally on the other side, i.e., on the northwestern or the left side of the Jiaojia Fault plane, there is also prospect for ore-finding, since the post-ore reactivation of the Jiaojia Fault has cut down or moved away the other half of the principal mineralized zone.

Jiaojia gold deposit was discovered by the Sixth Geological Brigade of the Shandong Bureau of Geology and Mineral Resources in 1967 on the Xincheng – Jiaojia Fault Zone. The Jiaojia Fault passes through the recent Jiaojia Mine on the northwest is a part of the Xincheng – Jiaojia Fault

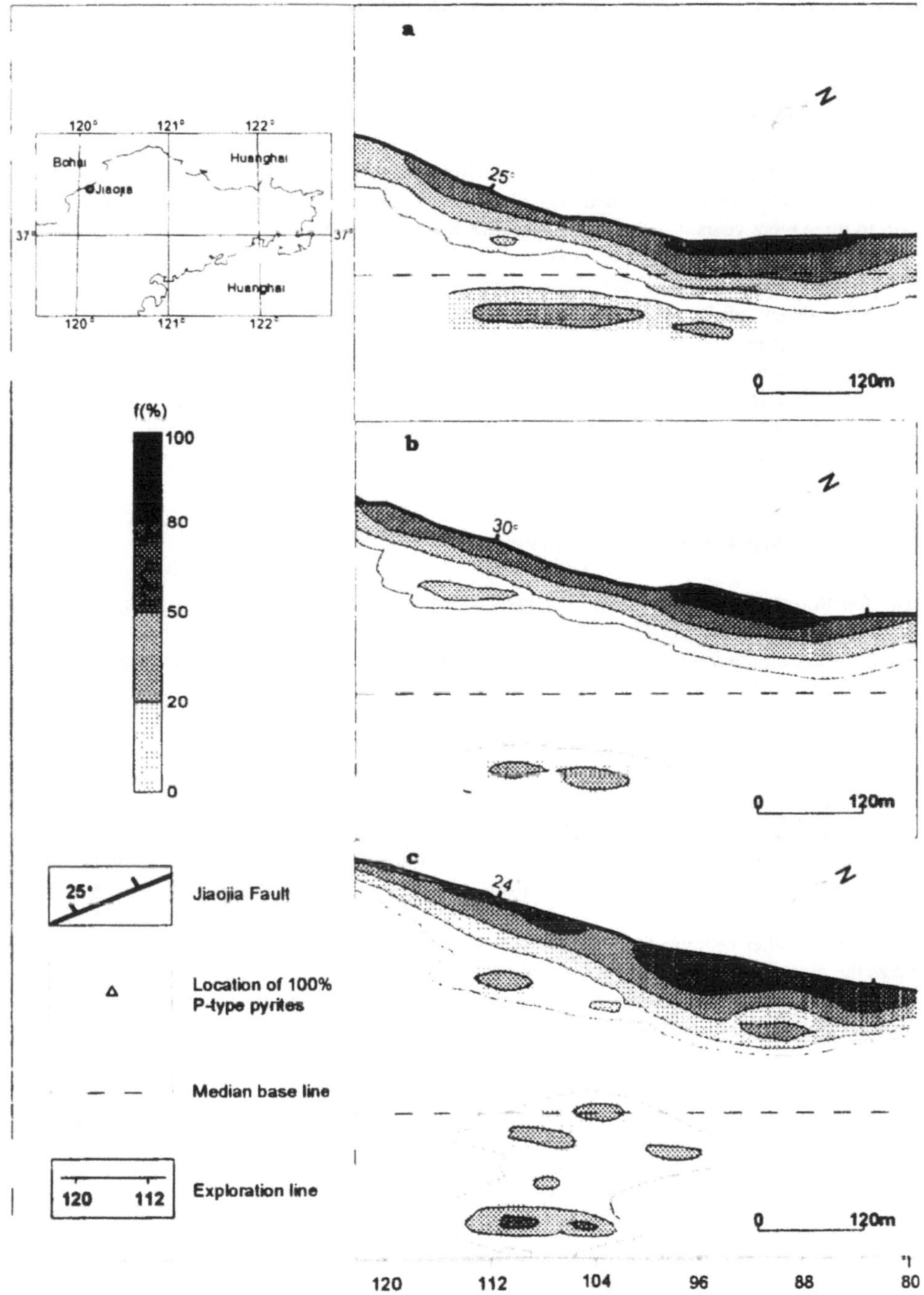

Fig 1 Mineralogical plane mappings of frequency of P-type pyrites (f) at three successive levels, Jiaojia Gold Mine, Jiaodong a at -110 m level, based on 42 samples and 920 measurements b at -150 m level, based on 40 samples and 850 measurements c at -190 m level, based on 45 samples and 1200 measurements Note: $f=P/(P+N) \times 100\%$

Zone. Through five years geological exploration and mine construction, the exploitation started in 1972.

The deposit is considered to be representative of the wall-rock alteration type of gold deposits in Jiaodong with the Neo-Archaean Jiaodong Group as overlying block and the Linglong Monzonitic Granite as the underlying block, but the study of genetic mineralogy and mineralogical mapping of the deposit were not started before 1995. Through about twenty five years mining, it was considered quite recently to be deficient of preserved reserve which permits exploitation only for two to three more years.

However the prospect of the mine is rather optimistic as is proved by mineralogical mapping by using frequency of P-type pyrites as checkup on crystallomorphological ballotage scores as mapping parameter. Both come to the same conclusion that there is good prospect down below - 190 m level in the depth and at least another twenty five years exploitation is available with the same mining rate as at present (Chen et al., 1996). Pyrite cubes in the Jiaojia gold deposit are mostly S-faced cubes developed from F-faced pentagonal dodecahedrons, indicating also high f_{S2} favourable for formation of large deposit [8, 11].

REGIONAL MINERALOGICAL MAPPING

The fourth mineralogical map (Fig. 2) represents the result of the regional mineralogical mapping of the northern Jiaodong Gold Ore Province, covering the terrane of the eastern half of the Penglai Municipality on the scale of 1:200,000 based on measurement of thermoelectric conduction types on 2373 grains of pyrite from 28 deposits and occurrences within 246 km^2.

The following information can be drawn from the mapping.

1. It shows distinctly the invisible zonation of the frequency of P-type pyrites.

2. The zonation shows that there are two distinct structural and mineralization trends in this terrane. One is northeast and the other is northnortheast. Both intersect in the central portion of the terrane.

3. There is neither f=0, nor f<10. On the contrary, there are both f=100 and f=90-100. The former takes the zigzag form appearing in the central portion of the map while the latter occurs not only in three seperate sectors in the middle of the map, but also occurs in the southwestern and the northwestern corners and especially in the elongate northeastern portion of the map.

4. The area of f=80-90 is larger than that of f=70-80 while the area of f=60-70 is still much larger than that of f=50-60.

5. The area of f<50 is only 4.9% of the mapped area, being less than 5% in total. That is to say, the area with f>50 is more than 95%. And the area of f<60 is about 12.07% of the mapped area, being less than 15%. Therefore, the area with f>60 is more than 85%.

6. The mapped frequency of P-type pyrites (f) shows that it is very high, being the highest so far known in Jiaodong. It illustrates why the superhigh-grade ores up to 20-30 wt% appear in the central portion of the terrane (C. D. Li, oral comm. 07-14-96) as prognosted and announced by the first author in 1990 before the mine cadres called together in the provisional concluding convention on the future prospect of the Heilangou Gold Mine in this region. It lies in the area where the

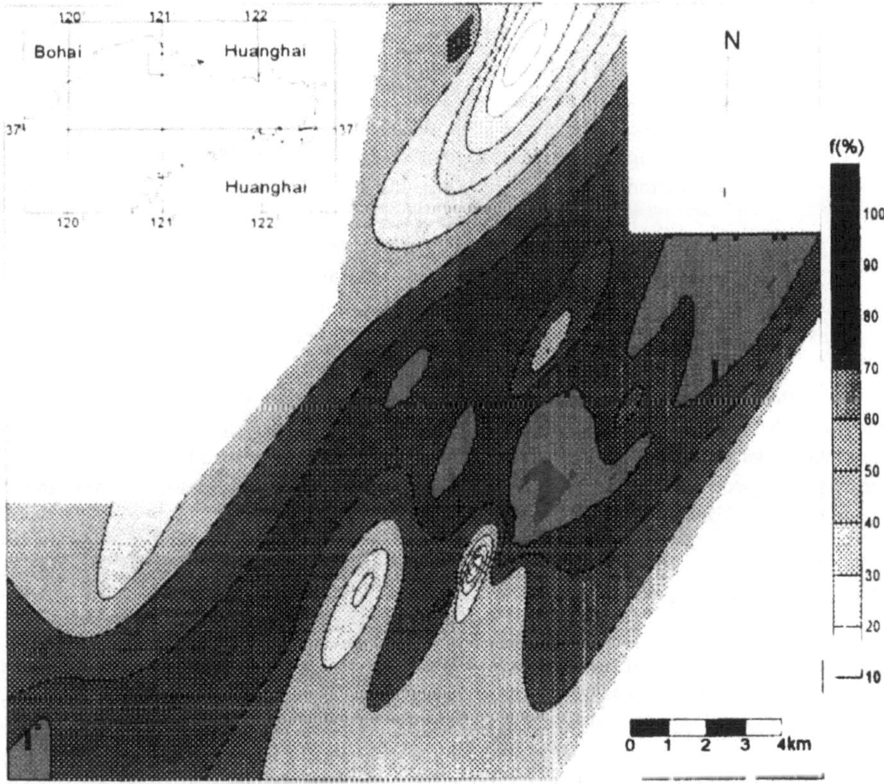

Fig 2 Regional mineralogical mapping of frequency of P-type pyrites (f), northern Jiaodong, based on 2373 measurements Blank area unmapped Note f=P/(P+N) × 100%

structural and mineralization trend turns from northnortheast to northeast from the southern half to the northern half of the terrane and with the occurrence of the zigzag-formed sector of 100% P-type pyrites in its center surrounded by 90-100% P-type pyrites on its periphery. It constitutes the "heart" of the high-grade gold mineralization of the whole mapped area.

7. The isolines are not closed on three sides of the map. Beside the central portion of the map, rich ores are expected to find in the northwestern, southwestern and northeastern parts of the terrane in ascending orders according to the distribution regularity of the frequency of P-type pyrites. As to the fourth side of the map, i.e., the southeastern side, it is known to be interrupted by a northnortheasterly trending fault called Xiaogujia Fault which marks the boundary between Penglai Municipality and Fushan District of Yantai Municipality.

The whole terrane is mostly covered by gold-hosting Guojialing Granodiorite isotopically dated 148.4+/-2.3 Ma (J_3) by Ar^{40}/Ar^{39} with hornblende, being younger than the Linglong Monzonitic Granite which is isotopically dated 170.6+/-2.3 Ma (J_2) by Ar^{40}/Ar^{39} with biotite [7]. However the occurrence of the overthrusting Neo-Archaean Jiaodong Group on its western side is similar. Dissimilarity in granitoids plays here important role in intensity of gold mineralization. For example the magnesiohornblende from Guojialing Granodiorite shows very intense Fe^{3+} Mossbauer spectrum indicative of very high f_{O2} with Fe^{3+} / Fe^{2+} up to 0.77, which is very favourable for the mobilization, transference and concentration of gold through later dynamometamorphism undergone to form high to superhigh grade of gold ores [7, 8, 14].

Though panning of placer gold was already very intensive in this terrane in Song Dynasty in 1026 A.D. or even earlier, it is however not known until 1990 that its large reserve with high- to superhigh-grade ore is closely related with the Guojialing Granodiorite through the theoretical study of Genetic Mineralogy [13]. The regional mineralogical mapping goes a step further practically along the direction of Prospecting Mineralogy.

Now the gold production in the central portion of this terrane has risen from 5035 liang (5058.7 oz.) in 1991 to 35500 liang (35667.2 oz.) in 1995 with the rate of 705% in increment within five years [20]. It is expected to produce more after more prospecting, exploration and exloitation of the deposits in this promising terrane as already disclosed by mineralogical mapping.

What is worth-mentioning is that the southwestern extension of this terrane goes beyond the boundary of Penglai Municipality into the northern part of Qixia County. Where it is covered by the southern extension of Guojialing Granodiorite and similar high frequency isolines of P-type pyrites, the prospect for high to superhigh grade of gold ores are promising as well and worth notice in northern Qixia County.

ACKNOWLEDGMENT

The authors are very grateful to JiaoJia Gold Mine and Penglai Administration of Control Gold General Corporation for their help and support in the field work. This study is supported by the grant from the Discipline Growth Point Foundation of Genetic Mineralogy, Department of Science and Technology, Ministry of Geology and Mineral Resources and the Scaling New Heights Foundation in Science and Technology, State Commission of Science and Technology, P.R.China.

REFERENCES

1. F. V. Chukhrov. Typomorphism — The most important problem of modern mineralogy. In : *Typomorphism of minerals.* F. V. Chukhrov, N. V. Petrovskaya and T. N. Shadlun (Eds). 1-14. Nauka Publishing House, Moscow (1969).
2. F. V. Chukhrov, A. I. Ginzbourg, N. V. Petrovskaya and T. N. Shadlun (Eds). *Typomorphism of minerals and its practical significance.* Nedra Publishing House, Moscow, 1-260 (in Russian) (1972).
3. N. P. Yushkin. *Topomineralogy.* Nedra Publishing House, Moscow, 1-228 (in Russian) (1982).
4. Chen Guangyuan, Sun Daisheng, Sun Chuanmin et al.. Genetic mineralogy of the Gongchangling Iron Mine, *Minerals and Rocks, Special Issue,* **4(2)**, 1-254 (in Chinese with English abstract) (1984).
5. Chen Guangyuan, Sun Daisheng and Yin Huian. *Genetic and prospecting mineralogy.* Chongqing Publishing House, Chongqing, 1-874 (in Chinese with English abstract) (1987).
6. Chen Guangyuan, Sun Daisheng and Yin Huian. *Genetic and prospecting mineralogy, second edition.* Chongqing Publishing House, Chongqing, 1-880 (in Chinese with English abstract) (1988).
7. Chen Guangyuan, Wang Jian and Huang Wenying. Hornblende favourable for indicating gold mineralization, *Proceedings ISGGE, Shenyang,* 688-690 (1989).
8. Chen Guangyuan, Shao Wei and Sun Daisheng. *Genetic mineralogy and gold deposits in Jiaodong region with emphasis on gold prospecting.* Chongqing Publishing House, Chongqing,

1-452 (in Chinese with English abstract) (1989).

9. Chen Guangyuan and Sun Daisheng. Six principal natures of mineral typomorphism. In: *Collected papers in memory of the 90th anniversary of Professor Feng Jinglan s birthday*. 121-133. Geological Publishing House, Beijing (in Chinese) (1990).
10. Chen Guangyuan and Sun Daisheng. Principles of mineral typomorphism — Its six essential natures, *Abs.* **1**, *15th IMA, Beijing*, 14-15 (1990).
11. Sun Daisheng and Chen Guangyuan. New advances in morphogenesis of pyrite, *Abs.* **1**, *15th IMA, Beijing*, 104-105 (1990).
12. Chen Guangyuan. Genetic mineralogy and mineralogical mapping, submitted paper in memory of the 70th anniversary of the Geological Society of China [unpublished], 1-5 (in Chinese) (1992).
13. Chen Guangyuan, Sun Daisheng, Zhou Xunruo et al.. *Genetic mineralogy and gold mineralization of Guojialing Granodiorite in Jiaodong region*. China University of Geosciences Press, Beijing, 1-230 (in Chinese with English abstract) (1993).
14. Chen Guangyuan, Sun Daisheng and Shao Wei. Jiaodong granitoids and gold mineralization, *Earth Science — Journal of China University of Geosciences*, **6**(1), 71-78 (1995).
15. Chen Guangyuan, Sun Daisheng, Huang Shaofeng et al.. *Genetic mineralogy of Lingshangou gold deposit in Jiaodong region*. China University of Geosciences Press, Wuhan, 1-150 (in Chinese with English abstract) (1995).
16. Chen Guangyuan and Sun Daisheng. Mineral typomorphism — the key problem of genetic mineralogy. In: *Retrospect of the development of geosciences disciplines in China, centennial memorial volume of Professor Sun Yunzhu (Y. C. Sun)*. Wang Hongzhen (Chief ed.). 116-125. China University of Geosciences Press, Wuhan (in Chinese) (1995).
17. Chen Guangyuan, Shao Wei, Sun Daisheng et al.. *Atlas of mineralogical mapping in Jiaodong gold province*. Geological Publishing House, Beijing, 1-128 (1996).
18. Chen Guangyuan. The development of modern genetic mineralogy in China. In: *Develop-ment of geosciences disciplines in China*. Wang Hongzhen et al. (Ed.). 93-98. China University of Geosciences Press, Wuhan (1996).
19. Li Shengrong, Chen Guangyuan, Shao Wei et al.. *Genetic mineralogy of Rushan gold orefield in Jiaodong region*. Geological Publishing House, Beijing, 1-116 (in Chinese with English abstract) (1996).
20. Mu Fanmin. Key mine needs to make key contributions, *China Newspaper of Geology and Mineral Resources*, No.2276 (in Chinese) (1996).

Proc 30th Int'l. Geol. Congr., Vol. 16, pp. 85-95
Huang Yunhui and Cao Yawen (Eds)
VSP 1997

THE SIGNIFICANCE OF MANGANESE OXIDES IN THE LATERITE PROFILE AT THE IGARAPÉ BAHIA GOLD DEPOSIT, CARAJÁS, BRAZIL

Weisheng ZANG
Quest International Resources Corp. P.O. Box 728, Cranbrook, B.C. Canada V1C 4J5

William S. FYFE
Department of Earth Sciences, University of Western Ontario London, Ontario, Canada N6A 5B7

Yuan CHEN
Department of Geology, Division of Science, Okanagan University College Kelowna, B.C., Canada V1V 1V7

Abstract

The Igarapé Bahia lateritic gold deposit in the Carajás region of the Amazon has developed on a primary Cu(Au) mineralization zone hosted by chlorite schists. The parent rocks of the chlorite schists include basalts, pyroclastic and clastic sedimentary rocks formed in an Archaean rift basin. The intense weathering resulted in accumulation of Fe-, Mn-, Al-oxides/hydroxides, and kaolinite in laterite profiles. The laterite profile is as much as 180 m thick, and from base to top, consists of saprolite, pallid, mottled, ferruginous zones and top soil. Gold orebodies mainly occur in the ferruginous zone.

Manganese oxides are present as a major constituent in some particular samples (MnO up to 18.06 wt%), although contents of MnO generally range from 0.14 to 5.86 wt %. Similar to Fe-oxides, these Mn-oxides also act as "scavengers" to control the geochemical behaviours of many elements, especially some of the first row transition metals, even gold, silver and rare earth elements. Electron microprobe analysis, SEM and XRD showed that the Mn-oxides are cryptomelane (a cryptomelane-hollandite solid solution) and lithiophorite. The former occurs in both upper and lower levels of the profile whereas the latter only in the lower levels (lower part of the mottled zone and the pallid zone). The cryptomelane in the upper levels occurs as micro crystallites or amorphous materials, showing that it is undergone leaching. By contrast, most Mn-oxides from the lower levels are well crystallized, where lithiophorite occurs as typical hexagonal plate crystals, and cryptomelane as needles or fabrics. In other reported laterite profiles, however, cryptomelane occurs near the unweathered rock, followed upward by nsutite, pyrolusite, a second generation of cryptomelane and then lithiophorite. The difference of Mn-oxides occurrence in this study is probably attributable to a two-stage evolution of the profile. The dissolution and formation of Mn-oxides are largely related to the water table change where an active oxidation/reduction horizon occurs. After the first stage of lateritization, an incision of the landscape with a well-developed drainage system resulted in the second stage evolution of the profile to form a leached pallid zone, and dissolution and precipitation of the Mn-oxides in the upper and lower levels, respectively.

Introduction

Over the past decade there has been a great increase in interest of the weathering of ores and regions of metal enrichment. We have become increasingly aware of the lack of knowledge of processes in such regions, and of the potential for secondary metal enrichment. The feature and development of gold grains in the Igarapé Bahia lateritic gold deposit in the Carajás region of the Amazon were described by Zang and Fyfe (1993[1]). In this paper we describe some features of

Mn-minerals. The Mn-oxides in the laterite profile may be present as a major constituent in some particular samples (up to 18 wt% MnO), although contents below 1.0 wt% are more common through the profile. These Mn-oxides could provide some evidence for evolution of this highly developed weathering profile. Similar to Fe-oxides, the Mn-oxides also contribute significantly to the surface activity of laterites due to their small crystallite size and large surface area. With such a high surface area, elements adsorbed on the surface can be present in major concentrations in the minerals. They act as "scavengers" to control the geochemical behaviours of many elements, especially the first row transition metals, even precious metal (Au and Ag) and rare earth elements.

Regional Geology

The Igarapé Bahia lateritic gold deposit is located in the Serra dos Carajás region, in the eastern part of the Amazon shield, Northern Brazil (Fig. 1). This region, covered by rain forest, consists of a WNW trending plateau at around 650 m altitude, surrounded by lowlands at around 250 m altitude. The geomorphological features suggest a stage of strong incision late in topographic evolution. The region occurs in an Archean continental rift (Lindenmayer, 1990[2]), where gold, copper, manganese and iron deposits occur, some very large.

The regional geology has been previously described in detail by Hirata et al. (1982[3]) and Lindenmayer (1990[2]). The Archean basement is represented by the Xingu complex composed of tonalitic to granodioritic polymetamorphic gneisses and migmatites, trondhjemites and gabbro-norites with subordinate supracrustal rocks in amphibolitic to granulitic metamorphic facies.The Salobo-Pojuca formation of late Archean age is made up of metamorphosed basic to intermediate metavolcanics with intercalations of clastic and chemical metasediments. It occurs in WNW trending belts enclosed by gneisses and migmatites of the Xingu complex. The Grao Para group is mostly made up of metamorphosed basalts, thick jaspilite beds and subordinated clastic sediments and felsic volcanics. Geochemical study of the volcanics indicates that they formed on a continental crust undergoing extension during late Archean. Coupled with the Salobo-Pojuca formation, they represent lateral facies changes of the volcano-sedimentary sequence deposited in a continental rift basin. The bedrock of the Igarapé Bahia lateritic gold deposit are considered to belong to the Rio Fresco formation. They are mainly composed of chlorite schists (probably metabasalts, metaclastic and pyroclastic rocks) and iron formation (Zang and Fyfe, 1995[4]).

The Laterite Profile

The lateritic profile is as much as 180 m thick. It can be divided into five horizons. From unweathered bedrock to top are saprolite zone, pallid zone, mottled zone, ferruginous zone and topsoil (Fig. 2). Composition of major oxides in the profile are given in Table 1. The unweathered rocks are mainly composed of quartz and chlorite in highly variable amounts. Other minerals include calcite, albite and micas, with sulfide-quartz veinlets and impregnated sulfide minerals (mainly chalcopyrite and bornite). The saprolite is as much as 40 m thick, with

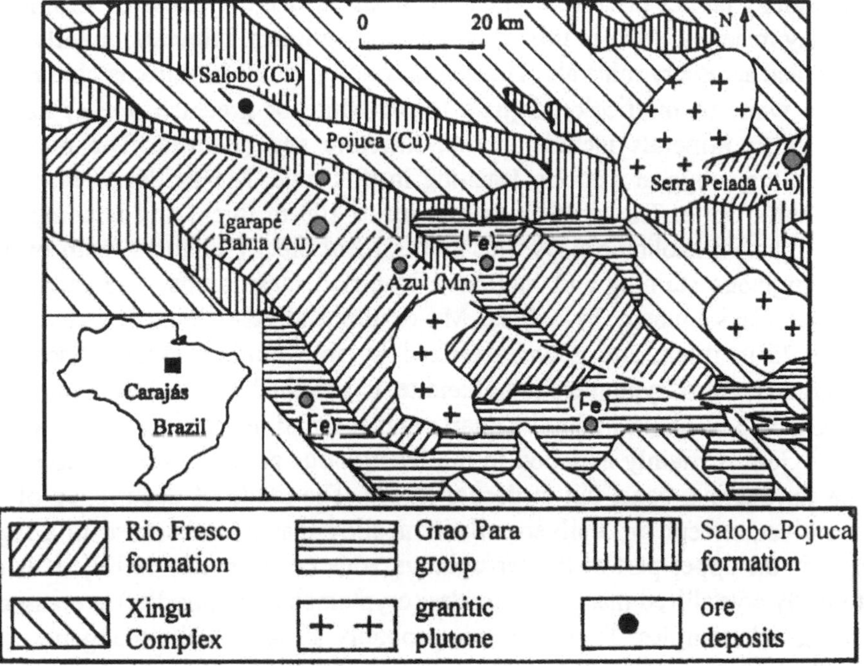

Fig. 1. Regional geology of the Serra dos Carajás area, Brazil (Modified from Hirata et al, 1982[3]).

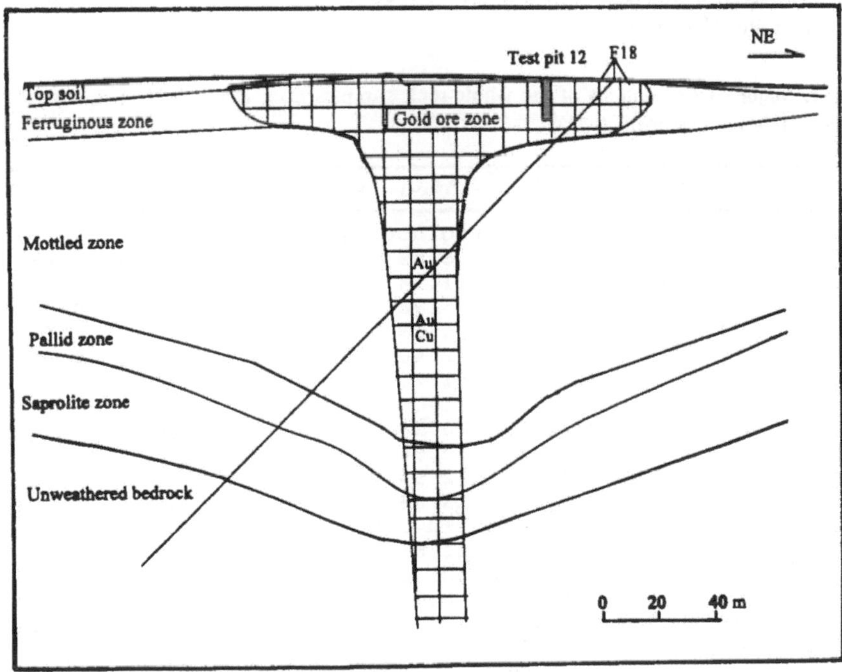

Fig. 2. Schematic cross section of laterite profile at the Igarapé Bahia lateritic gold deposit. Shaded area represents ore zone, the lower part of the "root" is dominated by copper mineralization.

some primary fabrics preserved. The mineral assemblage consists of kaolinite and other clay minerals. Also in this zone are relict primary minerals, mainly quartz, chlorite and accessory minerals. The pallid zone up to 20 m thick is characterized by pure white to multicolored (pale to off-white or greenish white) kaolinitic clays. Other minerals include quartz, ilmenite and goethite. The pure white kaolinite may suggest a strong leaching in this zone. Mn-oxides were observed in the pallid zone. The mottled zone is up to 100 m thick, and is composed of yellow to brown kaolinitic clays with mottled and blocky textures, with brown or red ferruginous staining and indurated iron-rich clay lumps. The minerals are kaolinitic clays, quartz, Fe-oxides, Mn-oxides and Al-oxides. The ferruginous zone is 25 m thick, and is characterized by iron oxide nodules, pisolites and ferruginous soil. Cryptomelane-hollandite is enriched in the lower part of this zone. Gibbsite, kaolinite and relict quartz and tourmaline occur in small amounts.

Gold is mainly enriched in the ferruginous zone. Silver is also enriched in the ferruginous zone, but at deeper levels. The mode of occurrence of silver is not clear, except for a silver-palladium alloy that was observed in a Ag-rich sample in the upper part of the ferruginous zone (Zang et al., 1992[5]). It is noted that poorly crystallized manganese oxides (cryptomelane-hollandite) are associated with Ag-rich samples. Thus, silver is probably sorbed by the manganese oxide minerals. Silver may also replace K^+ in the crystal structure of the minerals since Ag^+ (1.13 Å) has a similar radius to K^+ (1.33 Å). Anderson et al. (1973) reported that poorly crystallized manganese oxides sorbed significant quantities of silver. They suggested that surface exchange as well as structural exchange are involved in the uptake of silver.

Table 1. Major oxide contents of the whole-rock samples from Drillhole F18 and Test Pit 12 at Igarapé Bahia gold deposit (wt%)

sample	depth (m)	SiO_2	Al_2O_3	Fe_2O_3	TiO_2	MnO	MgO	P_2O_5
				ferruginous zone				
F18-1	0.0-0.2	5.34	6.08	79.30	0.54	0.14	0.68	0.54
F18-7	3.5-3.7	4.62	11.74	70.94	0.28	0.22	0.70	0.90
P-1	6.6-10.5	5.60	5.98	75.34	0.38	1.90	0.76	1.20
P-3	12.0-14.0	3.67	4.19	76.32	0.42	4.90	0.66	0.93
P-4	14.0-16.0	4.17	2.40	77.00	0.22	5.86	0.72	1.08
P-6	18.0-20.0	5.41	1.20	79.66	0.14	3.10	0.62	0.94
				mottled zone				
F18-16	24.7-25.0	21.81	20.28	44.52	1.68	0.21	0.36	0.53
F18-20	39.1-39.3	22.60	20.98	39.83	3.77	0.58	0.30	0.44
F18-27	70.2-70.4	25.04	22.53	39.22	1.30	0.66	0.33	0.28
F18-32	82.0-82.2	13.84	13.83	58.29	0.73	2.03	0.70	0.89
				pallid zone				
F18-41	118.8-119.1	42.35	22.85	20.35	4.10	0.44	0.22	0.13
				saprolite zone				
F18-43	125.5-125.8	13.25	8.97	39.03	0.50	18.06	0.88	0.53
F18-44	129.4-129.6	38.31	27.14	15.48	1.38	0.13	3.68	0.24
F18-50	148.5-162.0	47.54	13.88	22.93	0.92	0.14	8.60	0.19
				unweathered bedrock				
F18-55	217.8-226.3	86.12	5.02	4.74	0.11	0.03	1.36	0.06

Note: total iron as Fe_2O_3; CaO<0.02, NaO<0.01 and K_2O up to 0.14 in samples from pallid, mottled and ferruginous zones.

Mn-oxide Mineralogy

The Mn-oxides were identified from the ferruginous zone and the lower part of both the mottled zone and the pallid zone. They are readily recognized in hand-specimens by their distinctive dull black colour, finely banded or colloform texture. The XRD patterns of some Mn-oxide bearing samples indicate that the Mn-oxides are cryptomelane (a cryptomelane-hollandite solid solution) and lithiophorite. The Mn-oxides occurring in the upper levels of the profile (ferruginous zone) are cryptomelane whereas in the lower levels (lower part of mottled zone and pallid zone) are both cryptomelane and lithiophorite. Other differences are that gibbsite, hematite and cerianite are associated with the Mn-oxides from the upper levels, and that kaolinite is associated with the Mn-oxides from the lower levels. Goethite is associated with all the samples, but its concentration increases from the samples in the lower levels to those in the upper levels. This is also consistent with the general mineralogical trend of the profile which shows a progressive increase in Fe-oxides from the base to the top.

Morphology of Mn-oxide Minerals

Major feature of morphology of cryptomelane in the upper levels of the profile is that it occurs as microcrystallites or amorphous materials. The texture shows that it is undergone leaching. An amorphous cryptomelane nodule in a void shows a leaching feature (Fig. 3A). Cryptomelane with a relict texture is common in the samples from the upper levels of the profile (Fig. 3B, 3C).

By contrast, most Mn-oxides from the lower levels of the profile are crystalline, especially lithiophorite. The lithiophorite occurs as typical hexagonal platy crystals (Fig. 3D). Cryptomelane occurs as needles or fabrics, and locally concentrates as blocks (Fig. 3E). The cryptomelane also fills the fissures (Fig. 3F), together with Fe-hydroxides, and the texture shows that they are precipitating from the percolating solutions.

Chemical Composition of Mn-oxide Minerals

Cryptomelane was analysed with the electron microprobe in samples from different depths. The average values of each sample are given in Table 2. Unfortunately, lithiophorite was not analysed since Li can not be detected by electron microprobe. However, EDS spectrum shows that the mineral is typically a Mn-Al-oxide mineral, and no other significant elements were detected. The chemical formula ($LiAl_2Mn_3O_9 \cdot 3H_2O$) proposed by de Viliers and van der Walt (1945[7]) for lithiophorite is in accordance with the structure in which layers of MnO_6 and $(Al,Li)(OH)_6$ octahedra alternate along the c axis (Wadsley, 1952[8]). However, lithiophorite reported from laterites as well as other manganese deposits contains less quantity of Li, thus Parc et al. (1989[9]) proposed a chemical formula without Li, which is $Al_2Mn_3O_9 \cdot 3H_2O$. Since Li can not be detected with electron microprobe (and SEM) it is not known whether this mineral contains Li in the present study.

Cryptomelane and hollandite form a solid solution with a general chemical formula $A_{0-1}(Mn^{4+}, Mn^{2+})_8O_{16}$ where A is primarily Ba^{2+} in hollandite

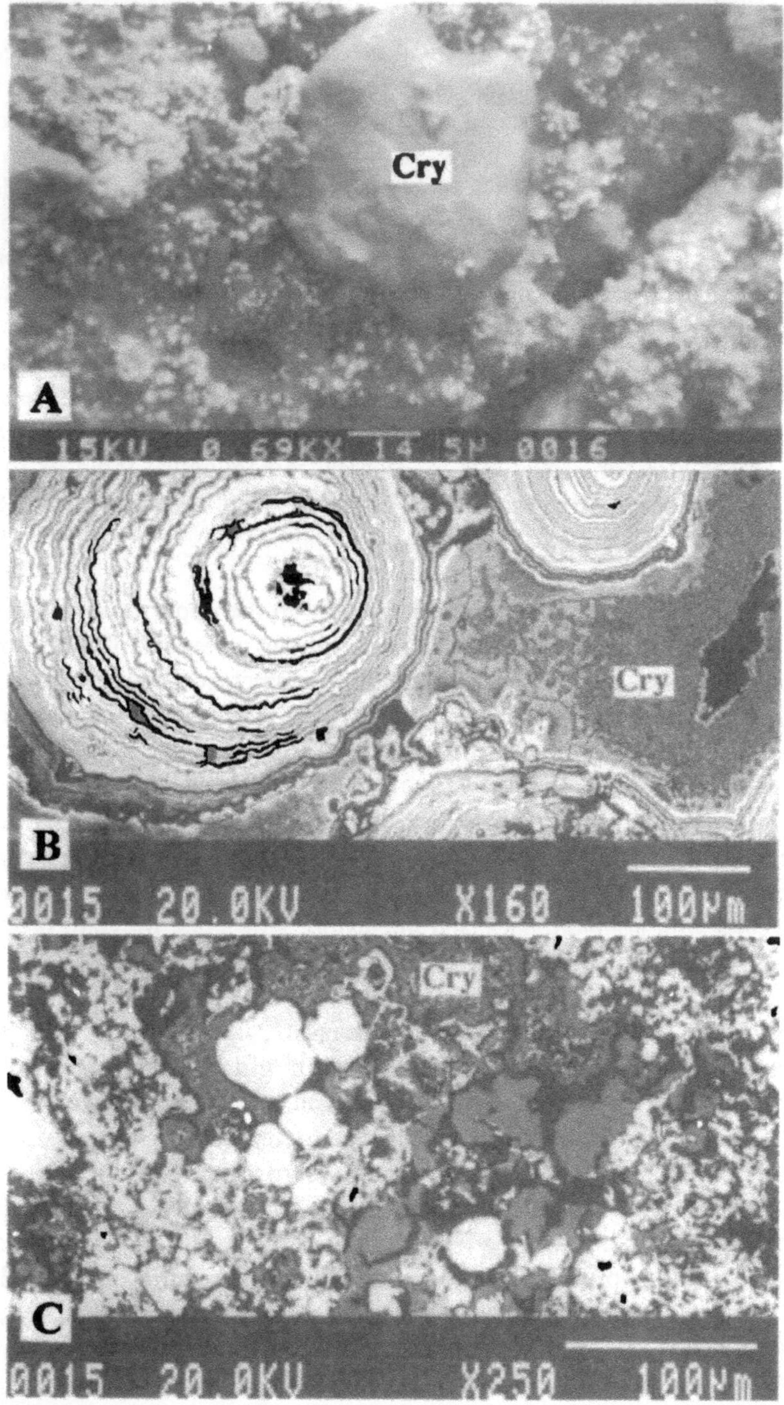

Fig. 3. SEM and back-scattered electron images of cryptomelane and lithiophorite from the laterite profile at the Igarapé Bahia gold deposit. A:SEM image of a cryptomelane nodule (Cry) in the void from the ferruginous zone. Background minerals are cerianite (white) and cryptomelane (grey). B, C: Back-scattered electron image of thin section, showing amorphous cryptomelane (Cry) with a relict texture, associated with Fe-oxides (white) from the ferruginous zone.

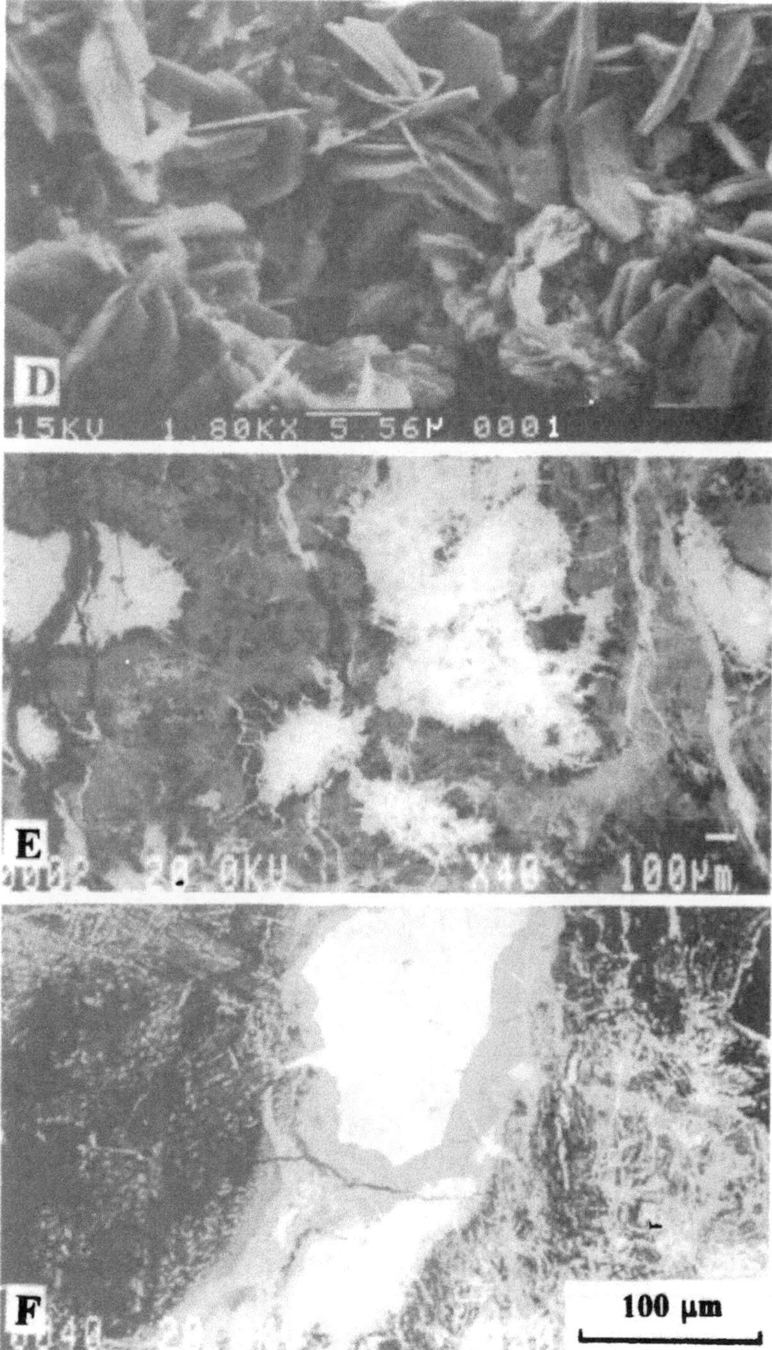

Fig. 3. Continued. D: SEM electron image of lithiophorite as typical hexagonal platy crystals with needle-like cryptomelane from the lower part of the mottled zone. E: Back-scattered electron image of thin section, showing cryptomelane fabrics (white) and weathered chlorite (grey) from the saprolite zone. F: Back-scattered electron image of thin section, showing crytomelane (white) and Fe-hydroxide (grey) filling a fissure in the saprolite zone.

and K^+ in cryptomelane. The cryptomelane consists of a framework of double chains of edge-sharing Mn-O octahedra containing large tunnels filled with K^+ ions. Most of the manganese is Mn^{4+}, but some Mn^{2+} substitutes for Mn^{4+} to balance the positive charge of tunnel cations (Hypolito et al., 1984[10]). In this study, cations were calculated on the basis of 8 cations (exclusive of K, Ba, Na, Ca). The average formulae of cryptomelanes from different depths of the profile are also given in Table 2. MnO and MnO_2 were calculated on the basis of charge balance. The calculation shows that MnO only occurs in samples from the lower levels, ranging from 0.08 to 4.64 wt% with an average value of 2.16.

Table 2. Average values of electron microprobe analyses of cryptomelane-hollandite solid solution from the Igarapé Bahia gold deposit

sample	B95	F18-12	P-4	F18-43
depth (m)	2.0	13.3-14.0	14.0-15.0	125.5-125.8
analyses	14	10	9	26
MnO_2	74.62	71.28	60.14	84.53
MnO	0.00	0.00	0.00	2.16
SiO_2	0.24	0.52	0.33	0.85
TiO_2	0.08	0.04	0.04	0.14
Al_2O_3	1.89	6.64	5.72	1.69
Cr_2O_3	0.02	0.00	0.00	0.01
Fe_2O_3	3.94	3.00	3.45	0.79
CuO	1.76	0.92	1.11	0.36
CoO	0.52	0.59	0.30	0.18
MgO	0.02	0.01	0.01	0.14
CaO	0.11	0.01	0.21	0.44
BaO	1.68	1.01	1.39	2.60
K_2O	3.14	2.74	2.01	3.43
Na_2O	0.07	0.14	0.02	0.38
Au	0.01	0.00	0.01	0.02
Ag	0.02	0.11	0.08	0.00
total	88.12	87.01	74.82	97.72

Average formulae:

B95 $(K_{0.54}Ba_{0.09}Na_{0.02}Ca_{0.02})_{0.67}(Mn^{4+}{}_{7.02}Fe^{3+}{}_{0.40}Al_{0.30}Cu_{0.18}Co_{0.06}Si_{0.03}Ti_{0.01})_8O_{16}$

P-4 $(K_{0.39}Ba_{0.08}Na_{0.01}Ca_{0.03})_{0.51}(Mn^{4+}{}_{6.35}Fe^{3+}{}_{0.40}Al_{1.03}Cu_{0.13}Co_{0.04}Si_{0.04}Ti_{0.01})_8O_{16}$

F18-12 $(K_{0.46}Ba_{0.05}Na_{0.04})_{0.55}(Mn^{4+}{}_{6.45}Fe^{3+}{}_{0.30}Al_{1.03}Cu_{0.09}Co_{0.06}Si_{0.07})_8O_{16}$

F18-43 $(K_{0.56}Ba_{0.13}Na_{0.09}Ca_{0.06}Mg_{0.03})_{0.87}$
$(Mn^{4+}{}_{7.16}Mn^{2+}{}_{0.32}Fe^{3+}{}_{0.09}Al_{0.26}Cu_{0.03}Co_{0.02}Si_{0.11}Ti_{0.01})_8O_{16}$

Composition of the cryptomelane indicate considerable incorporation of other ions. Usually SiO_2 and Al_2O_3 occur as impurities in the mineral, reflecting the presence of admixed kaolinite. This may explain the cryptomelane from the lower levels of the profile which has SiO_2 and Al_2O_3 contents at a reasonable ratio for kaolinite. However, in most cryptomelane samples from the upper levels of the profile, Al_2O_3 contents (up to 7.83 wt%) are unreasonably higher than SiO_2

contents (only up to 0.77 wt%). This suggests that Al may be incorporated into the cryptomelane structure. By contrast, Fe_2O_3 contents of the cryptomelane from the upper levels are greater than those from the lower levels of the profile. The Fe_2O_3 contents may largely reflect admixed goethite in the poorly crystallized cryptomelane. As described above, goethite replaces cryptomelane in the upper levels of the profile, and the replacement will inevitably result in admixture of these two minerals. The cryptomelane contains TiO_2 up to 0.34 wt%, but it occurs in the lower levels of the profile. Small amounts of Cr_2O_3 (up to 0.07 wt%) was also detected.

Cryptomelane contains K (K_2O 1.87-4.52 wt%) and Ba (BaO 0.66-4.93 wt%) as essential cations. It also contains small amounts of Mg (MgO up to 0.40 wt%), Ca (CaO 0.54 wt%), and Na (Na_2O 0.71 wt%). Contents of these cations are lower in the cryptomelane from the upper levels than that from the lower levels of the profile.

The cryptomelane from the upper levels of the profile contains more CuO and CoO than that from the lower levels. The former has CuO and CoO ranging from 0.70 to 2.09 wt% and from 0.15 to 0.68 wt%, respectively, whereas the latter ranging 0.11 to 0.71 wt% and 0.04 to 0.43 wt%. Taking into account Al_2O_3 contents in the cryptomelane, it is obvious that the CoO, CuO contents are higher in the samples with high Al substitution than those with low Al substitution. This phenomenon is in agreement with the proposal of Wagner et al. (1979[11]) that Al^{3+} substitution for Mn^{4+} promotes Co, Cu uptake. Also at lower levels Cu occurs in its own secondary minerals, such as malachite, azurite and native copper. The Co enrichment occurs because Co can be oxidized and much more strongly adsorbed onto Mn-oxide surfaces than other divalent ions (Murray and Dillard, 1979[12]; Taira et al., 1981[13]).

Silver and gold were also detected in the cryptomelane, but the latter may be from interferences during the analysis. However, Ag contents may reflect its intimate relation with Mn-oxides. In the profile, silver is enriched in the ferruginous zone, but it is enriched in the lower part, particularly in samples with more Mn-oxide minerals. Thus, silver is probably present as an adsorbed form on the manganese oxide minerals.

Discussion and Conclusions

The development of a pallid zone suggests that this laterite profile has been developed through two main stages: (1) development of a ferruginous zone and a mottled zone with saprolite over the unweathered rocks; (2) incision of landscape and modification of the profile to form the pallid zone and more saprolite. The morphology and geochemistry of the Mn-oxides support this hypothesis. At the first stage, Mn-oxides were formed in the upper levels from weathering of the parent rocks. At the second stage, however, with lowering of the weathering front these Mn-oxides have been leached, and then precipitated in the lower levels.

During the first stage Mn, K, Ba can be derived by weathering the hydrothermally altered bedrock in which Mn is mainly present in silicates.

Another Mn source is carbonate minerals (siderite and calcite). Potassium was mainly derived from muscovite, biotite and stilpnomelane. Base metals were undoubtedly derived from sulfides. Barium is probably associated with the primary mineralization zone as both barite and Ba-rich feldspars.

Noteworthy is the occurrence of lithiophorite. In the literature lithiophorite is frequently reported from upper levels of laterite profiles. Bernardelli and Beisiegel (1978[14]) have shown that weathering of the rhodochrosite-bearing parent rock at Azul, Carajás area, leads to the sequence that is cryptomelane nearest the fresh rock, followed upward by nsutite, pyrolusite, a second generation of cryptomelane, and then lithiophorite. Beauvais et al. (1987[15]) noted that cryptomelane and lithiophorite constitute the more stable forms of manganese in the superficial layers of the lateritic profiles in the same area. Very similar profiles have been described in other examples in Brazil (Horen, 1953[16]), and in Africa (Sorem and Cameron, 1960[17]). The distribution pattern of Mn-oxides can also be explained by activities of other components, such as K^+ and Al^{3+} ions. High activity of K^+ ions in the solutions released in the lower levels of the profile from weathering of bedrock will promote formation of cryptomelane, whereas high activity of Al^{3+} ions released by later dissolution of kaolinite in the upper levels will promote formation of lithiophorite in the presence of Mn^{4+} ions.

Lithiophorite, as a stable Mn-oxide mineral, should occur in the upper levels of this laterite profile if the above hypothesis is correct. However, lithiophorite is only identified in the lower levels in this study. This phenomenon is probably attributable to the two-stage evolution of the profile. The dissolution and formation of Mn-oxides are largely related to the water table where an active oxidation/reduction horizon occurs. At the second stage, in the strongly incised profile with a well-developed drainage system and a rapid water exchange this active horizon extended much further down. Among the Mn-oxides cryptomelane is more vulnerable to changing environments, as noted by Prajecus et al. (1990[18]), thus, it started to be dissolved in the upper levels.

Acknowledgements

The authors gratefully acknowledge the help of J. A. Forth for the preparation of polished thin sections, R. L. Barnett and D. M. Kingston for the assistance with electron microprobe analysis, R. Divison for the help with the SEM, Y. Cheng for the XRD analysis, and DOCEGEO for field support.

References

1. Zang, W., and Fyfe, W. S. 1993. A three-stage genetic model for the Igarapé Bahia lateritic gold deposit, Carajás, Brazil. Economic Geology, 88:1768-1779.

2. Lindenmayer, Z. G. 1990. Salobo sequence, Carajás, Brazil: geology, geochemistry and metamorphism. Unpublished Ph.D. dissertation, London, Ontario, Canada, University of Western Ontario, 405 p.

3. Hirata, W. K., Beisiegel, V. R., Ernardelli, A. L., Farias, N. F., Saueressig, R., Meireles, E. M., and Teixeira, J. F. 1982. Serra dos Carajás-Para state: iron, manganese, copper, and gold deposits: International Symposium of Archean and Early Proterozoic Geologic Evolution and Metallogeny, 1st, Salvador, Bahia, Brazil, 1982, Excursion Guide, p. 40-76.

4. Zang, W., and Fyfe, W. S. 1995. Chloritization of hydrothermally altered bedrock at the Igarapé Bahia gold deposit, Carajás, Brazil. Minerlium Deposita, 30:30-38.
5. Zang, W., Fyfe, W. S., and Barnett, R. L. 1992. A silver-palladium alloy from the Igarapé Bahia lateritic gold deposit, Carajás, Brazil. Mineralogical Magazine, 56:47-51.
6. Anderson, B. J., Jenne, E. A., and Chao, T. T. 1973. The sorption of silver by poorly crystallized manganese oxides. Geochimica et Cosmochimica Acta, 37:611-622.
7. de Villiers J. E., and van der walt, C. F. J. 1945. Lithiophotrite from the Postmasburg manganese deposits. American Mineralogist, 30:629-634.
8. Wadsley, A. D. 1952. The structure of lithiophorite (Al,Li)MnO_2. Acta Crystal.,5: 676-680.
9. Parc, S., Nahon, D., Tardy, Y, and Vieillard, P. 1989. Estimated sulubility products and fields of stability for cryptomelane, nsunite, birnessite, and lithiophorite based on natural lateritic weathering sequences. American Mineralogist, 74:466-475.
10. Hypolito, R., Valarelli, J.V., Giovanoli, R., and Netto, S. M. 1984. Gibbs free energy of formation of synthetic cryptomelane. Chimia, 38:427-428.
11. Wagner, G. H., Konig, R. H., Vogelpohl, S., and Jones, M. D. 1979. Base metals and other minor elements in the manganese deposits of west-central Arkansas. Chemical Geology, 27:309-327.
12. Murray, J. W., and Dillard, J. G. 1979. The oxidation of cobalt (II) adsorbed on manganese dioxide. Geochimica et Cosmochimica Acta, 43: 781-787.
13. Taira, H., Kitano, Y., and Kaneshima, K. 1981. Terrstrial ferromanganese nodules formed in limestone areas of the Ryukyu Island, Part I. Major and minor constituents of terrestrial ferromanganese nodules. Geochemical Journal, 15:69-80.
14. Bernarelli, A. L., and Beisiegel, V. R. 1978. Geologica economica da jazida de manganês de do Azul. Anais do XXXe Congresso Brazileiro de Geologia, Recife 4:1431-1444.
15. Beauvais, A., Melfi, A., Nahon, D., and Trescases, J. J. 1987. Pétrologie du gisement latéritique manganésifère d'Azul (Brésil). Mineralium Deposita, 22:124-134.
16. Horen, A. 1953. the manganese mineralization at the Merid mine, Minas Gerais, Brazil. Unpublished Ph.D. dissertation, Harward University, Cambridge, Massachusetts, 224 p.
17. Soren, R. K., and Cameron, E. M. 1960. Manganese oxides and associated minerals of the Nsuta manganese deposits, Ghona, West Africa. Economic Geology, 55:278-310.
18. Prajecus, B., Varga, R. A., Madgwick, L. A., Frakes, and Bolton, B. R. 1990. Effects of mineral composition on microbiological reductive leaching of manganese oxides. Chemical Geology, 88:143-149.

Proc. 30th Int'l. Geol. Congr., Vol. 16, pp. 97-108
Huang Yunhui and Cao Yawen (Eds)

MINERAL SURFACE COMPOSITION OF ROCKS AND ORES OF THE GOLDEN GIANT GOLD DEPOSIT OF HEMLO, ONTARIO, CANADA

Yuan CHEN

Department of Geology, Division of Science, Okanagan University College
Kelowna, B.C, Canada V1V 1V7

William S. FYFE

Department of Earth Sciences, University of Western Ontario
London, Ontario Canada N6A 5B7

Weisheng ZANG

Quest International Resources Corp., P.O.Box 728, Cranbrook, B.C.
Canada V1C 4J5

Abstract

The high sensitivity of the Auger Electron Spectroscopy (AES) makes it useful for searching trace fluid composition preserved on mineral surfaces and grain boundaries. In this study we used AES to investigate the surface composition of country rocks and ores of the Golden Giant gold deposit of Hemlo, Ontario.The Hemlo gold deposits are situated within a sequence of Archean metasedimentary and metavolcanic supracrustal rocks that are part of the Hemlo-Heron Bay greenstone belt within Wawa subprovince of Superior structural province, The greenstone belt has been subjected to a complicated history of deposition, magmatism, deformation, metamorphism and metasomatism,

The country rocks investigated include oligoclase-biotite-quartz schist, quartz-oligoclase-muscovite schist, quartz-oligoclase prophyritic breccia, quartz-eye muscovite schist and quartz-eye microcline granofels. The ores investigated include quartz-microcline siliceous ore, muscovitic ore, baritic ore and biotitic ore. In most samples chloride films with minor S and K are preserved on grain boundaries showing that Cl-bearing fluids were involved in formation of the rocks and ores. The country rocks close to the ore zones have high Cl content on mineral surfaces. The baritic ore contains distinctly high surface Cl (up to 6.3 wt%). The study of fluid inclusions in quartz veinlets from the Golden Giant deposit supports the AES results. The fluid inclusions are highly saline (19-29 NaCl wt%, equivalent). Analysis of fluid inclusion brine residues produced by evaporation shows that NaCl dominates, with variable amounts of $CaCl_2$ and carbonate species. It is surprising that the fluid inclusions have such low homogenization temperatures, ranging from 167 to 227°C. The data seems to support an epithermal model for some of the veins.

Introduction

Fluid inclusions trapped in minerals provide direct samples for investigation of the nature of fluids involved in a variety of environments. In recent years, however, there has been a great development in study of mineral surface composition with development of surface analytical techniques. Among them is Auger Electron Spectroscopy (AES). The high sensitivity of the technique to the light elements (except for H and He) makes it very useful for searching trace fluid composition preserved on mineral surface. Using AES, Mareschal et al. (1992[1]) reported for the first time the presence of graphite films and traces of Cl, S and Fe at grain boundaries from Kapuskasing gneisses, and appreciable Cl and graphite contents from deformed Brazilian iron formation of Minas Gerais. Frost et al (1989[2]) showed that thin graphite films could explain the high electrical conductivity of some granulite facies rocks from Wyoming.

If a rock has been involved in a fluid crystallization process, it is almost impossible that all the fluid is drained. Surface techniques will see the thin films (perhaps a few Å thick) representing the final fluids to pass along grain boundaries. Such fluids can be related to early or late, even surface unloading, processes. In this study, we used AES to investigate the surface composition of country rocks and ores from the Golden Giant gold deposit of Hemlo. This work is designed to see what types of fluids could be detected and what differences exist between ore-grade and barren rocks.

Fluid inclusions in quartz from the ore zone were also investigated. Semi-quantitative analysis of salt residues (decrepitates) from decrepitation of fluid inclusions was employed to obtain major element chemical composition besides thermometric analysis of fluid inclusions. This technique was proposed by Eadington (1974[3]), and has been employed subsequently to obtain inclusion compositions in a number of hydrothermal deposits (e.g. Heinrich and Cousens 1989[4]; and references therein).

Analytical methods

1. Auger electron spectroscopy (AES)

In an AES experiment, a high energy electron beam (1-10 kV) is focused on a sample's surface, resulting in inner shell ionizations of surface atoms. As ionized surface atoms relax, excess energy is disspated by either fluorescence, or the ejection of an electron with characteristic kinetic energies of the donor atom. The latter is known as the Auger process and the ejected electron is referred to as an Auger electron (Alford et al. 1979[5]). In this study, AES analysis was carried out using a Perkin-Elmer PHI model 600 scanning Auger microprobe. Survey scans were obtained by detecting and counting the number of Auger electrons with energies in the range of 0-1000 eV. Analysis was performed with a low energy electron beam (2.5 to 3 kV, 15 to 25 nA beam current). Semi-quantitative surface compositions are calculated using peak heights and sensitivity factors. The beam diameter is 0.04 μm and the penetrating depth is 10 to 20 Å. AES depth profiles are collected with reference to sputter time rather than depth. Sputter depth can be calculated using a sputter rate for a silica film that is 270 Å per minute. Rock samples were broken in order to expose fresh surfaces, which can be either grain boundaries or cleavage surfaces. The broken rock fragments with fresh surfaces were handled with clean tools to prevent surface contamination.

2. Electron microprobe analysis

Chemical composition of muscovite was determined with the JEOL JXA-8600 Superprobe, using an accelerating voltage of 15 kV. beam current of 10 nA, a beam diameter of 2-3 μm. The counting time for each element is 20 seconds. Calibration was against minerals, synthetic glasses and metals.

3. Fluid inclusion study

Doubly polished quartz wafers from the veinlets were studied on a Linkam TH600 heating-cooling stage that is calibrated against standard high-purity chemicals with known melting points as described by Macdonald and Spooner (1981[6]).

Homogenization temperatures of the fluid inclusions were measured at a heating rate of 0.5°C per minute. Salinity of the inclusion fluids was estimated by measuring temperatures of final ice melting of two-phase aqueous (L + V) inclusions or the melting temperature of daughter halite crystals. In order to initiate decrepitation the wafers were heated at a heating rate of 20°C per minute to 450°C. The decrepitated wafers were immediately mounted onto aluminium stubs, put under vacuum and coated with gold. The analytical work was performed on an ISI-DS 130 scanning microscope (SEM) equipped with energy-dispersive detection system (EDS). Semi-quantitative data of major element chemical composition were obtained with measurement of Kα peak counts on EDS spectrum by the computer system.

Geological setting

The Golden Giant gold deposit is part of the Hemlo gold deposit located in northwestern Ontario on the Northwestern shore of Lake Superior near Marathon (Fig. 1). The Hemlo Gold Deposit contains an undelimited reserve of more than 70,000,000 tonnes averaging 7.7 g/t of gold. It is also enriched in Mo, Sb, As, Hg, Tl, V and Ba. The Hemlo Gold Deposit is situated within a sequence of Archean metasedimentary and metavolcanic supracrustal rocks that are part of the Hemlo-Heron Bay Greenstone Belt within the Wawa Subprovince of the Superior structural Province. The deposit is located within a linear zone of heterogenous ductile and brittle deformation intruded by quartz and feldspar porphyritic trondhjemite dykes collectively referred to as the Hemlo Shear zone. The geology of the greenstone belt, and particularly of the Hemlo area has been described in numerous studies (e.g. Goad 1987[7]; Hugon 1986[8]; Muir 1982[9, 10]; Harris 1986[11], 1989[12]), and most authors agree that the greenstone belt has been subjected to a complicated history of deposition, magmatism, deformation, metamorphism and metasomatism.

The Hemlo gold deposit contains a diverse assemblage of minerals. Native gold is the principal gold mineral. Pyrite is the most abundant sulphide. Barite is the principal barium mineral and is found mainly within the Golden Giant orebody where it accounts for as much as 70% of some drill core samples (Harris 1986[11]). Microcline contains up to 16.6 wt% BaO. Other gangue minerals include quartz, sericite, V-muscovite, V-Sb-W-bearing rutile. Details of the mineralogy and geochemistry of the deposit and references to regional and local geology are reported by Harris (1989[12]). The main ore zone is a single ore body being mined by the Williams mine, the Golden Giant mine, and the David Bell mine. The ore-bearing rocks of the main orebody include quartz veins, white mica schist, microcline-rich gneiss, and granofels which strikes at about 110° to 115° and dip 60° to 70° northeast. Metamorphic grade in the vicinity of the Hemlo gold deposit is near the amphibolite-greenschist facies transition (Goad 1987[7]).

Rock and ore samples

In the present study, samples investigated include both gold ore and altered country rock. The former includes four types with different dominant mineral phases and the latter is composed of different types of mica schists, amphibolite and hydrothermal altered oligoclase-quartz porphyritic trondhjemite. Their classification and petrography were described by Goad (1987[7]).

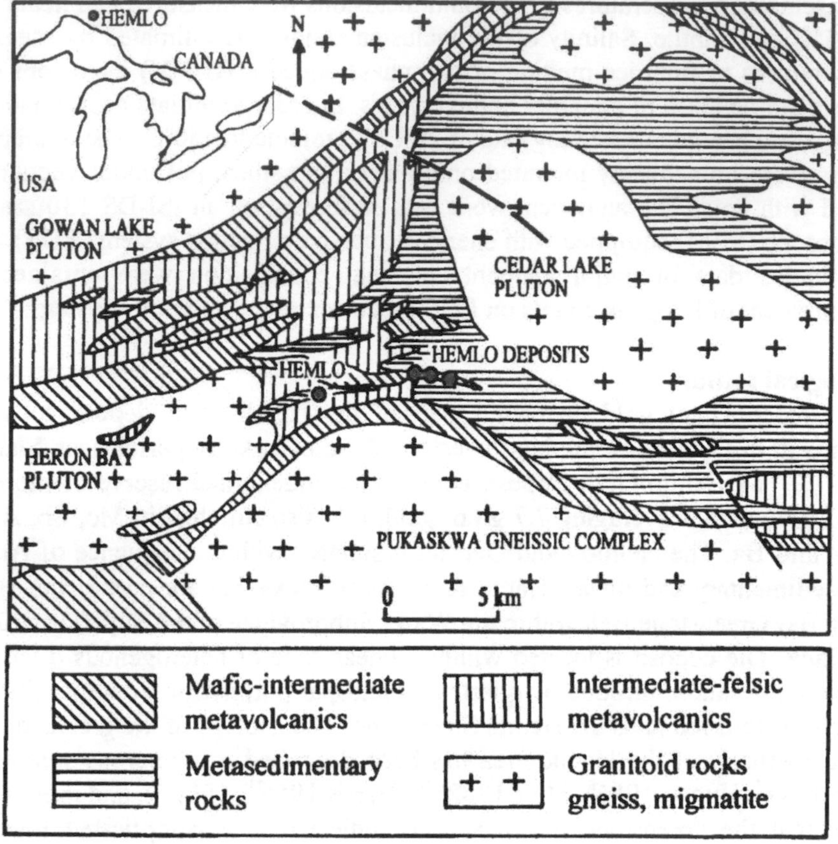

Fig. 1. Geological map of the Hemlo area (after Muir, 1982[9,10])

1. Altered country rock

Country rock investigated include four types: oligoclase-biotite-quartz schist (type I), quartz-oligoclase-muscovite schist (type II), lower quartz-oligoclase porphyritic breccias (type III), and quartz-eye muscovite schist and quartz-eye microcline granofels (type IV). Rock type I is located distal to the main ore zones relative to other types. It occurs in both the structural hangingwall and footwall of the Golden Giant deposit. In localities more proximal to areas of deformation and hydrothermal activity the schist contains garnet (up to 10%) and staurolite porphyroblasts. Quartz and/or calc-silicate veinlets, typically less than 2 mm wide, are common in the schist. This rock type was interpreted as a metamorphosed intermediate volcaniclastic rock. Rock type II is similar to rock type I, except for pervasive muscovite alteration. This rock type is interlaminated throughout the type I schist, especially in a 10 m thick interval immediately above the main ore zone. This rock commonly contains concordant and discordant micro-quartz veinlets which also contain barite and carbonate minerals. Rock type III is a clast-bearing rock with domains of oligoclase-quartz porphyritic trondhjemite in a micaceous matrix, which occurs in 3 horizons in the Golden Giant deposit. Only the lowermost horizon, that is referred to as the lower quartz-oligoclase porphyritic breccia, is right in contact with the ore zones. Rock type IV is also in contact with ore zones, and is mainly composed of plagioclase oligoclase-andesine, quartz and muscovite. The schist contains 2 to 7% lenticular quartz-eyes between 0.5 and 5 mm in size, oriented in the plane of schistosity. The quartz-eye microcline granofels occurs up to 10 m thick layers within the schist. The granofels is composed of microcline with variable amounts of quartz, albite to oligoclase, and some accessory minerals.

2. Gold ores

The ore zones at the Hemlo deposits are predominantly stratabound within country rock types III and IV. These rocks were interpreted as sheared and altered quartz-oligoclase porphyritic trondhjemite dykes (Goad 1987[7]). The ores within the Golden Giant deposit are mostly within the main ore zone which is 3-40 m thick, along a strike length of 100-650 m (Brown et al. 1986). The ores are mainly composed of quartz, microcline, muscovite (and V-muscovite), barite, biotite and disseminated and fracture filling opaque minerals. The opaque minerals include pyrite, molybdenite, native gold, stibnite, chalcopyrite, sphalerite, arsenopyrite, realgar, orpiment, cinnabar, ilmenite and magnetite.

Based on the major gangue minerals, Goad (1987[7]) classified four types of ores, including quartz-microcline siliceous ore (type I), muscovitic ore (type II), baritic ore (type III, with up to 60% barite) and biotitic ore (type IV). Ore type I is the dominant ore in the Golden Giant deposit, averaging approximately 10-15 g/t of gold (Brown et al. 1986[13]). Ore types II and III are lower grade than the ore type I, averaging 6 g/t of gold. Ore type II is mineralogically and texturally similar to the ore type I, but contains much more muscovite. Ore type IV is volumetrically less important than the other ores. Quartz veinlets (up to 5 cm thick) and micro-veinlets are common in these ore types, especially in ore types I and II. The veinlets are predominantly emplaced in the plane of schistosity but are also locally discordant. These veinlets also contain pyrite, barite, molybdenite, stibnite, realgar, cinnabar and native gold.

Results

1. Mineral surface composition

Country rock and ores were systematically scanned with AES, and results are given in Table 1 and Figure 2 . Chlorine was observed on mineral surface of quartz, muscovite, microcline from most country rock samples except for those from rock type I. The highest amounts of Cl preserved on country rock types II, III and IV are 1.5, 4.2 and 2.9 wt%, respectively. It demonstrates that the country rock closest to the ore zones have higher Cl contents on the mineral surface. Fluorine (2.8 wt%) has been observed on quartz surface in lower quartz-oligoclase porphyritic breccia (country rock type III). Appreciable amounts of Cl were also found on mineral surfaces (pyrite, quartz and barite) from quartz microcline siliceous ore (ore type I), and baritic ore (ore type III). Baritic ore contains higher Cl content (up to 5.6 wt%). Besides Cl, minor amounts of S and P have also been detected both from country rock and ores. Carbon was detected on all the sample surfaces, but only a few spectra clearly shows presence of graphite which is characterized by a carbon peak with a shoulder (Mareschal et al. 1992[1]). Most C on the mineral surface is probably adventitious carbon, as discussed by Hochella (1990[14]).

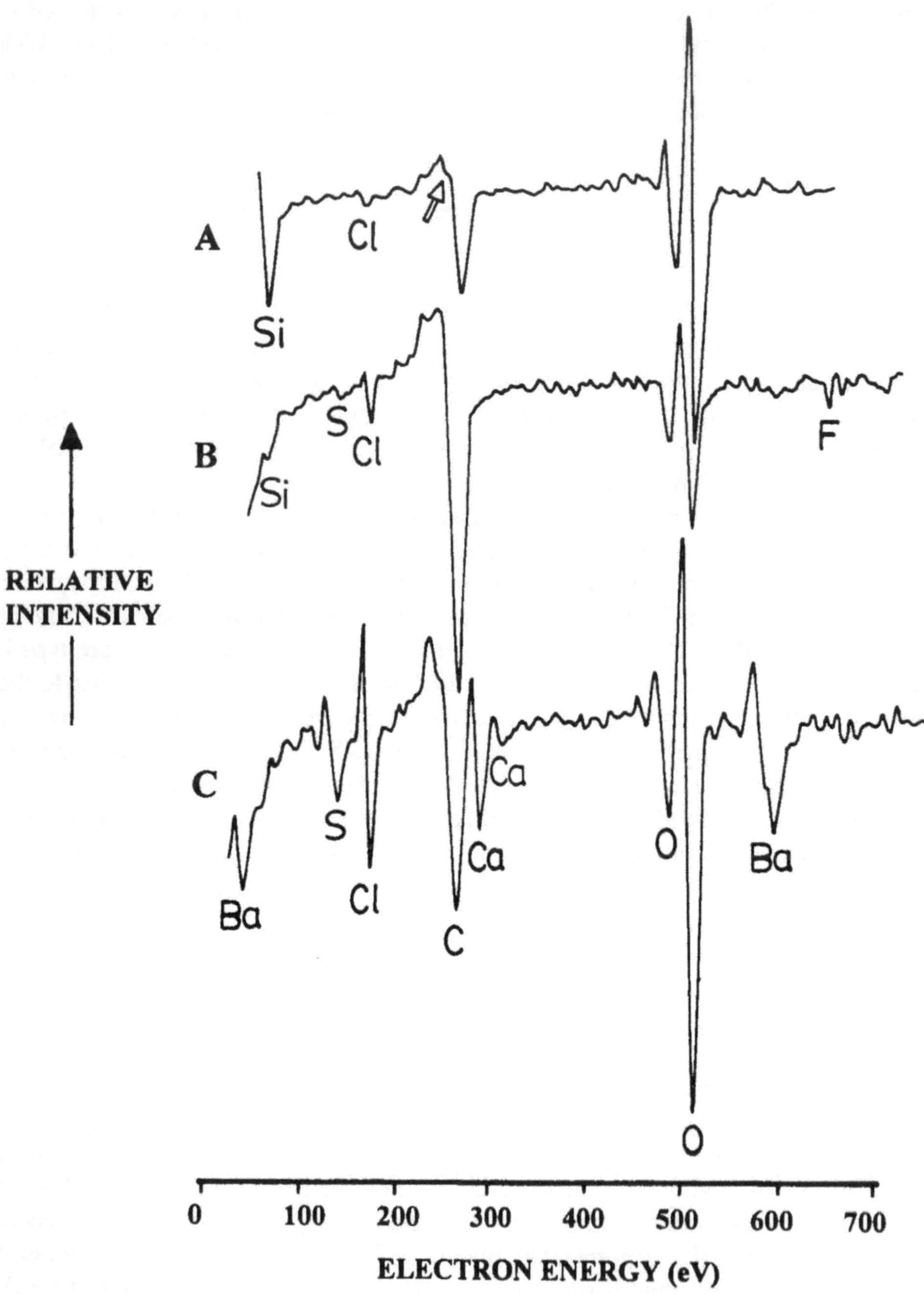

Fig. 2. Section of Auger spectra for quartz (A, B) from country rock (sample 27171) and barite (C) from baritic ore (sample 27142), showing the relative importance of Cl, S, F on quartz surface. Graphite is characterized by a carbon peak with a shoulder (arrow).

It should be noted that Cl contents on mineral surface are as high as 14.9 wt% after elimination of carbon from the results.

Table 1. Mineral surface composition of rocks and ores from the Golden Giant gold deposit (wt%).

	country rock						ore			
Type	II	III	III	IV	IV	IV	I	I	III	III
No.	27176	27183	27171	27146	27145	27187	27230	27166	27142	27142
Miner.	Ms	Qtz	Qtz	Ms	Ms	Mc	Py	Qtz	Brt	Brt
K	3.2			8.5	12.6	29.8			1.3	2.7
Ca	3.2	3.0		5.1	8.3	2.0		1.9	8.9	10.4
Ba									58.5	53.0
Al	5.2			0.9	4.4	2.5				
Fe							41.9			
Si	11.0	10.4	1.9	20.7	11.8	14.8		30.3		
Cl	1.5	1.2	4.2	0.8	2.9	0.9	0.7	1.0	5.6	5.3
F			2.8							
S		1.9	1.9	1.1	1.2	0.6	5.2		3.0	3.3
O	42.2	32.4	17.3	39.9	32.5	34.1	21.2	42.3	12.3	13.3
P		1.3					1.2			
C	33.6	49.8	71.9	23.0	26.3	15.3	29.9	24.5	10.5	11.9
	normalized data with elimination of carbon									
K	4.8			11.0	17.1	35.2			1.5	3.1
Ca	4.8	6.0		6.6	11.3	2.4		2.5	9.9	11.8
Ba									65.4	60.2
Al	7.8			1.2	6.0	3.0				
Fe							59.5			
Si	16.6	20.7	6.8	26.9	16.0	17.5		40.1		
Cl	2.3	2.4	14.9	1.0	3.9	1.1	1.0	1.3	6.3	6.0
F			10.0							
S		3.8	6.8	1.4	1.6	0.7	7.6		3.4	3.7
O	63.6	64.5	61.6	51.8	44.1	40.3	30.2	56.0	13.7	15.1
P		2.6					1.7			

NOTES: Ms=muscovite, Qtz=quartz, Mc=microcline, Py=pyrite, Brt=barite.

The Cl detected by AES probably occurs as films adsorbed on mineral surfaces, especialy quartz, microcline and pyrite that usualy do not contain Cl in their crytal structure. It is well known that muscovite contains Cl in its crystal structure. However, electron microprobe analysis (Table 2) demonstrates that the mucovite in this study contains Cl only up to 0.05 wt%, with most results below 0.01 wt%. These data confirm that Cl largely concentrates on mineral surfaces. The muscovite contains significant amounts of Cr_2O_3 (up to 0.30 wt%) and V_2O_5 (0.45 wt%). The Cr- and V-rich muscovite was also reported by Harris (1989[12]), and Pan and Fleet (1991[15], 1992[16]).

The highest Cl value (5.6 wt%) was detected on barite surface. AES depth profiles for barite clearly show that Cl content decreases with increasing sputter time (Fig. 3). The sputter depth can be converted to depth using a sputter rate for a silica film that is 270 Å per minute. Therefore, most Cl concentrated on the surface layer less than 135 Å, with the highest value near the surface of a few Å.

Table 2. Electron microprobe analyses of muscovite from the Golden Giant gold deposit (wt%).

sample No.	27145	27145	27145	27146	27146	27146
SiO_2	46.25	47.83	46.26	46.36	46.30	46.59
TiO_2	0.92	1.01	0.92	1.32	1.31	1.08
Al_2O_3	33.20	31.24	31.11	31.66	32.25	31.92
Cr_2O_3	0.26	0.02	0.30	0.07	0.02	0.02
V_2O_3	0.45	0.06	0.39	0.30	0.36	0.03
MnO	0.12	0.10	0.08	0.08	0.09	0.08
FeO	0.95	1.13	1.09	2.52	2.50	2.27
MgO	1.85	2.91	2.71	1.74	1.53	1.65
BaO	1.58	3.15	2.55	0.75	0.82	0.53
K_2O	10.91	10.19	10.97	11.35	11.25	11.18
Na_2O	0.50	0.24	0.25	0.32	0.39	0.37
F	0.00	0.00	0.00	0.00	0.00	0.00
Cl	0.00	0.01	0.00	0.00	0.00	0.01
F,Cl=O	0.00	0.00	0.00	0.00	0.00	0.00
Total	96.99	97.89	96.63	96.47	96.82	95.73
			numbers of ions on the basis of 22 oxygen			
Si	6.16	6.35	6.24	6.23	6.20	6.27
Al^{IV}	1.84	1.65	1.76	1.77	1.80	1.73
Al^{VI}	3.37	3.23	3.19	3.24	3.28	3.34
Ti	0.09	0.10	0.09	0.13	0.13	0.11
Cr	0.03	0.00	0.03	0.01	0.00	0.00
V	0.05	0.01	0.04	0.03	0.04	0.00
Mn	0.01	0.01	0.01	0.01	0.01	0.01
Fe	0.11	0.13	0.12	0.28	0.28	0.26
Mg	0.37	0.58	0.55	0.35	0.31	0.33
Ba	0.08	0.16	0.13	0.04	0.04	0.03
K	1.85	1.72	1.89	1.95	1.92	1.92
Na	0.13	0.06	0.07	0.08	0.10	0.10

2. *Fluid inclusions*

Primary fluid inclusions in some quartz veinlets from both country rock and ore ranges from 2 to 50 μm in length, randomly distributed through the quartz crystals. Morphologically, the inclusions form hexagonal negative crystals with a few inclusions in irregular form. At room temperature some inclusions are two-phase liquid/vapour inclusions. The volume of vapour bubble relative to the whole inclusion was estimated from 10 to 20%. Three-phase inclusions with cubic halite crystals are common in the studied samples. Some two-phase inclusions become three-phase ones with crystallization of halite crystals when the temperature is slightly lowered.

Homogenization temperatures for the fluid inclusions in minerals studied range from 167 to 227°C, with most temperatures below 200°C (Table 3). There is no large difference between country rock and ore samples. The inclusions with daughter minerals at room temperature suggest saturation of fluids with salts. This type of inclusions have salinity higher than 26 wt% NaCl in NaCl-H_2O system

(Crawford, 1981[17]). The temperature of daughter mineral melting for three-phase inclusions ranges from 126-161°C, suggesting a salinity of 28-29 wt% NaCl equivalent. The appearance of daughter minerals in two-phase liquid/vapour inclusions with temperature lowering indicates that the inclusions are almost saturated with salts at room temperature. The final ice-melting temperatures for two-phase inclusions range from -13 to -21°C, which suggest that some two-phase inclusions have salinity of 19-20 wt% NaCl equivalent. In general all inclusions in these sample are high salinity inclusions

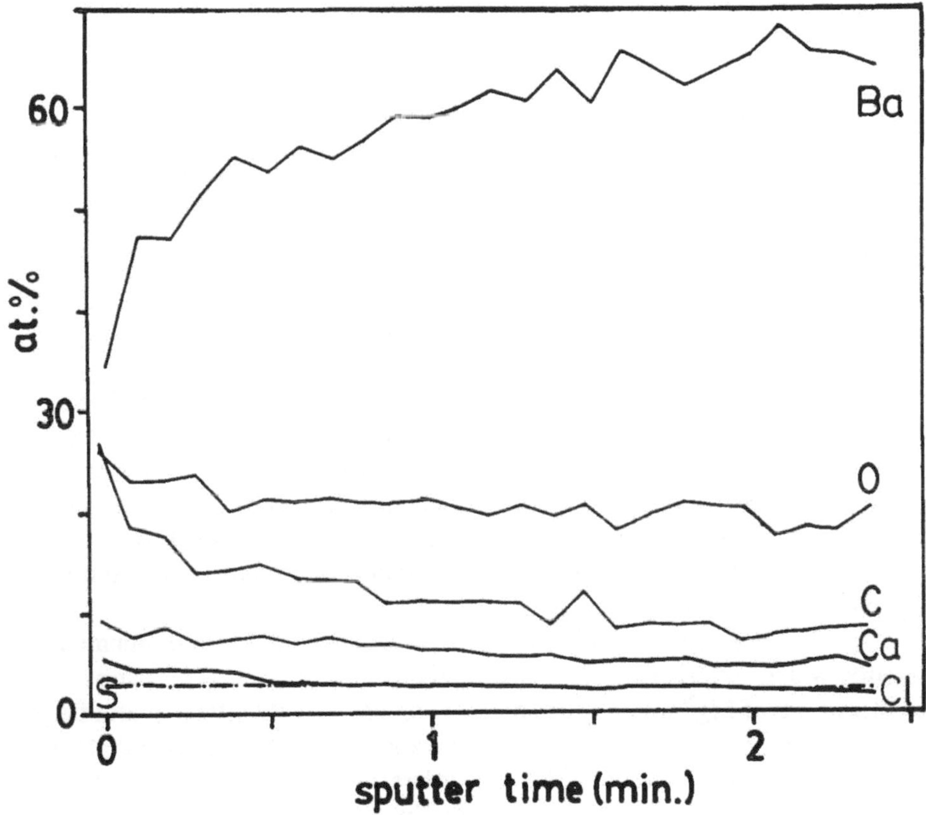

Fig. 2. AES depth profiles of elements on barite surface. Depth can be calculated using a sputter rate for a SiO_2 film that is 270Å per minute

Fluid released during decrepitation travelled to the polished surface along microfractures where volatile components, mainly H_2O and perhaps CO_2, were lost to the atmosphere and salt residues (decrepitates) were deposited. Decrepitates of the inclusions in quartz veinlets from country rock type I are typically characterized by amorphous Ca-rich mounds with cubic halite crystals scattered farther away. In contrast, most decrepitates of inclusions in quartz veinlets from ores are characterized by amorphous mounds, with only few tiny halite crystals observed, especially those from Sb-bearing inclusions.

Table 3. Decrepitate composition (wt%), homogenization temperature (T_h) and salinities of fluid inclusions in quartz veinlets from the Golden Giant gold deposit.

	country rock						ore			
Type	I		I		IV		I		I	II
Sample	27242		27213		27190		27158		27166	27151
Phase	xtl	amorp	xtl	amorp	amorp		amorp		amorp	amorp
Na	15.5	0.8	21.5	10.4	14.9	12.2	14.6	28.0	17.9	10.9
Ca	5.1	30.0	7.9	22.0	9.6	28.8	9.5	3.5	4.9	5.2
K	0.0	0.0	0.0	0.3	0.2	0.5	0.0	0.3	0.1	0.0
Mg	0.0	0.0	0.0	0.0			0.7	0.0	0.0	11.6
Si	30.8	5.1	11.5	9.7	1.4	3.9	16.0	16.4	22.5	2.2
Al	0.0	0.0	3.1	2.4	0.0	0.0	0.3	0.2	0.0	0.0
Fe	0.0	1.2	0.0	0.3	0.0	0.0	0.0	0.0	0.0	0.4
Ti	0.0	0.0	0.0	0.0	0.0	0.0	0.0	0.1	0.1	0.4
Cl	29.7	10.6	47.5	24.6	26.5	19.1	24.4	40.0	34.0	13.9
C	0.0	20.2	0.0	11.1	42.2	15.6	7.9	0.0	0.0	16.6
O	19.0	32.3	8.5	18.8	4.7	19.9	24.9	10.3	18.3	16.2
S	0.0	0.0	0.0	0.0	0.6	0.0	0.0	0.0	0.0	1.7
As			0.0	0.4			1.7	0.0	2.3	
Sb			0.0	0.0			0.0	1.2	0.0	21.1
T_h(°C)	204-217		182-193		209-227		167-172		188-199	183-194
S	26.5		28.0		22.0		29.0		20.0	19.5

NOTES: xtl=crystal, amorp=amorphous; S=salinity, NaCl wt% equivalent.

The semi-quantitative data of decrepitates are given in Table 3. In decrepitates from the country rock, the Ca-rich amorphous mounds are mainly composed of $CaCl_2$ and $CaCO_3$ with small amounts of NaCl. The crystals are dominated by NaCl. The decrepitates from the ores are more complicated than those from the country rock. Some decrepitates are dominated by NaCl, with or without $CaCl_2$. Some contain more C, suggesting a carbonate-rich fluid. In one sample, high Mg and Sb were detected.

These decrepitates indicate that anions in the fluids are dominated by Cl^-, together with CO_3^{2-}, HCO_3^-. Correspondingly cations are mainly Na^+ and Ca^{2+}, with small amounts of Mg^{2+} and K^+. The high Si contents largely reflect quartz fragments in the decrepitates. Small amounts of Al, Fe and Ti probably indicate that the fluids contain these elements as chlorides. It is interesting that the decrepitates from ores contain small amounts of S, As and Sb. The Sb in a particular sample even becomes major component. These elements reflect the nature of the fluids related to gold mineralization.

Discussion and conclusions

Various genetic models have been suggested for the origin of the mineralization of the Hemlo gold deposit. Some authors suggest the deposit was syngenetic or exhalative whereas some authors suggest epigenetic, structurally controlled or porphyry deposits (see Harris 1986[11]; and references therein). Most authors agree that this deposit is related to a deep shear zone system. In the syngenetic model, hot spring solution was suggested to be responsible for the metal-rich sediments (Patterson, 1986[18]), analogue to modern hot springs in New Zealand (Weissberg, 1969[19]). The metal-rich sediments could occur within porous units such as tuffs, fragments, and fault zones, resulting in alterations of muscovite, microcline, quartz and metal association of Au, Mo, As, Sb, Hg, and Ba. Other examples include those near Goldfield, Nevada, which also occur in tuffs (Boyle 1979[20]).

There is no doubt that fluids have played an important role in the mineralization processes. However, only a few studies have been done on the nature of fluids (e.g., Pan and Fleet 1992[21]). In this study, the chloride films observed on mineral surface both from ore zones and from country rock indicate that mineralization fluids are quite saline. The baritic ore contains distinctly higher surface Cl than the country rock. It is surprising that the fluid inclusions have such low homogenization temperatures. Further study is needed for a much wider range of ores and rock types. There are some clear problems involving fluid compositions and time and temperature regimes. Data presented here suggest an epithermal model but AES sees only the latest fluids to influence the rocks and ores.

In conclusion AES is a powerful tool to investigate mineral surface compositions, especially for searching the elements that indicate fluid nature in geological environments. This study demonstrates that traces of Cl, S and K are preserved on grain boundaries, cleavage surface of country rock and ores of the Golden Giant gold deposit of Hemlo. The results indicate that fluids were involved in formation of the rocks and ores. Fluid inclusions in quartz from the Golden Giant deposit are highly saline (19-29 NaCl wt%, equivalent). Analysis of fluid inclusion brine residues produced by evaporation shows that NaCl is dominant, with variable amounts of $CaCl_2$ and carbonate species. Homogenization temperatures of the fluid inclusions range from 167 to 227°C.

Acknowledgements

This research is supported by Ontario Geoscience Research Grant 437 to W.S. Fyfe. We thank R. E. Goad for collecting the samples, J. Forth for the preparation of polished thin sections, R.D. Davidson for the helping with SEM, and S. Ramamurthy for the assistance with AES analysis, R.L. Barnett for the assistance with electron microprobe analysis.

References

1. Mareschal, M., Fyfe, W. S., Percival, J, and Chan, T. 1992. Grain-boundary graphite in Kapuskasing gneisses and implications for lower-crustal conductivity. Nature, 357: 674-676.

2. Frost, B. R., Fyfe, W. S., Tazaki, K., and Chan, T. 1989. Grain-boundary graphite in rocks and implications for high electrical conductivity in the lower crust. Nature, 340: 134-136.

3. Eadington, P. J. 1974. Microprobe analysis of non-volatile constituents in fluid inclusions. Neues Jahrbuch für Mineralogie Monatshefte, 11: 518-525.

4. Heinrich, C. A., and Cousens, D. R. 1989. Semi-quantitative electron microprobe analysis of fluid inclusion salts from Mount Isa copper deposit (Queensland, Australia). Geochimica et Cosmochimica Acta, 53: 21-28.

5. Alford, N. A., Barrie, A., Drummond, I. W., and Herd, Q. C. 1979. Auger electron spectroscopy (AES) an appraisal. Surface and Interface Analysis, 1: 36-44

6. Macdonald, A. J. and Spooner, E. T. C. 1981. Calibration of a Linkam TH600 programmable heating-cooling stage for microthermometric examination of fluid inclusions. Economic Geology, 76: 1248-1258.

7. Goad, R. E. 1987. The geology, primary and secondary chemical dispersion of the Hemlo Au district metal occurrences, northwest Ontario. M.Sc. thesis, University of Western Ontario, London,

Ontario.
8. Hugon, H. 1986. The Hemlo gold deposit, Ontario, Canada: a central portion of a large scale, wide zone of heterogeneous ductile shear. Proceedings of Gold'86, an International Symposium on the Geology of Gold, Toronto, pp.379-387.
9. Muir, T. L. 1982. Geology of the Hemlo area, district of Thunder Bay. Ontario Geological Survey, Report 217, 65p.
10. Muir, T. L. 1982. Geology of the Heron Bay area, district of Thunder Bay. Ontario Geological Survey, Report 218, 89p.
11. Harris, D. C. 1986. Mineralogy and geochemistry of the main Hemlo gold deposit, Hemlo, Ontario, Canada. Proceedings of Gold'86, an International Symposium on the Geology of Gold, Toronto, pp. 297-310.
12. Harris, D. C. 1989. Mineralogy and geochemistry of the Hemlo gold deposit, Ontario, Canada. Geological Survey of Canada, Economic Geology Report 38, 88p.
13. Brown, P., Friesen, R., Kennedy, P. Kusins, R. and McNena, K. 1986. Golden Giant Mine geology. *In* The Hemlo gold deposits, Ontario. Joint Annual Meeting of Geological Association of Canada, Mineralogical Association of Canada, and Canadian Geophysical Union, Ottawa'86, Guidebook of Field Trip 4, pp. 37-43.
14. Hochella, M. F. 1990. Atomic structure, microtopography, composition, and reactivity of mineral surfaces. *In* Mineral-Water Interface Geochemistry, Reviews in Mineralogy, 23: 87-132.
15. Pan, Y. and Fleet, M. E. 1991. Barian feldspar and barian-chromian muscovite from the Hemlo area, Ontario. Cannadian Mineralogist, 29: 481-498
16. Pan, Y. and Fleet, M. E. 1992. Mineral chemistry and geochemistry of vanadian silicates in the Hemlo gold deposit, Ontario, Canada. Contributions to Mineralogy and Petrology, 109: 511-525.
17. Crawford, M. L. 1981. Phase equilibria in aqueous fluid inclusions. *In* Fluid Inclusions: Application to Petrology. *Edited by* L.S. Hollister and M.L. Crawford. Mineralogical Association of Canada, Short Course Handbook 6, pp.75-100.
18. Patterson, G. C. 1986. Regional field guide to the Hemlo area. *In* The Hemlo gold deposits, Ontario. Joint Annual Meeting of Geological Association of Canada, Mineralogical Association of Canada, and Canadian Geophysical Union, Ottawa'86, Guidebook of Field Trip 4, pp. 1-28.
19. Weissberg, B. G. 1969. Gold-silver ore-grade precipitates from New Zealand thermal waters. Economic Geology, 64: 95-108.
20. Boyle, R. W. 1979. The geochemistry of gold and its deposits. Geological Association of Canada, Bulletin 280, 584p.
21. Pan, Y. and Fleet, M. E. 1992. Calc-silicate alteration in the Hemlo gold deposit, Ontario: Mineral assemblages, P-T-X constraints, and significance. Economic Geology, 87: 1104-1120.

Proc 30th Int'l. Geol. Congr., Vol. 16, pp. 109-133
Huang Yunhui and Cao Yawen (Eds)

The Formation and Evolution of the Minerals in the Jinchuan Cu-Ni Deposit, Gansu Province, China

CAO YAWEN and CAI JIANHUI
Institute of Mineral Deposits, Chinese Academy of Geological Sciences, Beijing 100037, *China*
LI BING
China University of Geosciences, Beijing 100083, *China*

Abstract

The Jinchuan Cu-Ni Sulfide Deposit is the biggest magmatic Cu-Ni sulfide deposit in China. It occurs in the Jinchuan Ultramafic Massif which is located in central Gansu Province. The massif was formed through the intrusion of three stages. The lherzolite, plagioclase-lherzolite, olivine-websterite etc. were formed in the first and the second stages; and the sulfide dunite was formed in the third stage. The Cu-Ni orebodies of the Jinchuan Deposit mainly occur in the lherzolite and sulfide-dunite. The original magma of this massif is komatiitic. The main minerals of the massif and the orebodies are chrysolite, bronzite, endiopside, pargasite, plagioclase, Ti-bearing phlogopite, phlogopite, clinochlore, pyrrhotite, pentlandite, chalcopyrite, cubanite, mackinawite, magnetite, chromite, etc. The chrysolite, pyroxenes was often replaced by serpentine, magnesiocummingtonite and tremolite. The fabrics of the rocks in the massif and ores in the orebodies are similar. Based on the mineral intergrowth relationships the authors get to the conclusion that the evolution history of each intrusive can be divided into two stages: The first stage is the main solidifying stage. During this stage, about 70—80% of the volume of the intrusive was solidified, and chrysolite, pyroxenes, plagioclase, euhedral pargasite, etc. formed. The second stage is the ore forming and autometamorphism stage. In this stage, pyrrhotite, pentlandite, chalcopyrite, chromite(Chr^2), magnetite, etc. and some hydrosilicates such as Ti-bearing phlogopite, phlogopite, fibrous pargasite, tremolite, clinochlore, serpentine, etc. formed. There was a late stage liquation of Fe—Ni—Cu—Cr—S—O melt from silicate melt which was rich in water and alkalis in the residual magma after the main solidifying stage. The mineralization process of the Jinchuan Deposit is recognized as " crystallization differentiation + late stage liquation". The ore minerals of the Jinchuan deposit had a very special crystallization process. Simply, the process can be described as: chromite(Chr^2)→pyrrhotite→"pentlandite +magnetite solid solution (PMSS)"+iss; when PMSS exsolved, pentlandite and magnetite(Mt^2) formed; and chalcopyrite and cubanite formed when iss exsolved.

Keywords: Jinchuan, Cu-Ni deposit, mineral, formation

INTRODUCTION

The Jinchuan Cu-Ni Sulfide Deposit is the largest magmatic sulfide Cu-Ni deposit in China, and the third largest in the world. It is situated in central Gansu Province, China. A lot of explorative and geological work have been done since it was discovered in 1958. The geological characteristics, such as the tectonic setting, petrochemistry, the occurrence of the orebodies etc., have been introduced in tens of papers and research reports. Many geologists proposed that the main mineralization types of the deposit was related to the magmatic liquation[1, 2, 3]. Unfortunately, the mineralogical researches for the deposit is inadequate, and some common minerals in the deposit, e.g., Ti-bearing phlogopite, magnesiocummingtonite etc., are never reported in literatures. This paper presents the important results of a mineralogical study on the compositions, intergrowth relationships, and the formation and evolution of the common minerals in the deposit, excluding noble metal minerals (though they are abundant in the ores), trace, rare minerals and the minerals formed by weathering. Mineral abbreviations used in this paper are listed in Table 1.

Table 1. Abbreviations used

Name	Abbrev.	Name	Abbrev.
arfvedsonite	Ar	clinochlore	Cch
chromite	Chr	chalcopyrite	Cp
clinopyroxene	Cpx	cubanite	Cub
ilmenite	Il	mackinawite	Mac
marcasite	Mar	magnesiocummingtonite	Mc
magnetite	Mt	olivine	Ol
orthopyroxene	Opx	pargasite	Par
pentlandite	Pe	phlogopite	Phl
plagioclase	Pl	pyrrhotite	Po
pyrite	Py	serpentine	Serp
Ti-bearing phlogopite	Tph	tremolite	Tr
first stage intrusion	Σ^1	second stage intrusion	Σ^2
third stage intrusion	Σ^3	metal sulfides-oxides	MS

GEOLOGIC SETTING

Geology of the Jinchuan Ultramafic Massif

The Jinchuan Cu-Ni Deposit occurs in the Jinchuan Ultramafic Massif. The massif intrusived into the Archaeozoic Baijiajuzi Formation, in direct contacts with gneiss, ophicalcite, and zebra-isotropic migmatite. The isotopic ages of the massif is 1509—1526 Ma[1]. The massif is about 6500 meters long and 20—528 meters wide, extending downward for more than 1100 meters vertically, and has an outcropped

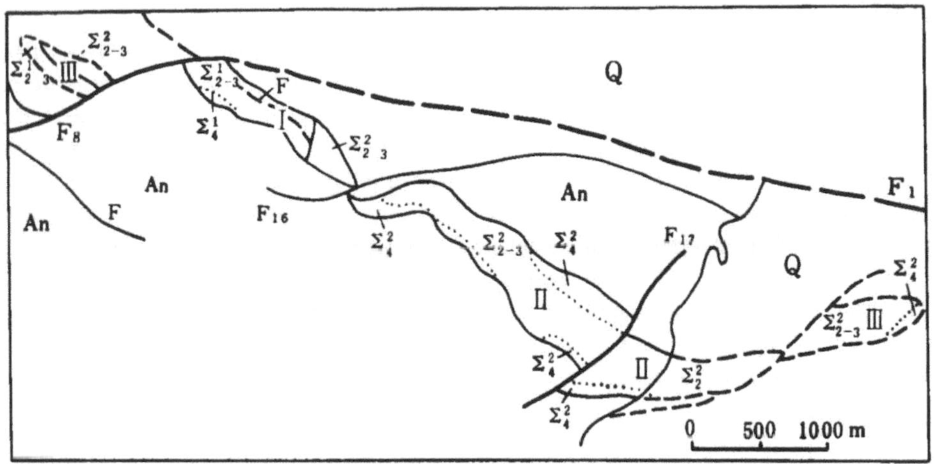

Figure 1. Geological map of the Jinchuna Ultramafic Massif, Gansu Province, China. Symbols: Q, Quaternary system; An, Archaeozoic Baijiajuzi Formation; F, fault; Σ^1_{2-3}, first stage lherzolite; Σ^1_4, first stage olivine-websterite; Σ^2_{2-3}, second stage lherzolite; Σ^2_4, second stage olivine-websterite; I, II, III, IV, number of mine areas. Geology modified after Tang Zhongli[1] and Jia Enhuan[2].

area of 1. 34 km^2(Fig. 1)[1,2]. The massif chiefly consists of lherzolite, plagioclase-lherzolite, olivine-websterite, and sulfide-dunite etc. It was formed through intrusion of three stages[1]. The third stage formed the sulfide-dunite, and the first and second stages formed the other rocks such as lherzolite, plagioclase-lherzolite, olivine-websterite, etc. The serpentine-tremolite-chlorite-gneiss occurs at the contact zone between the massif and the country rocks. Lherzolite is the most extensive rock in the massif, making up about 75% of the volume of the massif, and plagioclase-lherzolite, olivine-websterite, sulfide-dunite making up about 10%, 8%, 5. 6% of the volume of the massif respectively[3, 4].

Lherzolite is grayish black in color, and chiefly consists of olivine (chrysolite, abbreviated to Ol), clinopyroxene (mainly, endiopside, abbreviated to Cpx) and orthopyroxene (bronzite, abbreviated to Opx). There are a little pargasite (Par), phlogopite (Phl), oxides, sulfides, etc. occurring in the rock. Olivine is 1—3mm in diameter, and occurs as rounded crystalline grains. Cpx and Opx are subhedral, 1mm± in maximum dimension, in intergrowth with each other, and distribute at the intergranular openings of Ol, forming an interstitial structure. Rounded Ol grains can be included in pyroxenes, forming a poikilocrystallic structure. Pargasite is another primary mineral in the rock. It occurs as euhedral to subhedral stubby-prismatic crystals, 0. 5—1 mm in length, in intergrowth with pyroxenes. Oxide mineral is mainly chromite (Chr)(Fig. 3-F), included in Ol or pyroxenes, occurring as euhedral crystals, about 0. 02 mm in diameter. Metal sulfides are mainly pyrrhotite (Po), pentlandite (Pe), and trace chalcopyrite (Cp), distributing at the intergranulars of Ol, pyroxenes, or the centres of intergranular irregular zonal structure.

There is a special microstructure in lherzolite. Its characteristic is as follows. From Ol or pyroxenes crystals to the center of their intergranular opening, there is often a zonal structure. The first zone is the zone of fine grained, stubby-prismatic Par, Phl, and ferriferrous serpentine (Serp) that are in intergrowth; next to that is a zone of acicular or fibrous Par, which is in a parallel arrangement as comb structure; and then there is an assemblage of Par, Phl and clinochlore (CCh) in the center (Fig. 3-D). This structure can be called an intergranular irregular zonal structure. Droplike assemblages of sulfide minerals often occur at the center of this structure (Fig. 3-E). Because the boundary between the mineral assemblages of this structure and Ol or pyroxenes is regular, and the assemblages generally did not invade into Ol or pyroxenes, the origin of this structure is referred to the crystallization of the residual magma.

Ol was often replaced by serpentine. The replacement took place along the fractures of Ol, therefore separating Ol into fragments and forming a net-like structure. Pyroxenes was often replaced by tremolite (Tr), magnesiocummingtonite (Mc) or ferriferrous serpentine.

The mineral contents of lherzolite is as follows: Ol: 35—45%; Cpx: 4—8%; Opx: 1—3%; Par: 5—8%; Phl: 5—8%; Serp: 30—35%; Tr and Mc: 4—5%; Cch: 5—10%; Chr: $<$1%; sulfides: $<$1—2%. Based on the rock fabrics, the mineral contents of original rock are: Ol: 70—85%; pyroxenes: 8—15%; hydrosilicates: 5—10%; oxides+sulfides: 1—2%. The chemical compositions of this rock is given in table 2.

The petrologic characteristics of *olivine-websterite* is similar to that of lherzolite, and their difference is only in that the content of olivine is reduced and that of pyroxenes increased. Its petrochemical composition in Table 2 shows that the contents of SiO_2 and CaO are higher and ($FeO+Fe_2O_3$), MgO are lower than that of lherzolite.

Plagioclase-lherzolite generally occurs in the middle part of the massif as irregular schlirens next to lherzolite. It mainly consists of olivine (50—60%), pyroxenes (20—30%), and plagioclase (8—10%). The chromite, magnetite, Ti-bearing phlogopite and sulfides are in small amount. When sulfide was enriched, then the rock could become to lean ore.

The olivine in this rock also occurs as rounded grains, 1—2 mm in diameter, with Fo=83—85. Bronzite (Opx) is euhedral to subhedral, 2—3 mm in maximum dimension, and makes up 5—10% volume of the rock. Endiopside (Cpx) is subhedral to anhedral, 3—5 mm in maximum dimension, with a content about 15—20% in the rock. Plagioclase is subhedral platy crystals with 3—4 mm in maximum dimension. The later three minerals are in intergrowth and are distributed in the intergranular openings of olivine (Fig. 3-A). There are 2—3% brownish Ti-bearing phlogopite occurring in intergranular openings and in intergrowth with metal minerals such as pyrrhotite, chalcopyrite, pentlandite, ilmenite, etc. (Fig. 4-D).

Table 2. Average chemical composition (wt%) of the rocks in the Jinchuan massif

Rock type	lherzolite (10)*	olivine-websterite (1)	plagio-clase-lher-zolite(2)	sulfide-dunite (4)	average of first stage intrusives (3)	average of second stage intrusives (9)	average of the massif (30)
SiO_2	36.35	41.43	39.18	27.47	36.20	37.03	36.53
Al_2O_3	3.56	4.75	5.29	1.30	2.40	4.33	3.50
Fe_2O_3	6.23	5.34	3.91	8.35	7.31	5.37	6.29
FeO	9.17	6.71	11.48	18.38	6.17	10.68	8.26
TiO_2	0.28	0.48	0.39	0.15	0.25	0.31	0.31
CaO	2.34	4.26	3.93	0.71	1.24	3.06	2.43
MgO	29.66	27.88	26.36	26.19	32.95	27.85	30.19
K_2O	0.31	0.21	0.37	0.10	0.16	0.38	0.22
Na_2O	0.43	0.32	0.78	0.16	0.25	0.57	0.34
MnO	0.14	0.16	0.18	0.11	0.15	0.15	0.15
H_2O^+	8.02	6.58	5.51	8.84	9.46	6.98	8.04
CO_2	0.54	0.74	0.38	0.35	0.60	0.50	0.84
P_2O_5	0.06	0.10	0.07	0.04	0.07	0.05	0.06
Cr_2O_3	0.83	0.43	0.65	1.12	0.41	0.93	0.60
S	0.93	0.14	1.52	4.95	0.71	1.14	1.03
Cu	0.32		0.38	0.93		0.33	0.43
Co	0.03		0.02	0.07		0.03	0.04
Ni	0.52	0.12	0.50	2.27	0.27	0.6	0.45

The analyses performed by the Laboratory of Tianjin Institute of Geology and Mineral Resources.
* —presents the number of samples used for average.

Chromite can often be seen in this rock and it is similar to the ones in the lherzolite. The chemical composition of this rock is listed in Table 2, except that Al_2O_3 is slightly higher than that of lherzolite and olivine-websterite, all the other compositions are close to that of the later two rocks.

Sulfide-dunite makes up the intrusive of the third stage, mainly consisting of 75—85% olivine and 10—15% metal sulfide minerals and oxide minerals. Olivine is 50—90% serpentinized; and in the rich ore the olivine was altered to serpentine completely. Olivine generally occurs as rounded grains with a size of 1—3 mm (Fig. 3-C). The intergranular spaces among olivine grains were generally filled with metal sulfides and oxides, mainly including pyrrhotite, pentlandite, chalcopyrite, mackinawite, chromite, and magnetite etc., and formed a interstitial structure, or a disseminated structure to ores (Fig. 4-B). When metal minerals were enriched enough, olivine grains are surrounded by the metal minerals, and then the disseminated structure turns to a sideronitic structure. This structure can often be seen in rich ore. Pyroxenes are endiopside and bronzite. Bronzite makes up 70% volume of pyroxenes and together with endiopside have a content of 5—6% in the rock. Pyroxenes also occur in the intergranular spaces of olivine; and sometimes endiopside occurs as a corona surrounding bronzite (Fig. 3-B). The intergranular spaces among olivine grains could also be filled with hydrosilicates such as pargasite, phlogopite, clinochlore (Fig. 3-C) or the assemblages of brown Ti-bearing phlogopite and metal minerals. Chemically, this rock is of a highest ($FeO+Fe_2O_3$) content and lowest SiO_2, MgO, Al_2O_3 contents in contrast with the other rocks of the massif. Moreover, most of this rock can be used as ore.

The average chemical composition of the major rock types and different intrusives are given in Table 2. The results shows that the major rocks in the massif have the composition ranges of $SiO_2<42\%$, ($FeO+Fe_2O_3$): 11—33%, MgO: 26—33%, $Al_2O_3<6\%$, $CaO<5\%$, (Na_2O+K_2O)$<1\%$. This suggests that these rocks are the ultramafic rock which is of lean in Si and rich in Fe and Mg, and the original magma is komatiitic. Since the three intrusives have similar rock types and rock fabrics, they have similar evolution history.

Geology of the Deposit

The occurrence of orebodies: There are more than one hundred industrial orebodies in this deposit. Most of them occur in the central and lower part of the massif (Fig. 2), and a few small orebodies occur at the contact zone of the massif. Orebodies make up about 44% volume of the massif [4]. Based on the contact relationship between the orebodies and the country rocks and the genesis of the orebodies, the orebodies can be classified into two types, namely, type I and type II respectively.

Type I : There is no clear boundary between the orebodies and the country rocks; ores and the country rocks are similar in both minerals and fabric; the main difference between them is that in the orebodies ore minerals were enriched enough for an industrial utilizing, but not in the country rocks. The orebodies and country rocks formed synchronously. Orebodies of this type generally are bedlike or lentiform, several meters to over a thousand meters long along strike and tens to hundreds meters thick. Type I is the most important orebody type in the deposit, making up 96%± Cu and Ni reserves of the deposit. Interstitial and sideronitic structures are the most extensive structure in this type. Orebodies of this type exist in all intrusives of the three stage, and mainly occur in the intrusive of the third stage, because that the third intrusive itself is a huge orebody which makes up 85% Ni and Cu reserves of the total reserves. The orebodies in the intrusive of the first stage are very small in scale and in a very subordinate status.

Type II : There is clear boundary between the orebodies and the wall rocks; the ores of these orebodies consist almost completely of metal sulfides such as pyrrhotite, pentlandite and chalcopyrite etc. These orebodies generally are lentiform or veinlike, several to tens meters long and tens cm to about 20 m thick. They mainly exist at the bottom of the intrusive of the third stage or in wall rocks near the contact zone. Their genesis is referred to as a short distance moving of a sulfide-rich melt produced both by the magmatic crystallization differentiation and the liquation of the residual magma. The orebodies of this type make up about 4% Cu and Ni reserves

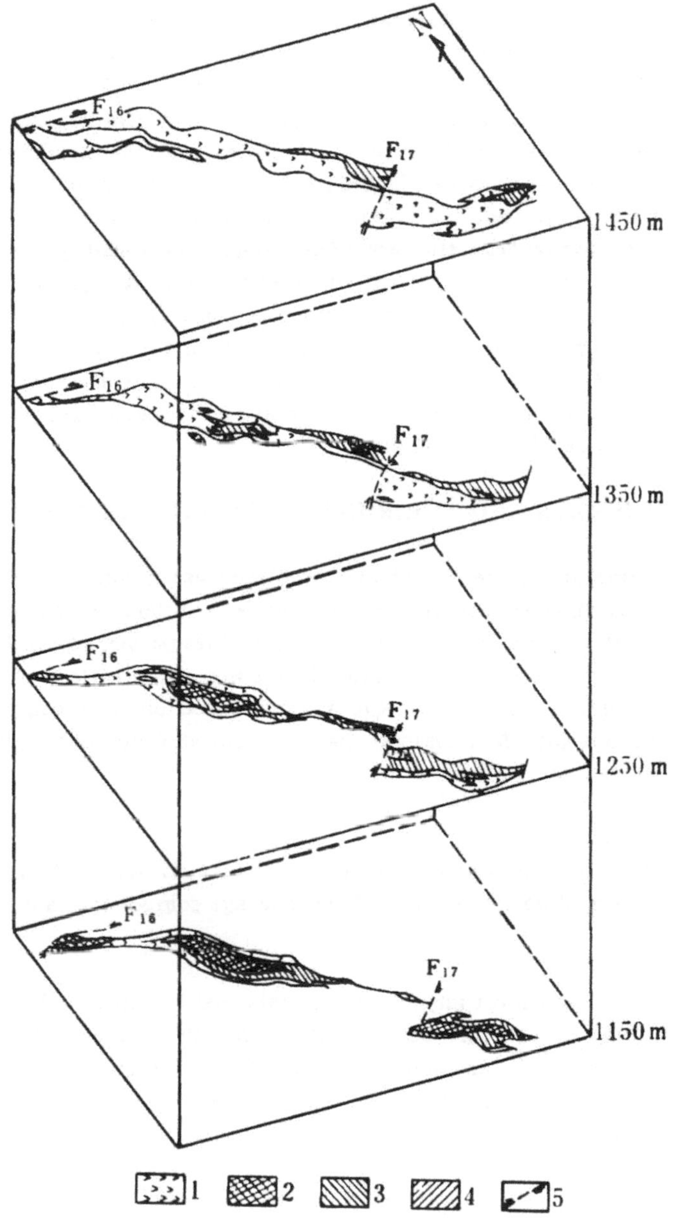

Figure 2. Sketch map of the mine area I of the Jinchuan Cu-Ni Sulfide Deposit. 1, lherzolite of second intrusive (Σ_{2-3}^{2}); 2, rich ore; 3, lean ore; 4, contact metamorphic lean ore; 5, fault. Geology modified after Tang Zhongli [1]

of the total ores.

Characteristics of the ores: Based on the ore structure, ores of the Jinchuan deposit can be divided into following three types. 1. Disseminated ore: It is characterized by its disseminated structure and often occurs in the orebodies of type I . Metal sulfides and oxides often filled in the intergranular spaces of olivine, pyroxenes, or plagioclase, etc. and scattered in the ore. The total content of metal minerals in the ore are 4—9%, which consist of about 2—5% pyrrhotite, 0. 5% chalcopyrite, 1—3% pentlandite, 0. 5—1% chromite, and 1% magnetite. Pyrrhotite occurs as anhedral grains and intergrew with chalcopyrite, chromite, and the assemblages of pentlandite+magnetite+mackinawite. 2. Sideronitic ore: It contains more metal minerals than disseminated ore, and the assemblages of metal minerals at the intergranular spaces of olivine, pyroxenes, etc. are linked together, forming a netlike appearance. Ores of this type generally contain about 8 — 10% pyrrhotite, 5% pentlandite, 2% chalcopyrite, 0. 3— 0. 5% cubanite, 0. 5% chromite, and 2% magnetite. Sideronitic ore commonly occurs in orebodies of type I in the third stage intrusives (Σ^3). 3. Massive ore: It almost consists of metal sulfides completely. The content ratios of Po:Pe:Cp=4. 3:1:1. There are a little euhedral magnetite in this ore, 0. 2—0. 5 in size, 1—4% contents.

COMMON MINERALS AND THEIR INTERGROWTH RELATIONSHIP

The common minerals in the massif and orebodies are olivine, pyroxenes, amphiboles, phlogopite, chlorite, serpentine, pyrrhotite, pentlandite, chalcopyrite, magnetite, chromite etc. Though differing in amount, these minerals can be found in almost all the rocks and ores in the massif. There are also a little amount of calcite, dolomite, trace platinoid minerals in the massif. Based on their composition, these minerals can be put into four classes, namely, silicates not containing H_2O, hydrosilicates, oxides, and sulfides.

Silicates not Containing H_2O

There are olivine, pyroxenes and plagioclase in this class. All of them formed in the main magmatic crystallization process. Their average composition and the ranges are listed in Table 3.

Olivine is a common and most important mineral, making up 70—85% volume (reconversion volume) of the Jinchuan massif. Generally, olivine occurs as rounded grains, 1—3 mm in maximum dimension (Fig. 3-C). Electron microprobe analyses results (Table 3.) show that the olivine contains 81 — 86 percent Fo, i. e. , it is chrysolite, and the composition variation range is narrow. It was commonly replaced by serpentine.

Pyroxenes make up about 10—15 reconversion volume of the massif. There are both clinopyroxene (Cpx) and orthopyroxene (Opx). Cpx:Opx ratios are 3:1—4:1 for the whole massif. The chemical composition analyses of Cpx (Table 3) show that it contains 46—66% En, 5—8% Fs, and 28—45% Wo. Most of Cpx are *endiopside*, and a little amount are diopside. The Opx contains 82—90% En, ranging 83—84%, so all the Opx in the Jinchuan massif are *bronzite*. Both Cpx and Opx are anhedral to subhedral stubby prismatic, 1—3 mm in maximum dimension, occur in the intergranular spaces of olivine, and intergrew with one another or with plagioclase, euhedral to subhedral coarse pargasite etc. Sometimes, the Cpx can include rounded olivine, and it can also be found that Opx embedded into olivine grains.

CAPTION FOR FIGURE 3 AND FIGURE 4

Figure 3:

A. Pl and Cpx are in intergrowth and filled the intergranular space of Ol and Opx. Transmitted light. Crossed polars

B. Opx has a corona of Cpx. Opx was replaced by $Serp^2$. Transmitted light. Crossed polars.

C. Ol grains are rounded, Phl, Par^2, and Cch filled the intergranular spaces. Transmitted light. One polar.

D. An intergranular irregular zonal structure. Transmitted light. One polar.

E. Droplike, irregular metal sulfide assemblages (MS) distribute in Par^2+Phl+Cch assemblage at the center of a interganular irregular zonal structure. Reflected light. One polar.

F. Par^1 intergrew with Cpx and MS. Transmitted light. One polar.

G. From Ol grain to the center of an intergranular space filled with sulfide and oxide assemblages (MS). The first zone is Par^1 and MS, and then is Phl+Par^2+Cch assemblage. There are droplike MS in the region of the latter assemblage near the MS, and there is no MS in Ol. Transmitted light. One polar.

H. Par^1 made up the transitional zone between Ol and MS in an interganular space. There are droplike inclusions in Par near MS. There is no MS in Ol. Transmitted light. One polar.

Figure 4:

A. A magnified micrograph of Figure 3-H. The inclusions consist of one, two, or more mineral phases. There are both sulfides and oxides in these inclusions. Reflected light. One polar.

B. MS filled the intergranular spaces of olivine grains, then formed the interstitial structure. There is no MS in those olivine grains that had been replaced by $Serp^1$ completely. Reflected light. One polar.

C. CP^1, Po^1, and Pe^1 are in intergrowth. All these minerals are anhedral. Irregular area of Cub within Cp^1 produced by exsolution on breakdown of iss. Reflected light. Crossed polars.

D. MS intergrew with Tph and they both occur at the intergranular spaces of Ol (replaced by $Serp^1$) and Pl. Flaky MS occurs along the cleavages of Tph. There is no MS in Ol or Pl. Reflected light. One polar.

E. Pentlandite (Pe^1) is separated into fragments by magnetite veinlets (Mt^2) and the boundaries between Mt^2 and Pe^1 are smooth. Mt^2 veinlets did not invade into pyrrhotite (Po^1) and silicates ($Serp^1$, originally, olivine). Reflected light. One polar.

F. Exsolution of graphic, flamelike, and dendritic mackinawite (Mac) and magnetite veinlets (Mt^2) within pentlandite (Pe^1). This is a magnified micrograph of a pentlandite fragment in Figure 4-E. Reflected light. Crossed polars.

G. A chromite (Chr^2) grain intergrew with Po, Cp, and Pe, consists of a homogeneous core and an outer zone. There are multiphase sulfide inclusions along the boundary of the core and the outer zone. Reflected light. One polar.

H. A grain of Chr^2. Its core is with a leopard texture, which has been resulted from the decomposition of an initially homogeneous phase. Reflected light. One polar.

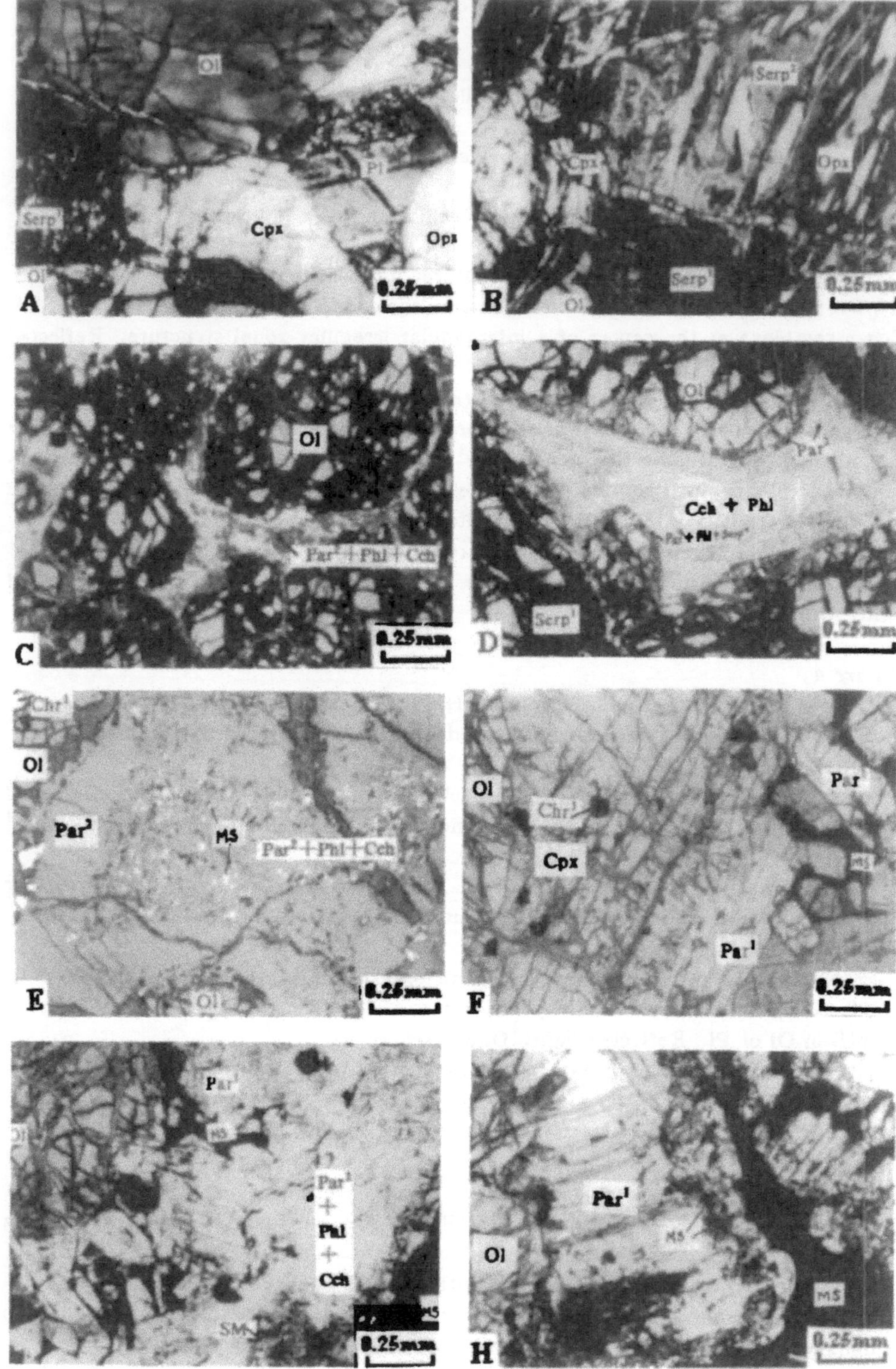

Ol
Pl
Serp
Cpx
Opx
Ol
0.25mm
A
Serp²
Cpx
Opx
Serp¹
Ol
0.25mm
B
Ol
Par²+Phl+Cch
0.25mm
C
Ol
Par²
Cch + Phl
Serp¹
0.25mm
D
Chr¹
Ol
Par²
MS
Par²+Phl+Cch
Ol
0.25mm
E
Ol
Chr¹
Cpx
Par¹
MS
Par¹
0.25mm
F
Par¹
MS
Par¹
+
Phl
+
Cch
SM
MS
0.25mm
Par¹
Ol
MS
MS
0.25mm
H

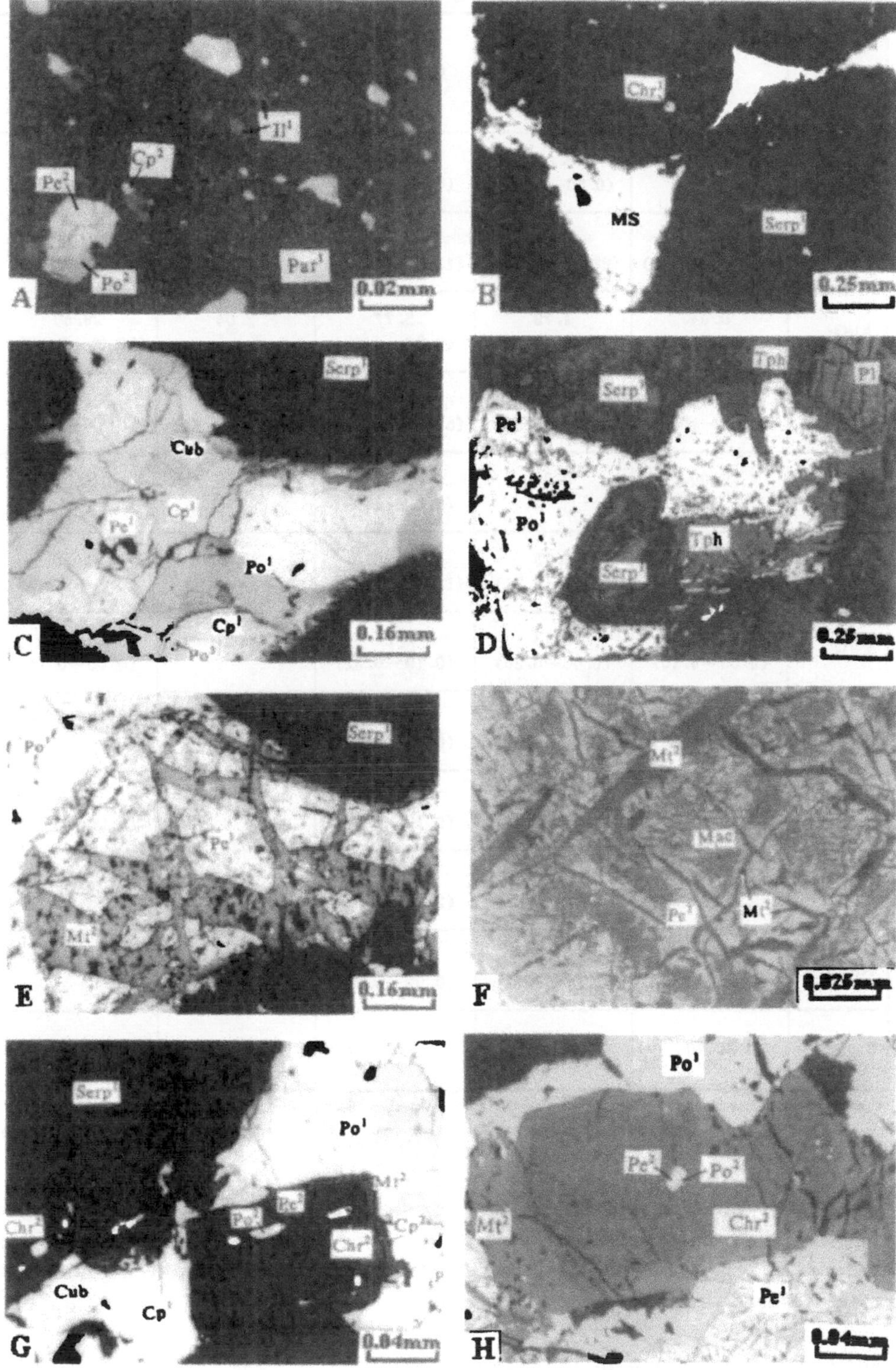
A
Il^1
Cp^2
Pe^2
Po^2
Par^1
0.02mm
B
Chr^1
MS
$Serp^1$
0.25mm
C
$Serp^1$
Cub
Cp^1
Pe^1
Po^1
Cp^1
0.16mm
D
Tph
Pl
$Serp^1$
Pe^1
Po^1
Tph
$Serp^1$
0.25mm
E
Po^1
$Serp^1$
Pe^1
Mt^2
0.16mm
F
Mt^2
Mac
Pe^1
Mt^2
0.025mm
G
$Serp^1$
Po^1
Mt^2
Po^2
Pe^2
Chr^2
Chr^2
Cp^2
Cub
Cp^1
0.04mm
H
Po^1
Pe^2
Po^2
Mt^2
Chr^2
Pe^1
0.04mm

Figure 4.

Table 3. Average composition (wt%) of silicates not containing H_2O

Mineral	Ol (10*)	Opx (bronzite) (9)	Cpx (endiopside) (5)	Cpx (diopside) (1)	Pl (2)
Na_2O	0.00	0.05[a] (0.00—0.16)[b]	0.60 (0.23—1.10)	0.30	5.05 (4.06—6.04)
MgO	43.94 (41.81—45.99)	30.66 (28.30—32.45)	19.90 (18.07—22.76)	16.69	0.02 (0.00—0.03)
Al_2O_3	0.04 (0.00—0.09)	1.79 (1.00—2.68)	3.77 (2.30—4.75)	2.04	29.59 (27.36—31.82)
SiO_2	39.92 (39.06—40.64)	55.03 (50.47—57.27)	51.83 (50.16—56.17)	52.41	51.67 (50.11—53.23)
K_2O	0.00	0.00	0.02 (0.00—0.09)	0.00	0.21 (0.00—0.42)
CaO	0.10 (0.00—0.15)	1.13 (0.06—2.71)	17.07 (13.46—19.53)	21.90	12.20 (10.77—13.63)
TiO_2	0.04 (0.00—0.10)	0.13 (0.00—0.20)	0.39 (0.19—0.52)	1.09	0.07 (0.00—0.14)
Cr_2O_3	0.13 (0.04—0.58)	0.37 (0.08—0.86)	0.83 (0.42—1.41)	0.64	0.03 (0.00—0.05)
MnO	0.27 (0.19—0.29)	0.24 (0.17—0.31)	0.10 (0.10—0.16)	0.10	0.02 (0.00—0.04)
FeO**	14.93 (13.00—16.30)	10.00 (8.08—14.52)	4.80 (2.86—5.63)	4.29	0.26 (0.25—0.26)
NiO	0.33 (0.00—0.55)	0.12 (0.00—0.32)	0.07 (0.02—0.14)	0.03	0.00
Total	99.70	99.52	99.38	99.5	99.12

* —numbers in brackets indicate the number of samples used for average. ** —all iron is expressed as FeO. a—average. b—composition varying range. Analysis Instrument: JEOL 733 electron microprobe; operating conditions were 15 kV, sample current 20 nA.

Occasionally the Opx may have a Cpx corona (Fig. 3-B). These fabrics indicate that the Opx formed slightly earlier than the Cpx. The Cpx was often replaced by tremolite or serpentine, but Opx was commonly replaced by magnesiocummingtonite or serpentine.

There are a small amount of feldspars, about 0.1 — 1% in the massif. Most feldspars are plagioclase. There is a little amount alkali feldspars in massive ore, and these alkali feldspars came from wall rocks. *Plagioclase* generally occurs in plagioclase-lherzolite, white color on hand specimen, 1—2 mm long, with a tabular form, and is in intergrowth with pyroxenes. Two electron microprobe analyses (Table 3) show that the An content of plagioclase is 48.6 to 56.0, so they belong to labradorite and andesine.

Hydrosilicates

Hydrosilicates are important components of the Jinchuan massif, making up about half of the volume of the massif by rough estimate. Amphibole, mica, chlorite, and serpentine are common minerals in this class. The average composition of these hydrosilicates are listed in Table 4 and Table 5.

Table 4. Electron microprobe analyses for amphiboles (wt%)

	Mc (4)	Tr (4)	Par[1] (4)	Par[2] (4)	Ar (1)
Na_2O	0.08 (0.00—0.23)	0.38 (0.16—0.72)	2.27 (2.13—2.64)	3.11 (2.69—3.52)	6.69
MgO	25.72 (25.04—27.20)	22.30 (21.66—23.10)	17.59 (16.14—18.34)	18.30 (17.53—19.29)	0.27
Al_2O_3	0.50 (0.11—0.72)	2.21 (1.75—3.17)	13.01 (12.82—13.48)	14.67 (12.74—16.12)	1.73
SiO_2	55.31 (53.98—56.20)	54.84 (53.28—56.25)	43.39 (42.17—44.91)	41.23 (40.57—42.74)	49.80
K_2O	0.00	0.00	0.05 (0.00—0.14)	0.37 (0.12—0.79)	0.00
CaO	5.10 (0.50—7.69)	13.56 (12.47—14.22)	13.24 (12.65—13.86)	10.89 (9.25—11.63)	0.02
TiO_2	0.03 (0.00—0.08)	0.09 (0.00—0.13)	0.47 (0.42—0.52)	0.13 (0.00—0.39)	0.00
Cr_2O_3	0.04 (0.00—0.08)	0.28 (0.05—0.74)	0.47 (0.27—0.60)	0.16 (0.05—0.25)	0.00
MnO	0.38 (0.29—0.51)	0.12 (0.07—0.18)	0.10 (0.08—0.11)	0.11 (0.09—0.14)	0.09
FeO	10.45 (7.22—14.98)	3.57 (3.13—3.91)	7.37 (6.13—10.42)	8.62 (7.84—9.38)	39.90
NiO	0.09 (0.06—0.13)	0.10 (0.05—0.18)	0.07 (0.00—0.16)		0.03
Total	97.7 (97.34—98.75)	97.45 (97.30—97.80)	98.03 (98.03—98.11)	97.59 (97.22—97.91)	98.53

Notes are the same as Table 3.

Table 5. Electron microprobe analyses for hydrosilicates other than amphiboles (wt%)

	Tph (2)	Phl (5)	Cch (2)	Serp[1] (6)	Serp[2] (3)	Serp[3] (3)
Na_2O	0.36 (0.00—0.72)	0.65 (0.12—1.73)	0.02 (0.00—0.03)	0.01 (0.00—0.08)	0.06 (0.03—0.08)	0.02 (0.00—0.05)
MgO	19.29 (18.27—20.31)	25.36 (23.91—27.55)	32.75 (32.12—33.37)	38.43 (34.28—42.42)	27.80 (17.88—33.84)	32.66 (29.60—34.91)
Al_2O_3	13.47 (12.17—14.76)	13.45 (12.76—14.37)	17.38 (17.13—17.62)	0.77 (0.00—4.00)	2.99 (2.70—3.55)	0.11 (0.09—0.17)
SiO_2	38.73 (38.19—39.27)	39.15 (37.77—41.44)	28.83 (27.62—30.03)	41.32 (36.36—44.12)	40.29 (35.17—47.01)	41.87 (38.26—46.01)
K_2O	10.00 (9.36—10.63)	8.98 (5.39—13.49)	0.00	0.00	0.01 (0.00—0.04)	0.00
CaO	0.00	0.02 (0.00—0.05)	0.05 (0.04—0.06)	0.10 (0.02—0.32)	0.52 (0.02—0.99)	0.10 (0.07—0.13)
TiO_2	8.12 (8.04—8.19)	0.19 (0.00—0.56)	0.06 (0.05—0.06)	0.09 (0.00—0.29)	0.17 (0.08—0.28)	0.04 (0.00—0.05)
Cr_2O_3	0.85 (0.68—1.01)	0.22 (0.00—0.68)	0.49 (0.31—0.67)	0.17 (0.04—0.71)	0.88 (0.61—1.03)	0.05 (0.03—0.06)
MnO	0.06 (0.03—0.08)	0.06 (0.03—0.11)	0.04 (0.00—0.08)	0.07 (0.00—0.19)	0.33 (0.00—0.85)	0.23 (0.15—0.30)
FeO	6.07 (5.79—6.35)	5.80 (4.40—6.87)	7.01 (5.73—8.29)	6.51 (2.09—16.24)	12.57 (11.46—14.54)	12.37 (6.55—15.75)
NiO	0.13 (0.05—0.21)	0.16 (0.14—0.22)		0.20 (0.00—0.34)	0.09 (0.04—0.14)	0.17 (0.06—0.30)
Total	97.09 (96.85—97.30)	94.04 (90.96—96.01)	86.63 (85.93—87.26)	87.68 (87.01—89.49)	85.71 (82.10—87.82)	87.62 (83.40—88.08)

Notes are the same as Table 3.

Members of amphibole group are magnesiocummingtonite, tremolite, arfvedsonite, and pargasite. The *magnesiocummingtonite* is in intergrowth with clinochlore, phlogopite, tremolite, and commonly occurs at the margin of the lherzolite intrusives in the eastern part of the mine area Ⅰ. The Magnesiocummingtonite is usually the metamorphic product of olivine or bronzite, but *tremolite* is the altered product of clinopyroxene. Both magnesiocummingtonite and tremolite occur as acicular assemblages, and their individual crystals are 0.5—1 mm long. Sometimes they formed pseudomorphs after the olivine, orthopyroxene, or clinopyroxene. Droplike, sheet-like or irregular ilmenite, magnetite, and multiphase metal sulfides can be found along the cleavages of those pseudomorphs. Euhedral prismatic *arfvedsonite* crystals can only be found in massive ores, and usually are blue in thin sections, in intergrowth with pyrrhotite or chalcopyrite. *Pargasite* extensively occurs in lherzolite and olivine-websterite. Based on the occurrences, pargasite can be divided into two types. The first type: Pargasite appears as euhedral prismatic crystals, 0.1—0.3 mm long, intergrowing with pyroxenes and metal sulfides (Fig. 3-F). Sometimes it replaced olivine. This type of pargasite is symboled as Par^1. It is worth to be noted that there are plenty droplike multiphase metal mineral solid inclusions in the Par^1 (Fig. 3-H, 4-A). The second type: pargasite occurs as fibrous, acicular, or micro-prismatic crystals, much more smaller than Par^1, and it is symboled as Par^2. Par^2 in intergrowth with clinochlore and phlogopite etc. could formed either a zone of the intergranular irregular zonal structure (Fig. 3-D) or a transition zone between olivine crystals and assemblages of metal sulfides and oxides (Fig. 3-G) Electron microprobe analyses results (Table 4) show that the Par^2 is a little richer in Na_2O, K_2O, and poorer in CaO and SiO_2 than Par^1.

Ti-bearing phlogopite and phlogopite are major members of mica group in the massif. Ti-bearing phlogopite is brown in color, 0.5—1 mm in maximum dimension, and generally appears as anhedral crystals. Intergrowing with sulfide and oxide mineral assemblages, Ti-bearing phlogopite occurs at the intergranular spaces of olivine, pyroxenes, and plagioclase (Fig. 4-D). Flaky, droplike, or irregular magnetite, multiphase sulfides can often be seen along the cleavages. Ti-bearing phlogopite extensively occurs in plagioclase-lherzolite, and can also be found in dunite sometimes. Phlogopite occurs in almost all rocks of the massif. It is tiny subhedral micaceous crystals, colorless, from 0.03 to 0.05 mm in maximum dimension, intergrowing with Par^2 clinochlore, as well as ferriferrous serpentine. It can often be seen that phlogopite intergrowing with other hydrosilicates replaced olivine or pyroxenes, or made up the irregular zonal structure.

Most of the chlorite in the Jinchuan massif are *clinochlore*. Clinochlore generally occurs as euhedral parallelogram crystals in intergrowth with Par^2, phlogopite, and ferriferrous serpentine. In the irregular zonal structures, clinochlore commonly concentrated in the central part and formed much bigger crystals than Par^2 or phlogopite.

Serpentine has the most extensively distribution in the massif. Based on the geneses and the composition serpentine can be divided into three types, designated $Serp^1$, $Serp^2$, and $Serp^3$. $Serp^1$ was formed through replacing olivine; TEM studies show that it consists of lizardite and antigorite, and the lizardite takes a major part. It can often be seen that there are some brown or greenish yellow spots in $Serp^1$ in thin sections, which are the intergrowthes of $Serp^1$ and hematite or ultrafine magnetite grains(?). $Serp^2$ was formed through replacing pyroxenes and it is richer in FeO and Al_2O_3 than $Serp^1$. The FeO content of $Serp^2$ is generally about 11%, so it is a ferriferrous variety of serpentine. $Serp^3$ has a bright green color in thin sections with a strong pleochroism, and is in intergrowth with phlogopite and clinochlore. All these serpentines were formed during the autometamorphism stage.

Oxides

Chromite and *magnetite* are common oxide minerals in the massif, but they are never very abundant. *Ilmenite* is rare. Chromite has two occurrences. One is the small (0.04—0.1 mm) euhedral chromite occurring in olivine or pyroxenes grains, designated as Chr^1. It is pale bluish grey under reflected light in polished section. The other is the chromite in intergrowth with metal sulfides occurring in the intergranular spaces of olivine, pyroxenes, and plagioclase etc., and it is symbolized as Chr^2. Chr^2 is much bigger than Chr^1 in grain size, generally ranging 0.1—0.8 mm. Under reflected light in polished sections, Chr^1 crystals are generally homogeneous in color, but Chr^2 crystals generally consists of a pale bluish grey core and a pale brownish grey outer zone, and there are often multiphase sulfide solid inclusions in the joint plane of the core and the outer zone (Fig. 4-G), and this occurrence of the sulfide inclusions indicates that this phenomenon was formed by crystallizing process. Electron microprobe analyses results (Table 6) show that the composition of the core of Chr^2 are similar to that of Chr^1, generally having 10—20 wt% Al_2O_3, belonging to Al—ferrochromite; but the outer zone of Chr^2 is much more richer in iron, with contents of about 60—70% FeO (see Table 6) (all iron is expressed as FeO.), belonging to Fe^{3+}-ferrochromite. Ilmenite lamellae can be found in the core of Chr^2. Under reflected light in polished sections, we can occasionally see that the core of Chr^2 has a leopard spot texture which was formed by the different reflected color of two minerals. One is a pale bluish grey mineral with similar composition to the Chr^1 and made up the matrix, and the other is a brownish grey one with similar composition (see Table 6) to the outer zone of Chr^2 and appears as irregular spots distributing in the matrix. This leopard texture must have been formed through the exsolution of the homogeneous core of Chr^2.

Magnetite has four occurrences, and they are designated as Mt^1, Mt^2, Mt^3, and Mt^4. Mt^1 appears as euhedral grains and is in intergrowth with metal sulfides occurring in massive ore and it may has a higher Cr content up to 9%. Mt^2 occurs as lamellae or irregular veins and is in close intergrowth with pentlandite, and only distribute in pentlandite grains (Fig. 4-E). Mt^3 is the small irregular magnetite occurring in hydrosilicates or distributing in the cleavages of hydrosilicates. Mt^4 is in intergrowth with pyrite and formed a myrmekitic texture, and they together replaced pyrrhotite. Electron microprobe analyses show that all the latter magnetite have nearly ideal magnetite composition.

Two occurrence types of *ilmenite*, designated Il^1 and Il^2, were found. Mainly occurring in the margin of the massif or in lherzolite, Il^1 appears as tiny droplike grains (about 0.005 mm in maximum dimension, sometimes intergrowing with sulfides) distributing in Par^1 (Fig. 3-H, 4-A); and it also occurs as subhedral tabular crystals intergrowing with Ti-bearing phlogopite and pyrrhotite. Droplike multiphase sulfide solid inclusions can often be found in the latter. Il^2 occurs as lamellae in the core of Chr^2. Electron microprobe analyses results (Table 6) show that Il^2 is richer in TiO_2 than Il^1.

Metal Sulfides

Over 30 metal sulfides have been found in the Jinchuan deposit. Pyrrhotite, chalcopyrite, pentlandite, cubanite, and mackinawite are the most common primary sulfides. The chemical composition of these minerals are listed in Table 7.

Table 6. Electron microprobe analyses for oxide minerals (wt%)

	Chr^{1} (12)	Chr^{2} core homogeneous (6)	Chr^{2} core, bluish grey (2)	Chr^{2} core, brownish grey (2)	Chr^{2} outer zone (2)	Il^{1} (6)	Il^{2} (1)
MgO	4.14 (1.14—9.77)	2.87 (0.00—4.76)	4.83 (4.58—5.07)	3.20 (1.11—5.29)	1.34 (1.29—1.39)	3.02 (1.83—3.98)	1.33
Al_2O_3	13.73 (3.13—19.71)	9.21 (1.65—15.24)	21.03 (20.42—21.63)	2.83 (2.23—3.43)	2.58 (1.64—3.51)	0.85 (0.02—2.88)	0.12
SiO_2	0.33 (0.21—0.66)	0.34 (0.20—0.68)	0.23 (0.19—0.26)	0.21 (0.15—0.26)	0.38 (0.34—0.42)	0.22 (0.15—0.25)	0.22
TiO_2	1.72 (0.46—3.27)	3.85 (1.28—11.28)	0.60 (0.40—0.79)	0.79 (0.78—0.90)	2.27 (2.21—2.33)	42.71 (45.34—57.33)	56.64
Cr_2O_3	37.37 (29.05—46.54)	38.72 (32.62—42.19)	33.07 (32.91—33.22)	17.56 (15.81—19.30)	26.84 (18.35—35.32)	1.05 (0.00—3.83)	0.27
MnO	0.41 (0.27—0.55)	0.67 (0.06—1.76)	0.48 (0.47—0.48)	0.26 (0.20—0.32)	0.63 (0.31—0.94)	1.27 (0.55—3.24)	5.51
FeO	40.54 (28.73—56.50)	42.49 (33.8—59.46)	38.60 (38.32—38.87)	69.26 (66.47—72.05)	61.96 (57.12—66.80)	42.20 (37.55—45.10)	34.53
ZnO	0.67 (0.27—1.27)	0.66 (0.05—1.46)	0.44 (0.35—0.53)	0.21 (0.19—0.22)	0.37 (0.2—0.54)		
NiO	0.22 (0.10—0.38)	0.12 (0.04—0.24)	0.10 (0.04—0.16)	0.5 (0.12—0.38)	0.13 (0.07—0.18)		

Notes are the same as Table 3.

Based on the occurrences the *pyrrhotite* can be sorted into following three categories. Po^{1} is the pyrrhotite occurring as anhedral to subhedral grains, 0.1—1 mm in diameter, intergrowing with pentlandite, chalcopyrite, etc. (Fig. 4-C). Lamellar twins can often be found in Po^{1}. Most of the pyrrhotite in the Jinchuan deposit occur in this form. Po^{2} is the pyrrhotite included in other minerals, occurring as solid inclusions alone or together with chalcopyrite and pentlandite in oxides such as Chr^{2}(Fig. 4-G, 4-H) or hydrosilicates such as Par^{1}(Fig. 4-A) or Ti-bearing Phlogopite. It generally has the grain size ranges of 0.01—0.05 mm. Po^{3} is the one formed by the exsolution of pentlandite or chalcopyrite (Fig. 4-C). The composition of the pyrrhotite of the three types are similar; the variation range of the composition are smaller than 2% (Table 7). Pyrrhotite was often replaced by pyrite+magnetite which was in intergrowth with a graphic texture.

Table 7. Electron microprobe analyses of metal sulfides

	Po (10)	Po[2] (2)	Pe (11)	Cp (10)	Cub (2)	Mac (6)
Cu	0. 05 (0. 01—0. 08)	0. 03 (0. 00—0. 06)	0. 60 (0. 00—3. 46)	34. 00 (33. 28—34. 84)	23. 37 (23. 06—23. 67)	5. 13 (0. 10—15. 28)
Fe	60. 95 (59. 58—62. 23)	59. 70 (59. 68—59. 71)	30. 91 (27. 89—36. 12)	30. 72 (29. 92—31. 92)	40. 93 (40. 61—41. 24)	38. 45 (21. 22—51. 46)
Ni	0. 09 (0. 00—0. 23)	0. 87 (0. 13—1. 61)	34. 44 (30. 55—36. 16)	0. 14 (0. 02—0. 48)	0. 07 (0. 05—0. 09)	19. 85 (11. 39—33. 55)
S	38. 01 (37. 39—39. 01)	38. 23 (38. 09—38. 36)	33. 17 (32. 31—33. 83)	34. 23 (33. 33—35. 25)	35. 04 (34. 86—35. 21)	35. 75 (34. 71—38. 95)
Co	0. 06 (0. 01—0. 09)	0. 09 (0. 07—0. 10)	0. 70 (0. 23—1. 39)	0. 07 (0. 03—0. 24)	0. 03 (0. 02—0. 03)	0. 55 (0. 21—1. 66)
Cr	0. 02 (0. 00—0. 09)	0. 54 (0. 14—0. 93)	0. 01 (0. 00—0. 04)	0. 18 (0. 00—1. 28)	0. 00	0. 00
Total	99. 18	99. 46	99. 83	99. 97	99. 44	99. 73

Operating conditions are the same as Table 3.

Pentlandite is the most important industrial mineral in the Jinchuan deposit. Similar to pyrrhotite, pentlandite have also three occurrences. Most of the pentlandite is symbolized as Pe^1, which occurs as irregular or rounded grains with a size range of 0. 1—2 mm, intergrowing with pyrrhotite and chalcopyrite (Fig. 4-C).

Pe^1 often occurs in the assemblage of several minerals. In this assemblage, pentlandite is a predominant mineral and made up the matrix of the assemblage; magnetite formed the irregular netlike veinlets or straight lamellae in the matrix. These magnetite veinlets and lamellae separated the pentlandite into fragments and they never invaded into the adjacent pyrrhotite or silicates (Fig. 4-E). The width of those veinlets and lamellae ranges from 0. 01 to 0. 16 mm and its genesis will be discussed later. The flame-like, graphic, or dendritic mackinawite, irregular chalcopyrite or pyrrhotite patches or spots are common within the pentlandite matrix, which seem to be formed from the decomposition of an initial solid solution (Fig. 4-F). The other two occurrences of pentlandite are: 1. It is in intergrowth with other metal sulfides such as pyrrhotite or chalcopyrite, occurring as inclusions within hydrosilicates (Fig. 3-H, Fig. 4-A) or chromite (Chr^2) (Fig. 4-G, 4-H). This pentlandite is symbolized as Pe^2. 2. It occurs as granular polycrystalline veinlets interstitial to pyrrhotite grains. This pentlandite is symbolized as Pe^3 and can only be found in the massive ore. Electron microprobe analyses results (Table 7) show that pentlandite generally have the S content around 33 wt% with a dispersion of 5 wt% Fe and Ni contents. There is no obvious relation between the compositional dispersion and the occurrences of pentlandite.

Chalcopyrite (symbolized Cp^1) generally occurs as anhedral grains in intergrowth with pyrrhotite and pentlandite (Fig. 4-C). There are also a little amount of chalcopyrite as solid inclusions within hydrosilicates, or as exsolution spots, patches within pentlandite. The chalcopyrite of these two occurrences are symbolized as Cp^2 and Cp^3 correspondingly. The composition of chalcopyrite is simple and has a very narrow variation (see Table 7).

Cubanite often occurs as irregular patches and lamellae within chalcopyrite produced by exsolution on breakdown of iss (Fig. 4-C) and its composition is also less changed.

Mackinawite is another common exsolution mineral in the Jinchuan deposit. Most of the mackinawite occur in pentlandite. The composition of mackinawite is of wide dispersion (Table 7); this may be caused by the complex exsolution process between it and pentlandite.

FORMING CONDITIONS OF ROCKS AND ORES

Nine geothermometers and geobarometers and some experimental phase diagrams are used to estimate the forming conditions of the rocks and ores of the Jinchuan massif. The authors have also carried out heating experiments and these experiments confirmed the estimates results.

Temperature

Estimated Temperatures by Geothermometers

Among numerous geothermometers available for spinel-pyroxenes-olivine assemblages, ones based on diopside-enstatite equilibrium relations[5, 6, 7], one based on $FeO-MgO-Al_2O_3-SiO_2$(FMAS), CFMAS (FMAS+CaO), and NCFMAS (CFMAS+Na_2O) experimental systems[8], and one based on solution model for Fe-Ti oxides[9] were used by the authors of this paper. These thermometers give 1024—1268℃ for pyroxene assemblage and 583℃ for ilmenite. If we consider the experimental results of Yoder and Tilley (1962)[10], then the olivine crystallization temperature may be 40℃ higher than pyroxene. Based on the stability of amphiboles from the comprehensive phase diagram of Gilbert (1982)[11], we can get that the forming temperatures for pargasite and tremolite may be 1070 ℃ and 900 ℃ correspondingly.

Heating Experiment for Natural Sample of the Jinchuan Cu-Ni Deposit

Three sideronitic ore and two massive ore specimen were used by the authors for the heating experiment to estimate the forming temperature of the rocks and the ores. These specimen were cut into slices with a thickness of 0. 5 mm and well polished. Heating processes were performed with a Leitz 1350 heating stage. A selected polished ore slice was put onto the heating stage, and was heat in a speed of 30℃/minute and the changes of the minerals in the slice was observed with a reflected light microscope during the heating process. The heating chamber was filled with argon gas so as to avoid the minerals from oxidizing during the process. The results of the experiments are summarized as follows. 1. Sideronitic ore (three sample, 6 experiments for each one): 900—1000℃, pentlandite (Pe) and intergrowing magnetite (Mt) veinlets melted together and changed to a homogeneous liquid phase; 1000—1080℃, pyrrhotite (Po) melted to liquid and the liquid mixed with the liquid of Pe+Mt; 1100℃, the homogeneous core of chromite (Chr^2) changed to two phases with a leopard spot texture (like Fig. 4-H); 1170℃, the outer zone of Chr^2 and silicates began melting. 2. Massive ore (two sample, two experiments for each one): 560—580℃, chalcopyrite (Cp) melted; 780℃, Po changed into two phases with a sheet-like texture, and then began to melt; 860—900℃, Pe and Po melted and the molten liquid mixed with Cp liquid to a homogeneous phase; 1000℃, euhedral Mt melted and mixed with Po+Cp+Pe liquid to one phase. So the forming temperature range of massive ore is lower than that of sideronitic ore. These results also present that for a mineral, its forming temperature is affected by the intergrowing minerals.

Considering the above results comprehensively, the forming temperature of the main minerals in the Jinchuan deposit are summarized as follows: Chromite and olivine, 1064 — 1038 ℃ (1215 on average); pyroxenes, 1024 — 1268℃ (1175℃ on average); chromite (Chr^2), 1100—1170℃; pargasite, 1070℃; tremolite 900℃; magnetite (Mt^1), 1000℃; pyrrhotite, 780—1080℃; pentlandite and intergrowing magnetite (Mt^2), 860—1000℃; chalcopyrite, 780℃.

Pressure

There is at the moment no reliable method of determining the equilibrium pressure of those mineral assemblages of the rocks in the Jinchuan massif. The Orthopyroxene geobarometer of Gasparik (1987)[8] and the empirical hornblende geobarometer of Hollister et al. [12] were used by us for estimating the forming pressures of pyroxenes and pargasite. They gave the pressure ranges of 9. 9—13. 5 Kb and 7—11Kb for Pyroxenes and pargasite correspondingly. Therefore, the pressure condition for the massif is estimated in the range of 7—13 Kb

Oxygen and Sulfur fugacities

Oxygen fugacity has been estimated based on the analyses of coexisting olivine, orthopyroxene and chromite (Chr^1) according to the methods of Wood (1991)[13] and Ballhaus (1990)[14], and they demonstrate that the oxygen fugacity $\lg fo_2$ =-6. 2~-4. 7 when olivine and chromite (Chr^1) were crystallizing, and that of pyroxenes is $lgfo_2$=-7.7~-5.7. The former may present the initial magma's fugacity, and the latter present the oxygen fugacity of residual magma after the orthopyroxene crystallized. Andersen's (1988) Fe-Ti oxides oxygen barometer[9] show that the oxygen fugacity for ilmenite is $lgfo_2$=-19.2. Because that the ilmenite is generally in intergrowth with hydrosilicates and sulfides in the Jinchuan Cu-Ni deposit, the result may present the oxygen fugacity when bulk sulfides formed. Based on Ernst's (1962) experimental results for arfvedsonite[15], the $lgfo_2$ of the arfvedsonite in massive ore is -17.5~-7.35, which may present the oxygen fugacity of the sulfide melt to form the massive ore. From analyses of pyrrhotites in different occurrences, according to Toulmin and Barton's (1964) method[16], sulfur fugacities for different pyrrhotite are estimated as follows: 1. Pyrrhotite (Po^1) in sideronitic and disseminated ores: $lgfs_2$=-2.4~0.2; 2. Pyrrhotite (Po^1) in massive ore: $lgfs_2$=-2.2~-0.3; 3. Pyrrhotite (Po^2) as inclusions in chromite (Chr^2): $lgfs_2$=0~0.5. The fugacity of Po^2 may present the sulfur fugacity of the initial sulfide melt and the fugacity of pyrrhotite for the massive ore may present that of a latter sulfide melt.

DISCUSSION

When and Where Did the Immiscible Sulfide-Oxide Melt Separate from the Silicate Melt?

For magmatic sulfide copper-nickel deposits, the ore mineralization happened at which stage of the magmatic evolution has been arguing since the end of last century. These copper-nickel sulfide ores are generally considered to have formed as a result of the separation of an immiscible sulfide-oxide melt (SOM) from a sulfur-saturated silicate melt shortly before, during, or after the emplacement[17]. But there are also quite a few researchers who believe that the ores were formed by the crystallization and metasomatism of the residual hydrothermal solution of the magma. For the Jinchuan Cu-Ni Deposit, most researchers suggest that the mineralization relates to the separation of an immiscible sulfide melt from the original magma before the bulk mineral crystallization. But the mineral intergrowth relationships do not support this hypothesis. The above description for mineral intergrowth relationships clearly show that the metal sulfides and oxides only occur at the intergranular spaces of olivine, pyroxenes, or plagioclase, and are in intimately intergrowth with hydrosilicates such as Ti-bearing phlogopite, pargasite, phlogopite, clinochlore, etc. (Fig. 3-E, F, 4-B, C, D). Quite a few spheric, droplike, flaky, or irregular sulfide and oxide multiphase inclusions occur in the hydrosilicates (Fig. 3-G, H, 4-A). These phenomena are generally considered as an important evidence of the existence of the immiscible SOM in the magma when these minerals were crystallizing. Oppositely, there are no multiphase inclusions in the earlier crystallized olivine, pyroxenes as well as plagioclase. These evidence indicated when olivine, pyroxenes as well as plagioclase were crystallizing, there was no separated SOM in the magma; and when pargasite, Ti-bearing phlogopite, phlogopite, clinochlore, etc. were crystallizing, there was coexisting immiscible SOM in the silicate melt. *The separation of the immiscible SOM from the silicate melt did not happen before or during the crystallizing episode of olivine, pyroxenes, as well as plagioclase, but after their crystallizing. In other words, the separation took place before or during the magmat-*

ic evolution stage when hydrosilicates except for serpentine ($Serp^1$ *and* $Serp^2$) *were forming*. Crystallization of the olivine, pyroxenes, as well as plagioclase resulted in the enrichment of Cu, Fe, Ni, S, as well as H_2O, K_2O, and Na_2O. This enrichment is an essential condition for the separation (or liquation) of the SOM from the residual magma. *It is just in this residual magma that an immiscible SOM separated from the silicate melt. So the mineralization process of the Jinchuan Cu-Ni Sulfide Deposit is recognized as: "crystallization differentiation + late stage liquation".*

Because olivine, pyroxenes made up 70 — 80% volume of any intrusive in the Jinchuan massif, after these minerals crystallized, 70—80% volume of an intrusive had been solidified. The residual magma distributed in the intergranular spaces of these crystallized minerals and can move through these spaces. *The separation (or liquation) took place just within these intergranular spaces.* The separated SOM can move down to the lower part or even the bottom of an intrusive because of the gravity effect. Therefore the orebodies generally occur at the lower part of the intrusives of the Jinchuan massif. If there was a dynamic force acting on an intrusive, it might squeeze the SOM out of the intergranular spaces and made the SOM move a short distance, and then formed the massive orebodies.

The Property of the Residual Magma

The results of mineral composition analyses show that the Mg/Fe ratio of the Mg-Fe silicates which formed by the crystallization of magma is about 5. 6; and the variation range of the Mg/Fe ratio is very narrow, generally, less than 1. The authors call this phenomenon as "Mg/Fe ratio certainty". Because the mineral content of the rocks in the intrusives is known, and what minerals crystallized before the separation of the immiscible SOM from the silicate melt is also known, the composition of residual magma of different rock types can be known by calculation. The calculation show that the residual magma contained 10—20 wt% SiO_2, 14—34 wt% H_2O, 1—2 wt% N_2O and K_2O, and increased the content of the minerogenic elements such as Cu, Ni, S,... by 3—5 times as compared with the original magma which had a composition ranges of 30—40 wt% SiO_2, 8—10 wt% H_2O, 0. 1—7 wt% S, 0. 5—3 wt% Ni, 0. 1—4 wt% Cu, 0. 5 wt%± K_2O and Na_2O. The temperature of the residual magma was about 1100 ℃ and the $\lg fo_2$ of it was around -7. 7.

The Genesis of the Hydrosilicates

The genesis of the hydrosilicates of the Jinchuan deposit can be classified into two typies. One is those formed through crystallization, including pargasite (both Par^1 and Par^2), Ti-bearing phlogopite, phlogopite, clinochlore, and serpentine ($Serp^3$). All these minerals crystallized from the residual silicate magma which was rich in H_2O and alkalis during, or shortly after (Par^1 might shortly before) the SOM separation. The other is those formed through replacing olivine and pyroxenes because of the autometamorphism of the intrusives. The minerals of this type include magnesiocummingtonite, tremolite, phlogopite, clinochlore and serpentine ($Serp^1$ and $Serp^2$). The former four minerals often intergrew with sulfide and oxide minerals and there are often multiphase sulfide and oxide mineral inclusions in these hydrosilicates, so they also formed during the mineralization period. There are generally no sulfide and oxide minerals in serpentine, so the serpentine ought to be formed after the mineralization.

The Origin of the Assemblage of Pentlandite and Magnetite, Mackinawite, and Chalcopyrite

As we pointed out before, most pentlandite always occur in the assemblages of magnetite veinlets, pentlandite, dendritic mackinawite, etc. (Fig 4-C, E, F). These magnetite veinlets are traditionally considered to have formed by the oxidation of pentlandite[18]. But the boundary between pentlandite and magnetite is smooth and straight, and there is no any characteristic of replacement. Moreover, these magnetite veinlets and lamellae distribute only in the pentlandite matrix, never invade the adjacent pyrrhotite grains and silicate grains (Fig 4-E). Therefore, this intergrowth relationship could not be the product of the oxidation of pentlandite. Sometimes, magnetite may take more than half of the volume of those assemblage grains, therefore there is no the possibility that the adjacent silicates such as olivine and pyroxenes give out so much iron to form magnetite through the process that they were replaced by serpentine. The heating experiment results show that at the temperature of 900 — 1000℃ the pentlandite and intergrowing magnetite veinlets in the sideronitic ore melted together and became a homogeneous liquid phase. Based on these evidence, the authors conclude that this assemblage formed through a decomposition of an initially solid solution of Cu—Fe—Ni—S—O. The solid solution exsolved to magnetite and a new solid solution of Cu—Fe—Ni—S firstly. The former formed the magnetite veinlets, and the latter exsolved further and formed the flame-like and dendritic mackinawite, chalcopyrite patches, and pentlandite.

The Evolution of the Sulfide-oxide Melt (SOM)

Because that the sulfides in the Jinchuan deposit always intergrew with oxides, the primary melt to form those minerals was certainly a sulfide-oxide melt (SOM). These sulfide and oxide minerals mainly include pyrrhotite, pentlandite, chalcopyrite, cubanite, mackinawite, magnetite, chromite, etc., so we infer that the components of the SOM should mainly be Fe, Cu, Ni, Cr, Al, Mg, S, and O. As chromite (Chr^2) is always euhedral and there are multiphase sulfide inclusions in it, it crystallized from the SOM firstly. The crystallization of Chr^2 depleted all the Cr, Al, and Mg. Then pyrrhotite, as well as a solid solution whose composition was equal to chalcopyrite + cubanite (iss), and a solid solution of Cu—Fe—Ni—S—O whose composition was equal to pentlandite + magnetite + mackinawite etc. condensed into three phases at the same time. With the decrease of temperature, iss exsolved to cubanite and chalcopyrite, and the solid solution of Cu—Fe—Ni—S—O changed to the assemblage of pentlandite + mackinawite + magnetite, etc. whose evolution has been discussed before. In brief, the process can be described as: chromite (Chr^2) → pyrrhotite + "pentlandite + magnetite solid solution (PMSS)" + iss. When PMSS exsolved, pentlandite and magnetite (Mt^2) were formed. Chalcopyrite and cubanite were formed when iss exsolved. This crystallization sequence is quite different from that of the Fe—Ni—Cu—S system described by Craig (1969) [19] and the Fe—S—O system described by Naldrett (1968)[20].

CONCLUSIONS

The main results of this study show that, first, the mineralization of the Jinchuan

Cu-Ni Sulfide Deposit related to both the mineralization differentiation and the liquation of residual magma and, second, the crystalline process of the minerals of sulfides and oxides are very complex and difficult to compare with those known experimental systems of Fe—S—O or Cu—Fe—Ni—S. The comprehensive paragenetic sequence for the minerals of the main rocks of the intrusives is summarized as follows. The first mineral to crystalize from the original magma was chromite (Chr^1). Subsequent to the onset of Chr^1, olivine, bronzite, endiopside as well as diopside, pargasite (Par^1), plagioclase (for plagioclase-lherzolite) began to crystallize orderly, and their main crystallizing episode of were overlapping. After these minerals crystallized, it happened that the liquation of SOM from the residual magma which distributed at the intergranular spaces of the mineral grains crystallized. Hydrosilicates such as pargasite (Par^2, and some Par^1), Ti-bearing phlogopite, phlogopite, clinochlore as well as ferriferrous serpentine crystallized from the separated residual silicate melt; and the SOM produced those metal minerals such as pyrrhotite, pentlandite, chalcopyrite, magnetite, chromite, etc. At last the residual hydrothermal solution react to olivine and pyroxenes, then the serpentine ($Serp^1$ and $Serp^2$) were produced. At the margin of the intrusives of the first and second intrusive stage, olivine and pyroxenes were replaced by magnesiocummingtonite, tremolite, as well as phlogopite, clinochlore, etc. with the mineralization.

Furthermore, for each intrusive of the Jinchuan massif, the evolution history can be divided into two stages: The first stage is the main solidifying stage. During this stage, chrysolite, pyroxenes, plagioclase, euhedral pargasite etc. formed, and about 70—80% of the volume of the intrusive was solidified. The second stage is the ore forming and autometamorphism stage. In this stage, pyrrhotite, pentlandite, chalcopyrite, chromite(Chr^2), magnetite etc. and some hydrosilicates such as Ti-bearing phlogopite, phlogopite, fibrous pargasite, tremolite, clinochlore, serpentine etc. formed.

REFERENCES

1. Tang Zhongli. Minerogenetic model of the Jinchuan copper and nickel deposit, *Geoscience* **4:4**, 55—64 (1990)

2. Jia Enhuan. Minerogenetic model and minerogenetic series of the Jinchuan copper-nickel deposit, *Geological Review* **32:3**, 276—286 (1986).

3. Jia Enhuan. Geological characteristics of the Jinchuan Cu-Ni sulfide deposit in Gansu Province, *Mineral Deposits* **5:3**, 27—38 (1986)

4. No. 6 geological team of the Geology and Mineral Resources bureau of Gansu Province. *Geology of the Baijiajuzi Copper-nickel Sulfide Deposit*, Geological Publishing House, 25—30 (1984).

5. B. J. Wood and S. Banno. Garnet-Orthopyroxene and Orthopyroxene-Clinopyroxe relationship in simple and complex systems, *Contr. Mineral petrol.* **42**, 109—124 (1973).

6. R. A. Wells. Pyroxenes thermometry in simple and complex systems, Contr. *Mineral Petrol.* **62**, 129—139 (1977).

7. B. T. Davis and F. R. Boyd. The join $Mg_2Si_2O_6$—$CaMgSi_2O_6$ at 30 Kilobars pressure and its application to pyroxenes from kimberlites, *Jour. of Geophysical Research*. 71, 3567—3576 (1966).

8. Tibor Gasparik. 1987, Orthopyroxene thermobarometry in simple and complex systems, *Contrib. Mineral. Petrol.* **96**, 357—370 (1986).

9. D. J. Andersen. Internally consistent solution models for Fe—Mg—Mn—Ti oxides: Fe—Ti oxides, *Am. Mineral.* **73**, 714—726 (1988).

10. H. S. Jr. Yoder and C. E. Tilley. Origin of basalt magmas: an experimental study of natural and synthetic rock systems. *Jour. Petrol.* **3**, 342—532 (1962).

11. M. C. Gilbert et al. Experimental studies of amphibole stability, In: *Reviews in Mineralogy*. **9B.**, 229—353 (1982).

12. L. S. Hollister et al., Confirmation of the empirical correlation of Al in hornblende with pressure of solidification of calc-alkaline plutons, *Am. Mineral.* **72**, 231—239 (1987).

13. B. J. Wood. Oxygen barometry of spinel peridotites, *Reviews in Mineralogy* **25**, 417—432 (1991).

14. C. Ballhaus. Oxygen fugacity controls in Earth's upper mantle, *Nature* **348**, 437—440 (1990).

15. W. G. Ernst. Synthesis, stability relations, and occurrence of riebeckite and riebeckite-arfvedsonite solid solutions, *Jour. Geol.* **70**, 689—736 (1962).

16. P. Toulmin and P. B. Barton. A thermodynamic study of pyrite and pyrrhotite, Geochem. Cosmochim. Acta **28**, 641—671(1964).

17. A. J. Naldrett. *Magmatic Sulfide Deposits.* Clarendon Press, Oxford University Press (1989).

18. No. 6 Geological Team of Gansu Bureau of Geology and Mineral Resources. *The plates of ore structures and textures of copper-nickel ore deposit in Jinchuan.* The Gansu People's Publishing House, 74—76 (1983).

19. J. R. Craig & G. Kullerud. Phase relations in the Cu—Fe—Ni—S system and their application to magmatic ore deposits, *Econ. Geol. Mon.* **4**, 344—358 (1969).

20. A. J. Naldrett. 1968, Melting relation over a portion of Fe—S—O system and their bearing on temperature of crystallization of nature sulfide-oxide liquids, Carnegie Inst. Washington Year Book **66**, *419—427 (1968)*.

Proc. 30th Int'l. Geol. Congr., Vol 16, pp. 135-149
Huang Yunhui and Cao Yawen (Eds)

Reaction Space: Transition from Eclogite to Amphibolite Facies in An Ultra-high Pressure Metamorphic Terrane from Zhucheng, Shandong, E. China

LAI XINGYUN, SU SHANGGUO
Department of Geology, China University of Geosciences, Beijing, 100083, China

Abstract

Eclogites and the related rocks are widely distributed over an ultra-high pressure metamorphic terrane in SE Shandong Province, eastern China. The petrography and mineral chemistry of the rocks preserved the transitional hitory from eclogite to amphibolite facies. The reaction space proposed by Thompson (1982) was applied to investigate the metamorphic processes. In a closed system (N'C'M'A'S'), the reaction space was conducted based on the two basis reaction vectors: *α)* 2di+ *tk* + 2*mc* =py and *β)* py+2qz+ *tk* + 2*pl* =2ab+ 2*mc*. Advancement of retrograde altaration of eclogite along the vector di+ *tk* + *pl* + qz=ab leads to the formation of symplectic intergrowth of clinopyroxene plus albite around omphacite. Abundance of quartz in eclogites limits the volume and shape of the reaction polytope, hence the retrograde metamorphic production, e. g. symplectites, are more easily developed in quartz-basaltic eclogites other than olivine-basaltic ones. If the system (N'C'M'A'S'H') is open to water, formation of the hornblende is attributed to the hydration of eclogites during retrograde metamorphism. Its reaction space is defined by the four independent net transfer reactions: *A)* ab= *ed* +4qz; *B)* 2ab+ 2*mc* =py+2qz+ *tk* + 2*pl*; *C)* ab=di+ *tk* + *pl* +qz; *D)* 4tr +5qz+ 3*ed* = 3py+11di+ 3*pl* +4H_2O. The transfer of garnet and clinopyroxene into amphibole is mainly controlled by the reaction vector *D*. The other basis reaction vectors *(A, B, C)* are responsible for changes of mineral compositions and modes. Furthermore, least-squares approximation was used to estimate the mineral abundances.

Keywords: eclogite, amphibolite, reaction space, closed system, open system, least-squares

INTRODUCTION

The idea of reaction space was first put forward by J. B. Thompson [14, 15] as a useful tool to aid in understanding mineral compositional and modal changes in metamorphic process. Although disturbed by both algebraic and geometric methods included in reaction space, a number of geologists successfully have applied this new tool to the analysis and solution of geological problems in the last decade. For example, Thompson [15] described the reactions in amphibolite, greenschist and blueschist and Poli [12] investigated the transition from amphibole eclogite to greenschist facies in the Austroalpine domain (Oetztal Complex) based on the principle of reaction space. It is well known that the transition from garnet feldspar amphibolite to eclogite is a very wide P-T field extending from somewhere about 5kbars, where the garnet-amphibole pairs starts to appeare in basic rocks, to 10-20kbars, where plagioclase decomposes, then up to higher pressure, where garnet and omphacite coexist [12]. Because the stable field of plagiclase is controlled primarilly by bulk composition. which ranges from about 5kbars to 15kbars [13]. some very low pressure eclogites may contain sodic plagioclase. Whether or not amphibole is present in eclogites depends to a large extent on the water contained in rocks and metamorphic temperature. It is very common that Na-riched amphibole is stable together with garnet and omphacite in eclogite facies. Consequently. unlike the metamorphism of pelitic schist where the paragenesis changes significantly as a function of metamorphic conditions (*P*, *T*, *X*), the metamorphism of mafic rocks is characterized by continuous changes in mineral compositions and modes other than by the introduction of new phases [15].

This paper is to try to interpretate the evolutionary history from eclogite to garnet amphibolite in Zhucheng area. Shandong Province. eastern China on basis of reaction space.

GEOLOGICAL SETTING AND PETROLOGY

Area of SE Shandong Province in eastern China is well known for its numerous occurences of coesite (or its pseudomorph)-bearing eclogites [4. 17]. Eclogite and associated rocks are exposed in the Su-Lu ultra-high pressure metamorphic (UHP) terrane. which is generally believed to be the eastern extension of the world well-known Dabie Mountains UHP metamorphic belt in central China. and is offset by the gigantic sinistral Tanlu Fault [19]. The Su-Lu terrane is represented by a Proterozoic metamorphic complex composed of granitic gneisses. schists. amphibolites. and small amounts of metasedimentary rocks including marbles and quartzites. Eclogites and serpentinized peridotites occur as blocks or lenses on a centimetre to several hundreds meter scale in granitic gneisses. A number of gneisses and schists immediately surrounding eclogite blocks have preserved high-pressure metamorphic relics [7]. So far. most previous reseachers have proposed that eclogite blocks were formed at about 220Ma [6. 8. 9]. The samples of eclogites and garnet amphibolites were collected from outcrops in Taohang. Zhucheng County in southern Shandong Province (see Fig. 1 and explanation of Lai in [7]). The margins of most eclogite blocks are commonly altered to amphibolites. A typical rock sample of 30cm long. 20cm wide and 15cm high is used as a main object of the study. Five thin sections in succession were cut down from the core to rim of the sample. It is easily observed that fresh eclogite. which dominantly constitutes the sample core. is replaced gradually by garnet amphibolite. which occurs near the rim. and then replaced by epidote amphibolite in the outermost rim. Based on the micro-structures of the eclogites and related rocks. three kinds of mineral parageneses. which may represent three different retrograde metamorphic stages. can be identified.

1) Fresh eclogite. consists mainly of garnet. omphacite. quartz and minor rutile. Although no coesite occurs in the present thin sections. its occurrence as inclusions in garnet is documented by Zhang [18]. Omphacite and rutile are other phases that may present as inclusions in garnet. Tripple junction microstructure between garnets and omphacites indicates that thermodynamic equilibrium has been approached in eclogite stage.

2) Coronitic assemblages of pyroxene phases. they are characterized by coronitic assemblages consisting of symplectitic intergrowth of Ca-clinopyroxene and albite around omphacite. A number of omphacite pseudomorphs are replaced by the very fine-grained

intergrowth. The corona is belived to be produced by the reaction of omphacite + SiO_2 → symplectic clinopyroxene + albite [16] in the period of exhumation of the eclogite blocks.

3a) Garnet amphibolite. this rock commonly occurs around fresh eclogites and is dorminantly composed of garnet, amphibole, plagioclase, quartz and small amounts of sphene and opaques. Intergrowth of bluish green amphibole and plagioclase is widespread along garnet-rims and fractures. Amphibole varies from nanometer-scale crytocrystals to milimeter-scale subhedral columnar gains. Rutiles are surrounded by sphene corona. With exception of clinopyroxenes contained within garnets as inclusions, no pyroxene phases have been preserved in matrix of amphibole and plagioclase. The retrograde reactions from the eclogite to the amphibolite facies can be written as Grt+Qtz→Cpx+Plg, Cpx+Plg+H_2O→Hb+Qtz [18].

Table 1. Representative mineral analyses of eclogites, symplectites, and garnet amphibolites from Zhucheng area, Shandong

Min.	Gt1	Gt2	Gt9	Gt13	Cpx7	Cpx5	Hb11	Hb14	Pl153	Pl19
Stage	①	②	③	③′	①	②	③	③′	②	③
SiO_2	41.88	39.36	40.09	38.75	56.93	55.07	40.04	40.58	68.27	64.48
TiO_2	0.00	0.00	0.09	0.07	0.00	0.20	0.00	0.30	0.06	0.06
Al_2O_3	22.57	22.19	22.67	23.98	11.00	11.77	21.99	19.66	19.60	22.50
Cr_2O_3	0.00	0.06	0.15	0.00	0.16	0.13	0.07	0.04	0.00	0.01
FeO	18.60	20.76	18.22	18.45	3.09	3.00	11.34	12.93	0.00	0.13
MnO	0.41	0.46	0.45	0.44	0.00	0.00	0.21	0.19	0.00	0.02
MgO	9.85	9.81	8.69	8.27	8.51	8.95	10.29	9.72	0.00	0.00
CaO	6.15	7.07	8.60	8.99	12.49	13.30	10.44	11.21	0.23	2.47
Na_2O	0.08	0.06	0.11	0.06	6.78	6.66	3.01	2.97	11.58	10.26
K_2O	0.04	0.05	0.00	0.00	0.04	0.10	0.00	0.38	0.06	0.08
Total	99.57	99.81	99.07	99.01	99.00	99.18	97.98	97.97	99.8	100.01
Si	3.123	2.979	3.036	2.946	2.024	1.961	5.767	5.896	2.988	2.840
Ti	0.000	0.000	0.005	0.004	0.000	0.005	0.000	0.033	0.002	0.002
Al	1.983	1.980	2.023	2.149	0.461	0.494	3.678	3.367	1.011	1.168
Cr	0.000	0.004	0.009	0.000	0.004	0.004	0.008	0.005	0.000	0.000
Fe^{3+}	0.000	0.072	0.000	0.000	0.000	0.034	0.765	0.371	0.000	0.000
Fe^{2+}	1.160	1.242	1.154	1.173	0.092	0.055	0.581	1.200	0.000	0.005
Mn	0.026	0.029	0.029	0.028	0.000	0.000	0.025	0.023	0.000	0.001
Mg	1.095	1.107	0.981	0.937	0.451	0.475	2.177	2.105	0.000	0.000
Ca	0.491	0.573	0.698	0.732	0.476	0.507	1.587	1.745	0.011	0.117
Na	0.012	0.009	0.016	0.009	0.467	0.460	0.828	0.837	0.983	0.876
K	0.004	0.005	0.000	0.000	0.002	0.005	0.000	0.070	0.003	0.004

3b) Garnet epidote amphibolite. Compared with garnet amphibolite, mineral assemblage of this rock is characterized by an appearance of epidote. The fine-grained epidote are distributed around the yellowish green amphibole crystals. Micro-structure of this rock shows that epidotes have derived from the decomposition of amphiboles. Albite-riched plagioclase intergrows along with amphiboles. Intergrowth of amphibole and epidote is thought to result from the breakdown of garnet: Grt+Na+H_2O→pargasite+

epidote [16].

MINERAL CHEMISTRY

Mineral compositions were determined by the electronic probe method at the Analysis and Survey Centre of Changchun University of Earth Sciences in China. Lots of symplectic minerals are too fine to be precisely determined by the availabe techinque, hence we only obtained some limited data of representative minerals. The mineral compositions are listed in Table 1.

Garnet

The garnets essentially consist of almandine (Alm), pyrope (Pyr), grossular (Gro), while spessartine (Sps) is generally <0.01%. The fresh eclogite garnets are more enriched in pyrope content, Pyr=0.40%, Alm=0.42%, Gro=0.18%, and Sps=0.01%, than any others. From eclogite through symplectic intergrowth of Cpx + albite to amphibolite facies, garnet compositions increase in Gro content, but decrease in Pyr content, and remain essentially unchangeble or slightly decrease in Alm content (Fig. 1). The amphibolite garnet is characterized by the highest in Gro+And+Uva and lowest in Pyr, i. e. Pyr = 0.33%, Alm = 0.41%, Sps = 0.01%, and Gro = 0.26%. What's more, most of garnets from amphibolites are strongly zoned with relatively Pyr-rich core, which pressumably were formed during the eclogite stage, and Pyr-poor and grossular-rich rim possibly testifing to the amphibolite of epidote amphibolite facies overprint in the period of retrograde metamorphism.

Amphibole

Amphibole occurs only in garnet amphibolites and epidote amphibolites and is mainly composed of hornblende. While changing from bluish green and fine-grained to yellowish green and coarse-grained crystals, amphibole decrease from 0.413 to 0.255 in Na (M4), from 1.445 to 1.137 in Al^{VI} and ranges from 0.415 to 0.582 in Na (A). This trend may reflect decrease of P-T during epidote amphibolite facies overprint. Furthermore, most of amphiboles are distributed along the fractures and rims of garnets, which indicates that fluid infiltration play an important role in the formation of amphiboles.

Clinopyroxene

clinopyroxene of eclogite is omphacite with jadeite content ranging from 0.45 to 0.52. Because of very fine in size, compositions of symplectic pyroxene around omphacite haven't been gained in this study. However, lots of evidence from previous studies have shown that the symplectic clinopyroxene is either Ca-clinopyroxene such as diopside, salite and so on [6, 13, 18] or sodic clinopyroxene which varies greatly in term of jadeite content, ranging from a few percent to as high as 26% [16, 18]. It is obvious that a symplectic intergrowth of clinopyroxene plus albite represents a retrograde assemblage after omphacite. A common feature is that these symplectic clinopyroxenes contain a distinctly lower jadeite component compared to eclogite omphacites.

Plagioclase

Plagioclase is a common retrograde mineral, which intergrows with either clinopyroxene or amphibole. Its composition ranges from $An_{1.2}$ to An_{20}. Plagioclase coexisting with clinopyroxene generally contains higher albite end-member, which is close to pure

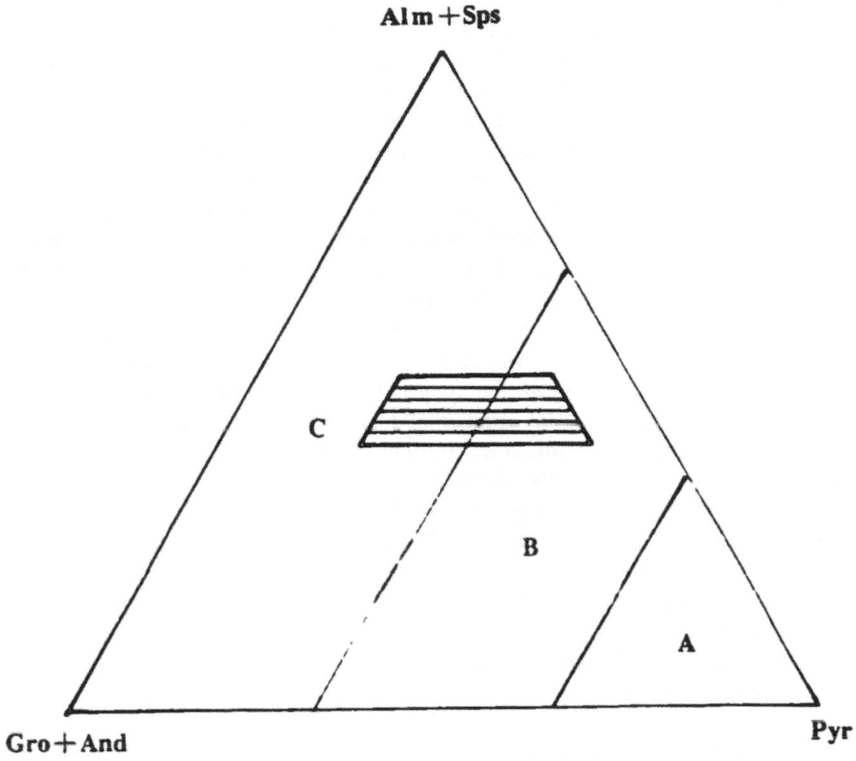

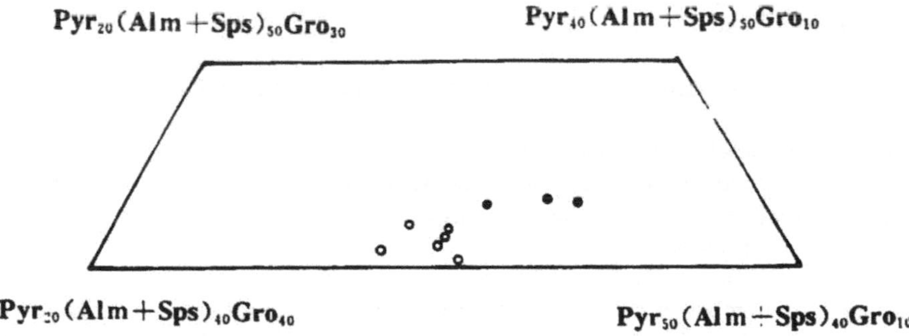

Figure 1 Compotional plots of analysed garnets in field of type B eclogites (●) [2]and garnet amphibolites (○). The shaded parts of the above triangle is enlarged below to show the plots in detail.

albite. but that associated with amphibole and epidote has lower albite, which is close to oligoclase.

REACTION SPACE FOR A CLOSED SYSTEM

Phases and components

As presented by Thompson (1982) [14, 15], one can determine the chemical reactions between phases in petrological system defined by a number of additive components (ad) or "end members" and exchange components (*ex*) or "vectors". Retrograded eclogite with symplectic intergrowth of clinopyroxene plus albite around omphacite is characterized by a four-mineral assemblage: garnet, clinopyroxene, plagioclase and quartz. Three of the four minerals are complex solid-solutions, and they lie (to an excellent approximation) within the composition space defined by the eleven oxides: SiO_2, TiO_2, Al_2O_3, Fe_2O_3, FeO, MnO, MgO, CaO, Na_2O, and K_2O. Following the recommendation presented by Poli (1990) [12], this eleven oxides system can be turned into a condensed system: N′=Na2O+K2O, C′=CaO, M′=MgO+FeO+MnO, $A'=Al_2O_3+Fe_2O_3+TiO_2$, $S'=SiO_2$. Because of absence of hydrous phase in the four-mineral assemblage, it is reasonable to suppose that "dry" eclogite and related rocks lie within a closed system without gain or loss of water to the environmental medium (ENV). We thus have the number of chemical system, $C_s = 5$. The additive components involved the four-mineral assemblage are shown as follows:

diopside	di	$Ca(Mg,Fe,Mn)Si_2O_6$
pyrope	py	$(Mg,Fe,Mn)(Al,Fe^{3+},Ti)Si_3O_{12}$
albite	ab	$(Na,K)(Al,Fe^{3+},Ti)Si_3O_8$
quartz	qz	SiO_2

and the exchange components include:

tk	$(Al,Fe^{3+},Ti)_2(Mg,Fe,Mn)_{-1}Si_{-1}$
pl	$Ca(Al,Fe^{3+},Ti)(Na,K)_{-1}Si_{-1}$
mc	$(Mg,Fe,Mn)Ca_{-1}$

Any phase of the four-mineral assemblage can be described in term of the additive and exchange components:

CPX	di	*tk*	*pl*	*mc*
GAR	py	*mc*		
PLG	ab	*pl*		
QTZ	qz			

The sum of the independent components of phases, $\Sigma_\varphi C_\varphi$, is 9

Reactions involved in the closed system

In the condensed system, number of reactions, Nh, is $C - C_s = 9 - 5 = 4$. It is easily obtained that two pure echange reactions can be written as follows:

$$mc\,[GAR] = mc\,[CPX] \quad (1)$$

$$pl\,[CPX] = pl\,[PLG] \quad (2)$$

Therefore, number of the net transfer reactions, Nt, is equal to $Nh - Nx = 2$, where Nx is the number of pure exchange reactions. According to [13], all net transfer reactions related to the system are shown as:

α) $2di + tk + 2mc = py$ (3)

β) $py + 2qz + tk + 2pl = 2ab + 2mc$ (4)

$di + tk + pl + qz = ab$ (5)

$py + pl + qz = ab + di + 2mc$ (6)

Only two of them are independent, the other two can be obtained by linear combination of the two independent reactions. We shall use reactions (3) and (4) to define the basis vectors for a two-dimensional net transfer space, and mark them with α and β, respectively. The changes of additive components, Δn_{ad}, and exchange components, Δn_{ex}, can be written in units of oxygen equivalent and mole units of exchange components per oxy-unit transfered, respectively:

$$\Delta n_{di} = -\alpha \tag{7a}$$
$$\Delta n_{ab} = \beta \tag{7b}$$
$$\Delta n_{py} = \alpha - 3\beta/4 \tag{7c}$$
$$\Delta n_{qz} = -\beta/4 \tag{7d}$$

and

$$\Delta n_{tk} = -\alpha/12 - \beta/16 \tag{8a}$$
$$\Delta n_{pl} = -\beta/8 \tag{8b}$$
$$\Delta n_{mc} = \beta/8 - \alpha/6 \tag{8c}$$

It is necessary to determine a reference-point where $\alpha_0 = 0$ and $\beta_0 = 0$, hence $\alpha = \alpha - \alpha_0 = \alpha$ and $\beta = \beta - \beta_0 = \beta$. Starting from the reference—point, namely coordinate origin of the reaction space, any variation in the phase abundances can be calculated according to (7) and (8), where $\Delta n_{ad} = n_{ad} - n^0_{ad}$, $\Delta n_{ex} = n_{ex} - n^0_{ex}$. By combining (7) and (8), we can gain a set of equations discribing the variation of mineral compositions:

$$X_{tk}[CPX](n^0_{di} - \alpha)/3 = X^0_{tk}[CPX]n^0_{di}/3 - \alpha/6 - \beta/8 \tag{9a}$$
$$X_{pl}[CPX]\,(n^0_{di} - \alpha)/3 + X_{pl}[PLG](n^0_{ab} + \beta)/4 = X^0_{pl}[CPX]n^0_{di}/3 + X^0_{pl}[PLG]n^0_{ab}/4 - \beta/8 \tag{9b}$$
$$X_{mc}[CPX](n^0_{di} - \alpha)/3 + X_{mc}[GAR](n^0_{py} + \alpha - 3\beta/4)/6 = X^0_{mc}[CPX]n^0_{di}/3 + X^0_{mc}[GAR]n^0_{py}/6 + \beta/8 - \alpha/6 \tag{9c}$$

where $X_{ex}[PHA]$ is the amounts of the exchange vector per formula unit (p. f. u.) of the phase.

Reaction polytope

The advanvcement of any reaction is limited within a reaction space, i. e. reaction polytope. The shape and volume of a reaction polytope are defined by the amounts of reaction phase and exchange-capacity of mineral solid-slolution. The exchange-capacities, of which $[ex^+]$ and $[ex^-]$ account for maximum and minimum possible amounts of the exchange vectors in minerals, respectively, are summerized as follows:

$X_{tk}[CPX] = 0$		*for tk* −	(10*a*)
$X_{tk}[CPX] = 1$		*tk* +	(10*b*)
$X_{pl}[PLG] = -1$	$X_{pl}[CPX] = 0$	*pl* −	(10*c*)
$X_{pl}[PLG] = 0$	$X_{pl}[CPX] = 1$	*pl* +	(10*d*)
$X_{mc}[GAR] = -3$	$X_{mc}[CPX] = 0$	*mc* +	(10*e*)
$X_{mc}[GAR] = 0$	$X_{mc}[CPX] = 1$	*mc* +	(10*f*)

Substituting the exchange-capacity limits, equations (10), in (9), we can obtain the equation of the "line" of the reaction polytope in a two-dimensional space:

$$(1 - X^0_{tk}[CPX])n^0_{di}/3 = \alpha/6 - \beta/8 \qquad [tk+] \tag{11a}$$
$$X^0_{tk}[CPX]n^0_{di}/3 = \alpha/6 + \beta/8 \qquad [tk-] \tag{11b}$$
$$(1 - X^0_{pl}[CPX])n^0_{di}/3 - X^0_{pl}[PLG]n^0_{ab}/4 = \alpha/3 - \beta/4 \qquad [pl+] \tag{11c}$$
$$(1 - X^0_{mc}[CPX])n^0_{di}/3 - X^0_{mc}[GAR]n^0_{py}/6 = \beta/4 \qquad [mc+] \tag{11d}$$
$$(3 + X^0_{mc}[GAR])n^0_{py}/6 + X^0_{mc}[CPX]n^0_{di}/3 = \beta/\beta - \alpha/6 \qquad [mc-] \tag{11e}$$

As for the equation of the "line" on which a additive phase is run out can be established by setting n_{ad} in equation (7) at zero:

$$n^0_{di} = \alpha \tag{12a}$$
$$n^0_{py} = -\alpha + 3\beta/4 \tag{12b}$$
$$n^0_{ab} = -\beta \tag{12c}$$
$$n^0_{qz} = \beta/4 \tag{12d}$$
$$n^0\,qz = \beta/4 \tag{12d}$$

One eclogite (Sample P11-2GS1) composition was determined by wet analysis in the Chemical Analysis Lab. of China University of Geosciences (Beijing). It consists of 46.66wt% SiO_2, 0.28% TiO_2, 15.14% Al_2O_3, 3.48% Fe_2O_3, 7.82% FeO, 0.24% MnO, 7.47% MgO, 10.39% CaO, 3.84 Na_2O, and 0.55% K_2O. We suggest that the eclogite consisting of garnet, omphacite, quartz and minor rutile lies at the origin, reference-point, of the reaction space. It is easily to obtain the amounts of each phase at the origin (n^0_{ad}) by recalculating the rock composition in terms of terms of the component diopside, pyrope, grossular, quartz and jadeite. The Colombi's NTROCK program [3] was used in this recalculation. The number of additive component in oxy-units and exchange component in mole p.f.u. at the origin are shown as follows:

$n^0_{py} = 0.47$, $n^0_{di} = 0.51$, $n^0_{qz} = 0.02$, $n^0_{ab} = 0.0$;

$X^0_{di}[CPX] = 0.61$, $X^0_{pl}[CPX] = 0.61$, $X^0_{mc}[GAR] = -0.96$, $X^0_{mc}[CPX] = 0.0$

We must state that n_{ab} and $X_{mc}[CPX]$ was setted at 0 in the recalculating process. Substituting the n^0_{ad} and $X^0_{ex}[PHA]$ in (11) and (12), we can construct a two-dimensional

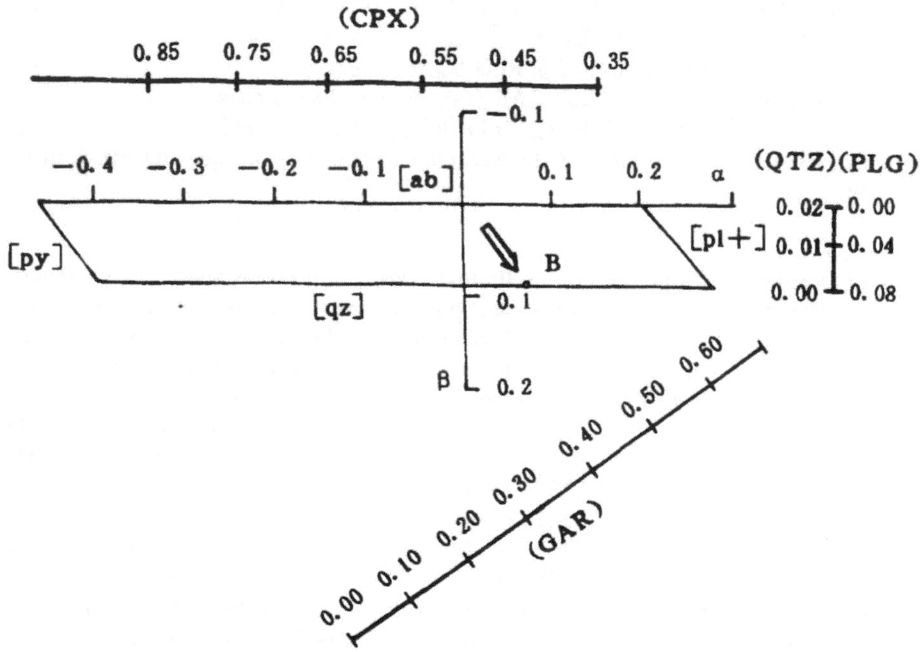

Figure 2 Reaction polytope for the eclogites and retrograded eclogites from Shandong, E. China. Coordinate axes α and β correspond to degrees of advancement on net transfer reactions (3), 2di+ *tk* + 2*mc* =py, and (4), py+2qz+ *tk* + 2*pl* =2ab+2mc, respectively. Lines of the polytope correspond to exhaustion of pyrope (py), albite (ab), quartz (qz) from the assemblage CPX+GAR+PLG+QTZ, and loss of exchange capacity of *pl*+. The eclogite lies at the origin. The arrow represents the reaction vector (5), i.e. di+ *tk* + *pl* +qz=ab. Starting from the eclogite position (at origin), the arrow points to an assemblage containing sodic clinopyroxene + garnet + plagioclase, which made up the symplectites around ompacite in retrograded eclogite. No changes of garnet asboundance and composition take place along the vector (5). Coordinate (GAR), (QTZ), (PLG), and (CPX) indicate the amounts of these minerals in units of oxygen equivalent.

reaction space, which is bounded by [ab], [py], [qz], and [pl^+] line (Fig. 2). The three mineral assemblage, Grt-Omp-Qtz, eclogite lies at the origin. The arrow within the reaction polytope represents the reaction vector (5). i. e. di+ *tk* + *pl* +qz=ab. which slope is 3/4.

Any advancement of reaction along the arrow means no changes of composition and mode of garnet occurs. the amounts of quartz and jadeite content of omphacite gradually decrease. but plagioclase increase. This reaction pathway can accounts for the formation of symplectic intergrowth of clinopyroxene and albite around omphacite in the retrograded eclogite. The reaction ends with the exhaustion of quartz. At the end, point B. of the reaction. the retrograded eclogite should be composed of 0. 08 oxy-units albite. 0. 47 garnet. and 0. 45 clinopyroxene. and the clinopyroxene composition in point B should be sodic clinopyroxene with jadeite content of 0. 6 mole p. f. u. because α and β are 0. 06 and 0. 08. respectively. However. most of the symplectic clinopyroxenes around omphacites are Ca-clinopyroxene and don't contain so high jadeite content as metioned above. A reasonable explanation may be that retrograde metamorphism starting from eclogite stage initially involved localized reactions which occured either within textureal / mineralogical domains defined by the original (precursor) mineral grains or at the boundaries between such domains. In addition. reaction coronas around relict omphacite or textural domains indicate disequilibrium during retrograde alteration of eclogites. At most. some domainal or mosaic equilibrium has been reached only within local domains. The realistic reactions can produce different effects compared with the result by inference based on reaction space analysis. The reaction space is built up on complete. not partial. equilibrium taking place in the petrological system.

As for eclogites. amounts of quartz is an important factor to affect the stabilities of mineral during retrograde metamorphism. It is more difficult for the reaction (5). di + *tk* + *pl* +qz=ab. to occur in olivine-basaltic eclogites than in quartz-basaltic ones. This accounts for that the symplectic intergrowth of clinopyroxene and albite can well developed in quartz-rich eclogites. and around omphacites adjacent to quartz as well.

Changes of garnet composition and aboundance wasn't considered in the above-mentioned discussions. Actually. this is not the case. The ratio of grossular content in garnet increased gradually during the retrograde metamorphism. If we assume that the exchange component *mc* in clinopyroxene didn't change significantly. i. e. orthopyroxene has not been formed. during the retrograde metamorphism. increase of amounts of grossular means that garnet aboundance will decrease in accordance with the equation (9c). In case of the grossular content decreases. a reaction vector will be terminated at a point to the left of the point B. and the slope of the reaction pathway will be greater than 3/4.

REACTION SPACE FOR AN OPEN SYSTEM

Reactions involved in an open system

Unlike the coronatic assemblages of eclogites. garnet amphibolites derived from eclogites consist of five minerals: garnet. plagioclase. quartz. clinopyroxen. and amphibole. Occurence of amphibole requires participation of water from environment (ENV) owing to absence of hydrous phase in "dry" eclogites. As a result the chemical system is N'C'M'A'S'H' (H'=H_2O) rather than N'C'M'A'S'. and the number of the chemical system. C_s. is 6. Based on the closed system. an additive component tr Ca_2 (Mg. Fe. Mn)$_5$ (Si)$_8O_{22}$ (HO)$_2$. and an exchange component *ed* (Na. K) (Al. Fe^{3+}) Si_{-1}

should be supplemented in the open petrological system. This is because that an amphibole composition space is defined by a four-independent components: tr, *tk*, *pl*, and *ed*. For the purpose of being simple and clear, we neglect the composition displacement along exchange vector mc in pyroxene and amphibole composition space. Adapting the similar method used in the closed system, we conclude that the total number of independent phase components, $\sum_{\varphi} C_{\varphi}$, is 8+5=13, and the total reaction numbers, $Nh = C - C_e = 13 - 6 = 7$.

The independent pure exchange reactions are summarized as follows:

$$pl[\mathrm{CPX}] = pl[\mathrm{PLG}] \quad (11)$$

$$pl[\mathrm{AMP}] = pl[\mathrm{PLG}] \quad (12)$$

$$tk[\mathrm{CPX}] = tk[\mathrm{AMP}] \quad (13)$$

$Nx = 3$, and $Nt = Nh - Nx = 7 - 3 = 4$.

Following recommendation of Thompson [14], we choose four independent net transfer reactions as the basis vectors of the rection space open to water:

A) $\mathrm{ab} = ed + 4\mathrm{qz}$ (15)

B) $2\mathrm{ab} + 2mc = \mathrm{py} + 2\mathrm{qz} + tk + 2pl$ (4a)

C) $\mathrm{ab} = \mathrm{di} + tk + pl + \mathrm{qz}$ (5a)

D) $4\mathrm{tr} + 5\mathrm{qz} + 3ed = 3\mathrm{py} + 11\mathrm{di} + 3pl + 4H_2O$ (16)

The reaction (4a) and (5a) occur in the closed system, (16) in the open system, and (15) in either closed or open system. A reference-point was choosed as an origin of the reaction space, where a garnet amphibolite is stable. This garnet amphibolite consists of 0.1619 oxy-units tr, 0.4622 py, 0.1755 di, 0.1928 ab and 0.0076 qz, and $X^0_{tk}[\mathrm{AMP}] = 0.170$, $X^0_{pl}[\mathrm{AMP}] = 0.465$, $XA^0_{ed}[\mathrm{AMP}] = 0.511$, $X^0_{tk}[\mathrm{CPX}] = 0.222$, $X^0_{pl}[\mathrm{CPX}] = 0.099$, $X^0_{pl}[\mathrm{PLG}] = -0.011$, $X^0_{mc}[\mathrm{GAR}] = -0.698$

Mineral composition and mode of retrograded eclogites are determined by the following equations:

$$n_{ab} = n^0_{ab} - A - B - C \quad (17a)$$

$$n_{di} = n^0_{di} + 0.75C + 0.6226D \quad (17b)$$

$$n_{tr} = n^0 tr - 0.9057D \quad (17c)$$

$$n_{py} = n^0_{py} + 0.75B + 0.3396D \quad (17d)$$

$$n_{qz} = n^0_{qz} + A + 0.25B + 0.25C - 0.0943D \quad (17e)$$

and

$$n_{tk} = n^0_{tk} + 0.625B + 0.125C \quad (18a)$$

$$n_{pl} = n^0_{pl} + 0.125B + 0.125C + 0.0283D \quad (18b)$$

$$n_{ed} = n^0_{ed} + 0.125A - 0.0283D \quad (18c)$$

$$n_{mc} = n^0_{mc} - 0.125B \quad (18d)$$

Combination of (17) and (18) in the way used in (9) can give the relationship between n^0_{ad} and X_{ex} [PHA]:

$$X_{tk}[\mathrm{CPX}]n^0_{di}/6 + X_{tk}[\mathrm{AMP}]n^0_{tr}/24 = X^0_{tk}[\mathrm{CPX}]n^0_{di}/6 + X^0_{tk}[\mathrm{AMP}]n^0_{tr}/24 + 0.0625B + (0.125 - 0.75X_{tk}[\mathrm{CPX}])C + (0.9057X_{tk}[\mathrm{AMP}]/24 - 0.6226X_{tk}[\mathrm{CPX}]/6)D \quad (19a)$$

$$X_{pl}[\mathrm{CPX}]n^0_{di}/6 + X_{pl}[\mathrm{PLG}]n^0_{ab}/8 + X_{pl}[\mathrm{AMP}]/24 = X^0_{pl}[\mathrm{CPX}]n^0_{di}/6 + X^0_{pl}[\mathrm{PLG}]n^0_{ab}/8 + X^0_{pl}[\mathrm{AMP}]n^0_{tr}/24 + X_{pl}[\mathrm{PLG}]A/8 + (0.125 + X_{pl}[\mathrm{PLG}])B/8 + (0.125 - 0.75X_{pl}[\mathrm{CPX}]/6 + X_{pl}[\mathrm{PLG}]/8)C + (0.0283 + 0.9057X_{pl}[\mathrm{AMP}]/24 - 0.6226X_{pl}[\mathrm{CPX}]/6)D \quad (19b)$$

$$X_{ed}[\mathrm{AMP}]n_{tr}^{0}/24 = X_{ed}^{0}[\mathrm{AMP}]/24 + 0.125A - (0.0283 - 0.9057X_{ed}[\mathrm{AMP}]/24)D \qquad (19c)$$

$$X_{mc}[\mathrm{GAR}]n_{py}^{0}/12 = X_{mc}^{0}[\mathrm{GAR}]n_{py}^{0}12 - (0.125 + 0.75X_{mc}[\mathrm{GAR}]/12)B - 0.3396X_{mc}[\mathrm{GAR}]D/12 \qquad (19d)$$

The exchange-capacity limits related to amphibole are: *tk* = 2 (*tk* +), *tk* = 0 (*tk* −); *pl* = 2 (*pl* +), *pl* = 0 (*pl* −); *and ed* = 1 (*ed* +), *ed* = 0 (*ed* −). The others are same as those used in the closed system. Therefore, we have obtained a set of equations representing the four-dimensional "planes" of the reaction space in an open petrological system:

$$n_{ab}^{0} = A + B + C \qquad [ab] \qquad (20a)$$

$$n_{tr}^{0} = 0.9057D \qquad [tr] \qquad (20b)$$

$$n_{py}^{0} = -0.75B - 0.3396D \qquad [py] \qquad (20c)$$

$$n_{di}^{0} = -0.75C - 0.6226D \qquad [di] \qquad (20d)$$

$$n_{qz}^{0} = -A - 0.25B - 0.25C + 0.0943D \qquad [qz] \qquad (20e)$$

$$4n_{di}^{0} + 2n_{tr}^{0} - 40X_{tk}[\mathrm{CPX}]n_{di}^{0} - X_{tk}^{0}[\mathrm{AMP}]n_{tr}^{0} = 1.5B - 0.6793D \qquad [tk+] \qquad (21a)$$

$$-4X_{tk}^{0}[\mathrm{CPX}]n_{di}^{0} - 4X_{tk}^{0}[\mathrm{AMP}]n_{tr}^{0} = 1.5B + 3C \qquad [tk-] \qquad (21b)$$

$$4n_{di}^{0} + 2n_{tr}^{0} - 4X_{pl}^{0}[\mathrm{CPX}]n_{di}^{0} - 3X_{pl}^{0}[\mathrm{PLG}]n_{ab}^{0} - X_{pl}^{0}[\mathrm{AMP}]n_{tr}^{0} = 3B \qquad [pl+] \qquad (21c)$$

$$-3n_{ab}^{0} - 4X_{pl}^{0}[\mathrm{CPX}]n_{di}^{0} - 3X_{pl}^{0}[\mathrm{PLG}]n_{ab}^{0} - X_{pl}^{0}[\mathrm{AMP}]n_{tr}^{0} = -3A + 0.6792D \qquad [pl-] \qquad (21d)$$

$$n_{tr}^{0} - X_{ed}^{0}[\mathrm{AMP}]n_{tr}^{0} = 3A + 0.2264D \qquad [ed+] \qquad (21e)$$

$$-X_{ed}^{0}[\mathrm{AMP}]n_{tr}^{0} = 2A - 0.6792D \qquad [ed-] \qquad (21f)$$

$$-X_{mc}^{0}[\mathrm{GAR}]n_{py}^{0} = -1.5B \qquad [mc+] \qquad (21g)$$

$$-3n_{py}^{0} - X_{mc}^{0}[\mathrm{GAR}]n_{py}^{0} = 0.75B + 1.0189D \qquad [mc-] \qquad (21h)$$

$$4X_{pl}^{0}[\mathrm{CPX}]n_{di}^{0} + 3X_{pl}^{0}[\mathrm{PLG}]n_{ab}^{0} + X_{pl}^{0}[\mathrm{AMP}]n_{tr}^{0} - 4X_{tk}^{0}[\mathrm{CPX}]n_{di}^{0} - X_{tk}^{0}[\mathrm{AMP}]n_{tr}^{0} - X_{ed}^{0}[\mathrm{AMP}]n_{tr}^{0} = 3A - 1.5B - 1.3584D \qquad [sh] \qquad (21i)$$

where [*sh*], (21i), stands for the stoichiometrical condition for both amphibole and pyroxene where $X_{pl} = X_{tk} + X_{ed}$.

Determination of the origin and least-squares approximation

To establish the reaction space for the five-mineral assemblage, it is necessary to select a reference-point, i. e. the origin. This can be done by considering an amphibolite which contains garnet, clinopyroxene, amphibole, plagioclase, and quartz. The mole fraction of exchange vectors, X_{ed}^{0}, in every mineral phase can be calculated in accordance with Thompson's suggestion [14, 15]. However, mineral modes, n_{ed}^{0}, are unknown. Least square analysis is a very usful method to solve this problem. Bryan etc. [1] applied this method to estimate proportions in petrographic mixing equations, afterwards, Morris [10] used it in magma fractional crystallization processes. We modified the version of Morris' basic program and utilized it in estimation of mineral abundances.

Given mineral chemical compositions, we can estimate mineral modes by using least-squares approximation. A number of minor phase such as rutile, magnetite are taken into consideration during the calculation. Observed values, estimated values, and mineral weight proportions and oxygen-equavelent proportions obtained from least-squares approximation for the fresh eclogite (ecl.), symplectite of retrograded eclogites (sym.) and garnet amphibolite are list in the Table 2. Reaction polytope and metamorphic processes.

By substituting n_{sm}^0 and X_m^0 [PHA] in (20) and (21). we can establish the reaction polytope. However, this reaction polytope is bounded within a four-dimensinal space, whose basis coordinate axes are represented by four indepent reaction vectors: *A*, *B*, *C*, and *D*. In order to get an audio-visual polytope form. we project the reaction space in a three or two-dimensional space which is defined by arbitray three or two of the four basis vectors. All face equations of the reaction polytopes were calculated by means of Clombi's NTPLOT program [3]. Fig. 3 is obtained from projection along basis vector A onto the space *B-C-D* and *C-D*, and Fig. 4 from projection along *B* onto the *A-C-D* and *A-D*.

Table 2. Observed (V. OBV) and fitted (V. EST) compositions, squared residulas (S. R.), and mineral witght proportions and oxy—equevalent proportions (in brackets) obtained from least—squares approximation for eclogite (Ecl.) and garnet amphibolites (Gt—Amp.). "Origin" stands for the garnet amphibolite lies at the reference—point (origin) of the reaction space.

Sample		Ecl.	Gt—Amp. (Origin)	Gt—Amp.
P11—6GS	V. OBS	V. EST	V. EST	V. EST
SiO_2	48.56	48.56	48.56	48.56
TiO_2	0.45	0.45	0.45	0.45
Al_2O_3	17.14	17.85	16.94	17.72
FeO	14.08	12.53	14.09	13.10
MnO	0.25	0.25	0.34	0.24
MgO	6.53	9.15	7.69	8.29
CaO	10.45	8.33	9.75	9.29
Na_2O	2.36	2.53	2.70	1.36
K_2O	0.18	0.04	0.30	0.23
S. R.		14.29	2.00	6.74
CPX		0.361 (0.375)	0.180 (0.176)	0.0 (0.0)
GAR		0.610 (0.60)	0.479 (0.462)	0.275 (0.260)
QTZ		0.021 (0.025)	0.006 (0.008)	0.158 (0.188)
PLG		0.0	0.174 (0.193)	−0.047 (−0.051)
AMP		0.0	0.166 (0.162)	0.604 (0.603)

In case of knowing mineral modes (Tab. 2) and compositions (Tab. 1). we can estimate the advancement along each basis vector based on the set of equations (20) and (21) (Figs. 3 and 4)

The coordinates of various assemblages are shown in Table 3.

We projected in Fig. 3 and 4 the eclogite ((①)). symplectitic assemblage of retrograded eclogite ((②)). which was obtained from the point B in Fig. 2. and garnet amphibolite ((③)). It is easily observed that a trend from eclogite ((①)) to symplestite stage ((②)) is nearly parallel to the vector *C*. i. e. di+ *tk* + *pl* =ab. This is the case metioned in the closed system and this vector corresponds to the well-known reaction jadeite+quartz→albite [5]. It is this reaction that controlled omphacite mode and composition during the retrograde alteration from eclogite to symplectic intergrowth of Ca-clinopyroxene and albite. The advancement along *C* should be favoured by a decrease in pressure as a re-

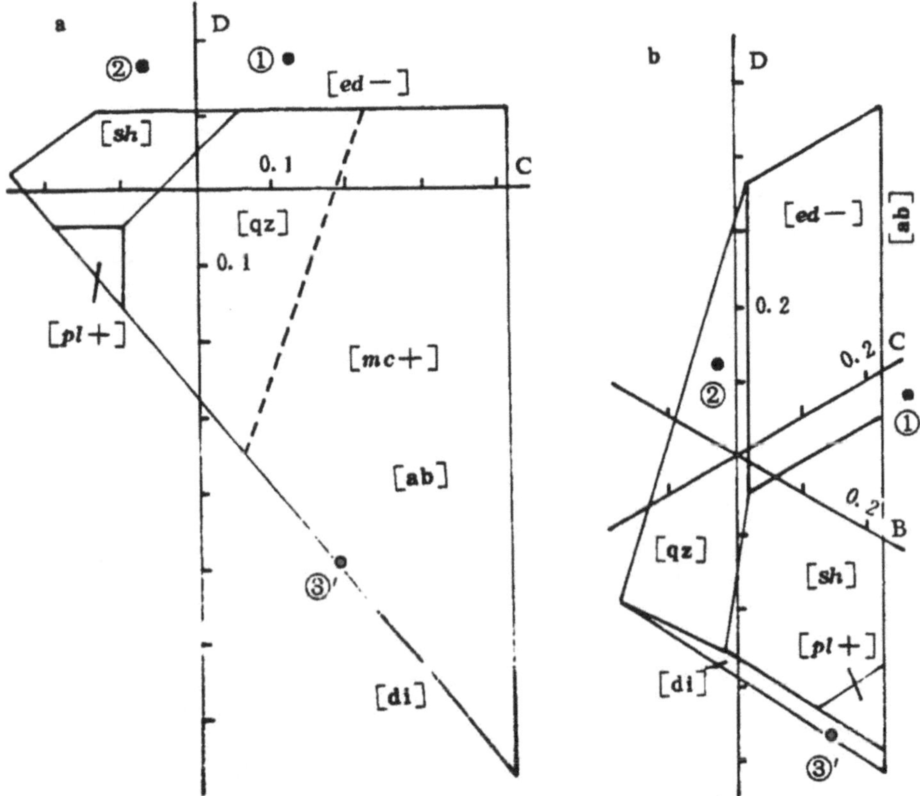

Figure 3 Reaction polytope for eclogite and garnet amphibolite from Shandong Province, E. China. Coordinate axes *B*, *C*, and *D* correspond to degrees of advancement on the net transfer reactions 2ab + 2*mc* = py + 2qz + *tk* + *pl*, ab = di + *tk* + *pl* + qz, and 4tr + 5qz + 3*ed* = 3py + 11di + 3 *pl* + $4H_2O$, respectively. Faces of the polyhedron stand for elimination of clinopyroxene (di) and quartz (qz) from the five-mineral assemblage GAR + CPX + AMP + PLG + QTZ, or loss of exchange capacity. The light labels represent hidden faces. The Fig. 3a was obtained from the projection of the three-dimensional polyhedron *A* − *C* − *D* on the plane *C* − *D*. Number ①, ②, ③′ correspond to the eclogite, symplectitic intergrowth assemblage contained within the retrograded eclogites, and garnet amphibolite shown in the Tables 1, 2, and 3.

sult of exhumation of the eclolgites. The transition from eclogite (①) or retrograded

Table 3 The advancement along the basis vectors (*A*, *B*, *C*, and *D*) of the eclogite (ecl.), the symplectite (sym.) and garnet amphibolite (gt-amp.)

Stage	Sample	A	B	C	D
①	ecl.	0.028	0.103	0.118	0.178
②	sym.	−0.012	0.026	−0.069	0.178
③	Gt−amp.	0	0	0	0
③′	Gt−amp.	−0.3651	−0.049	0.170	−0.487

eclogite (②) to ganet amphibolite (③′) is mainly controlled by the reaction vector *D*. 4tr + 5qz + 3 *ed* = 3py + 11di + 3 *pl* + $4H_2O$. This reaction vector approximately corresponds to the reaction Grt + Omp + H_2O = Hb + Plg (An + Ab) + Qtz proposed by Oh

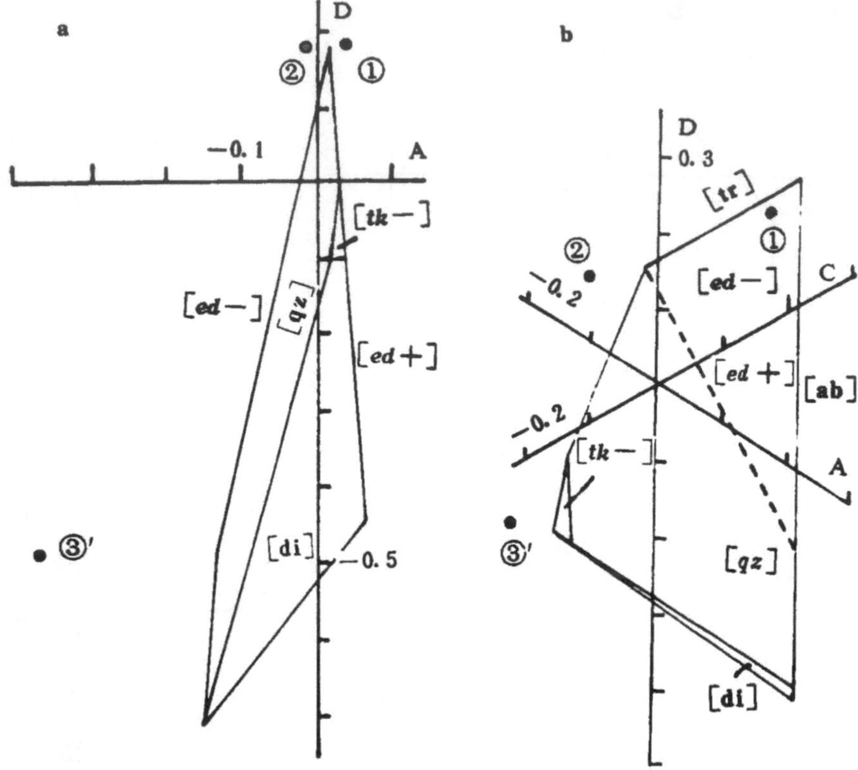

Figure 4 Projecton of reaction polytope from a four—dimensional space defined by basis vectors *A*, *B*, *C*, and *D* on three— and two— dimensional space defined by the vectors *A* — *C* — *D* and *A* — *D*, respectively. The coordinate *A* represents the basis reaction vector ab= *ed* +4qz. See the explanation of the Fig. 3 for the other signs.

(1992) [11]. which as a boundary equilibrium curve separates eclogite from amphibolite facies. and it is a significant hydration reaction that transforms the eclogites into garnet amphiblotes. The amounts of omphacite is an important constraint to control the reaction procedure. For example. the point (③′) in Fig. 3 represents the state that the reaction was terminated owing to the exhaustion of clinopyroxene. hence pyroxene phase was absent from most garnet amphibolites. but a number of omphacites was still present as inclusions contained in garnets. So far. lots of previous studies [16. 18. 19] and petrographical evdences indicate that the reaction vector *D* is an important net transfer reaction to accout for the formation of hornblende and plagioclase after breakdown of garnet and omphacite during retrograde metamorphism. It is possible for the reaction vector *B*. 2ab+2 *mc* =py+2qz+ *tk* + 2*pl*. in reverse to cause the transfer of a grossular content in garnet into plagioclase and the elimination of quartz. The reaction *A*. ab= *ed* +4qz. can account for edenite substitution in amphibole. A decreasing trend of edenite subsitution was found in respond to the retrograde alteration of amphiboles. This means that the reaction *A* proceeded towards the left hand side (Fig. 4).

Acknowledgements

This paper was completed with the financial support of the Metamorphic Geological Dynamic Open Laboratory of Changchun University of Earth Sciences and the Nature Science Foundation of China. The authors thank Prof. You Zhengdong of China University of Geosciences (Wuhan) for his patient and helpful reviews of an early version of the manuscript and Prof. Wang Renming of Peking University for his engagement and powerful supports.

REFERENCES

1. W. B. Bryan, L. W. Finger, and F. Chayes. Estimating proportions in petrographic mixing equtions by least-squares approximation, *Science*, **163**, 163—164 (1969)
2. R. G. Coleman, D. E. Lee, L. B. Beatty, and W. W. Brannock. Eclogites and eclogites: their dfferencies and similarities, *Bull Geol. Soc. Am.* **76**: 483—508 (1965)
3. A. Colombi. Rspace: a set of programs to define completely the reaction space of J. B. Thompson, Jr., *Computers & Geosciences*, **15**, 403—440 (1989)
4. M. Enami and Q. Zang. Quartz preudomorphs after cocsite in eclogites from Shandong Province, east China, *American Mineralogist*, **75**, 381—386 (1990)
5. T. J. B. Holland. The experimental deterniation of activities in disordered and short range ordered jadeitic pyroxenes, *Contrib. Mineral Petrol.*, **82**, 214—220 (1983)
6. B. M. Jahn, J. Corichet, B. Cong, Tzen-Fu Yui. Ultra-high Nd eclogites from an ultra-high pressure metamorphic terrane of China, *Chem. Geol.* (manuscript) (1995)
7. X. Y. Lai, S. S. Su, and Z. Chen. High-pressure relics in gneisses surrounding the eclogite in Zhucheng area, Shandong, China, *Chinese Science Bulletin*, **40**, Suppl., 83—85 (1995)
8. S. Li, Y. Z. Chen, N. J. Liu, D. L. Zhang, Z. M. Zhang, Q. D. Zhang, D. M. Zhao. U-Pb zircon ages of eclogite and gneiss from Jiaonan Group in Qingdao area, *Chinese Sci. Bull.*, **38**, 1773—1777 (1993)
9. S. Li, Y. Xiao, D. Liu, Y. Chen, N. Ge, Z. Zhang, S. S. Sun, B. Cong, R. Zhang, S. R. Hart. Collision of the North China and Yangtze Blocks and formation of coesite-bearing eclogite: timing and processes, *Chem. Geol.*, **109**, 89—111 (1993)
10. P. A. Morris. MAGFRAC: a basic program for least-squares approximation of fractional crystallization, *Computers & Geosciences*, **10**, 437—444 (1984)
11. C. W. Oh. The petrogenetic relationship among high-P/T metamorphic facies including the eclogite and epidote-amphibolite facies in modal basltic system, *Jour. Geol. Soc. Korea*, **28**, 298—313 (1992)
12. S. Poli. Reaction space and P-T paths: transition from amphibole eclogite to greenschist facies in the Auxtroalpine domain (Oetzal Complex), *Contrib. Mineral petrol.*, **106**, 399—416 (1991)
13. A. E. Ringwood. Composition and petrology of the earth's mantle, *McGraw-Hill, U.S.A.* (1975)
14. J. B. Thompson. Reation space: an algebric and geometric approach, Mineralogical Soc. *America, Reviews in Mineralogy*, **10**, 35—52 (1982)
15. J. B. Thompson, J. Laird, and A. B. Thompson. Reactions in amphibolite, greenschist and blueschist, *Jour. Petrology*, **23**, 1—27 (1982)
16. X. M. Wang, J. G. Liu and S. Maruyama. Coesite-bearing eclogites from the Dabie Mountains, central China: petrogenesis, P-T paths, and implications for regional tectonics, *J. Geology*, **100**, 231—250 (1992)
17. J. Yang and D. C. Smith. Evidence for a former sanidine-coesite-eclogite at Lanshantou, eastern China, and the recognition of Chinese Su-Lucoesite-eclogite province, 26 *Terrastraces*, **1**, 74 (1989)
18. R. Y. Zhang, B. Cong, J. G. Liu. Su-Lu ultra-high pressure metamorphic terrane and expanation of its origin, *Acta Petrologica Sinica*, **9**, 211—255 (1993)
19. R. Y. Zhang, J. G. Liu and B. Cong. Petrogenesis of garnet-bearing ultramafic rocks and associated eclogites in the Su-Lu ultra-high-P metamorphic terrane, eastern China, *J. Metamorphic Geol.*, **12**, 169—196 (1994)

Proc 30th Int'l. Geol. Congr., Vol 16, pp 151-157
Huang Yunhui and Cao Yawen (Eds)
VSP 1997

Alternative Analysis of Room Temperature IR-spectra of Quartz

G.V.NOVIKOV, D.G.KOSHCHUG, H.RAGER
Institute of Experimental Mineralogy, RAS; Dept. of Geology, Moscow State University; Philipps University, Marburg, Germany.

Abstract

Analysis of any spectroscopic data comprises the determination of line parameters (position, halfwidth, line shape, intensity), and their errors. Depending on complexity of the spectrum some preliminary transformation of the initial spectrum may be needed to elucidate the approximate position of the individual components. In this work it is proposed to generate the model of the experimental spectrum with the help of original mathematical procedure which results in a new spectrum with better spectral resolution. The procedure is applied to room temperature IR-spectra of quartz. It is shown that spectral resolution is improved by a factor of 3. The validity of room temperature results is confirmed by low temperature measurements.

Keywords: IR-spectroscopy, spectra deconvolution, quartz

INTRODUCTION

Analysis of any spectroscopic data comprises the determination of line parameters (position, halfwidth, line shape, intensity) and their errors. Usually the analysis is carried out by a special programs based on the approximation of experimental spectrum by the sum of components. Depending on the complexity of spectrum some preliminary transformation (like first and second derivation, Fourier transformation) of the initial spectrum may be used to elucidate the number and approximate position of the individual components.

To demonstrate a new approach for numerical spectra analysis infrared absorption spectra of hydroxyl in quartz have been chosen.

During the last 30 years infrared (IR) spectroscopy was widely used to identify different types of hydroxyl vibrations in quartz [1, 2, 5]. The Al substitution for Si in this material is often associated with an incorporation of additional impurities like H^+, Li^+, Na^+, K^+ etc. which are located as charge compensators in the structural channel running along *z*. The interaction of OH groups with these ions gives rise to typical absorption lines in the range 3300 to 3600 cm^{-1}. They are indicative for the kind and concentration of impurities and, therefore, can be taken for the characterization of the quality of quartz. Moreover, a precise characterization is important for many technical applications where quartz is used as raw material. Low temperature spectra of quartz are interpreted in [1, 2] and the lines corresponding to the main impurities are distinguished. Some problems in detailed interpretation of low temperature spectrum of quartz are not solved so far. At room temperature most of lines are more broadened than at low temperature and the distance between some lines is smaller than their halfwidth. Any attempts to fit the room temperature spectra of quartz will give no correct results because of considerable correlation between parameters of strongly overlapped lines.

As it will be shown in this work this correlation can be diminished using some original mathematical procedure [4] and the analysis of room temperature IR-spectra of quartz gives physically reliable results.

EXPERIMENTAL

The samples are natural quartz single crystals of high quality from the North Ural (Russia). The overall content of impurities is less than 10^{-2} w. %. All samples contain small amounts of Al, Li, Ge, OH and possibly Na, K. For the measurements 2 single crystal plates with a thickness of approximately 2 to 3 mm were prepared. The direction of the optical z axis is within the plate for sample 1 and declined with 45° to the plate for sample 2.

The measurements were performed at room temperature using the Bruker Fourier Transform Infrared Spectrometer IFS-88. For comparison liquid nitrogen measurements were also carried out using a Carl Zeiss cryostat and an UR 20 Infrared Spectrometer. All measurements were performed using unpolarized light. Because the IR absorption due to hydroxyl vibrations is most interesting only the range from 3100 to 3700 cm^{-1} is considered.

NUMERICAL ANALYSIS OF SPECTRA

The analysis of experimental spectrum includes in general two steps. At first one has to fix the model of the experimental spectrum: namely, to determine the number of components in the spectrum, their relative arrangement and their intensities. In case of such a complex spectra as spectra under consideration this step is the most important. The second step is more ore less technical - to improve the proposed model to fit the experimental spectrum.

In this work it is proposed to generate the model of experimental spectrum with the help of original mathematical N-procedure which transforms the experimental spectrum $Y(x)$ according to the equation:

$$Y(x, p, \Delta x) = Y(x) - p[Y(x + \Delta x) + Y(x - \Delta x)].$$

Here, Y is the intensity, x is the wavenumber, $p = 2 * [-2 - \delta^2 + \sqrt{(\delta^4 + 5\delta^2 + 4)}] / \delta^2$, and $\delta = 2\Delta x/G$. Δx and G are the procedure parameters, which have to be chosen by the user. These values should be chosen thus to give after the N-transformation an optimum spectral resolution at acceptable signal-to-noise ratio. In general best results are obtained at G approximately equal to the experimental halfwidth of the components in the initial spectrum. The Δx value usually is close to $G/2$ or less, if the signal-to-noise ratio is not too low which depends on the noise level of the spectrometer (or other registration equipment). Decrease of Δx improves the spectral resolution but at the same time the signal-to-noise ratio falls.

Fitting procedure is carried out in accordance with [4]. The original computer program allows to vary any line parameters: position, halfwidth, intensity, line shape, background intensity. The results of calculations include the values of variable parameters, their dispersion and general correlation factors [3].

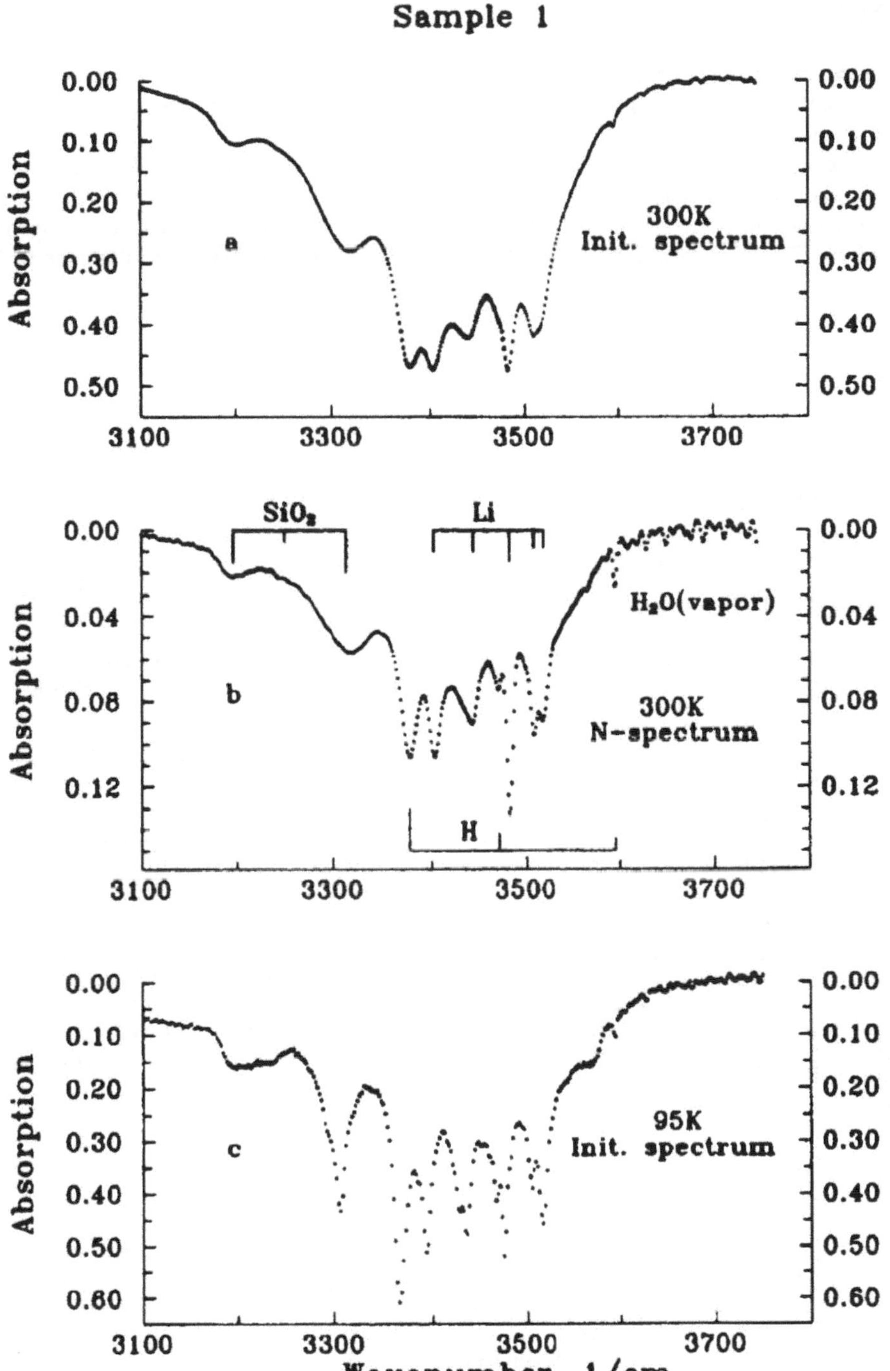

igure 1. IR absorption spectra of sample 1: a - initial room temperature spectrum, b - spectrum after N-transfor-ation, c - low temperature spectrum.

RESULTS AND DISCUSSION

The experimental room temperature spectra are shown in Fig. 1a and 2a for samples 1 and 2, respectively. In each spectrum three groups of known lines are distinguished. They are shown schematically in Fig. 1b and 2b:

1. The broad absorption distributions in the range 3200 to 3300 cm^{-1} are due to combination vibrations of the quartz matrix.
2. The narrow lines in the range 3300 to 3600 cm^{-1} are due to different types of hydroxyl vibrations. The lines caused by single OH groups are marked by "H"; the vibrations of OH perturbed by the interaction with Li are marked by "Li".
3. The very narrow weak lines which appear at wavenumbers higher than 3600 cm^{-1} are due to water vapor inside the spectrometer.

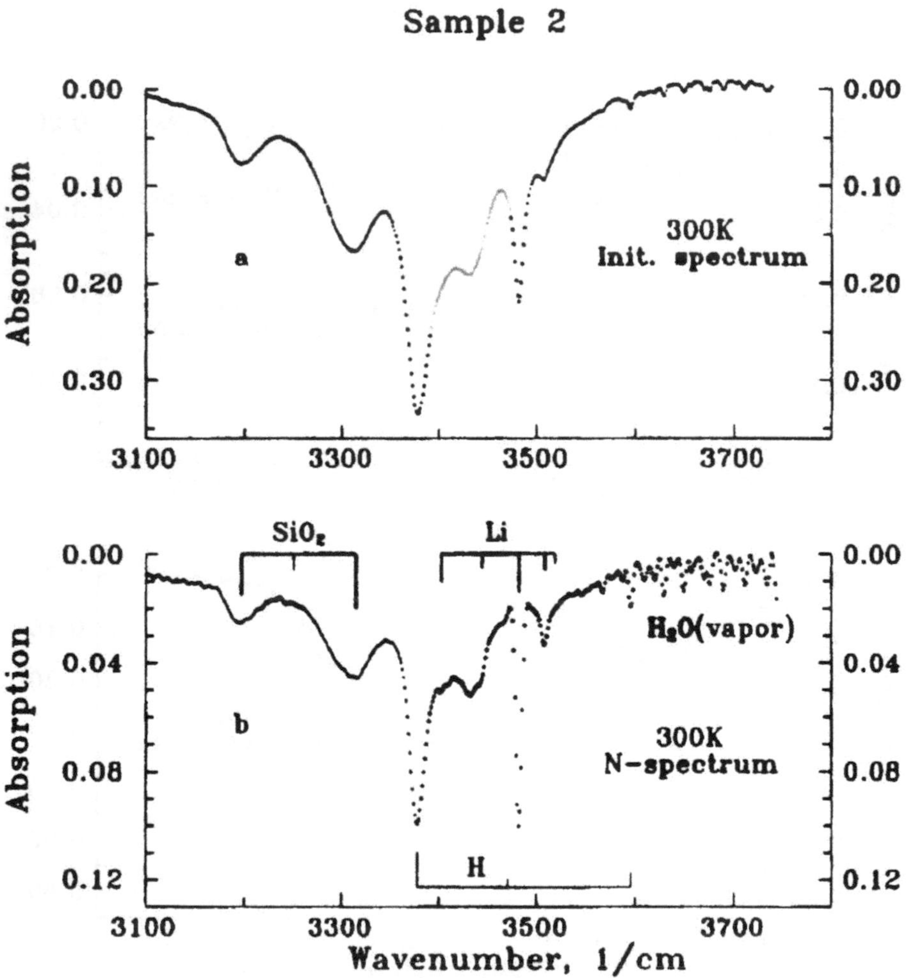

Figure 2. IR absorption spectra of sample 2: a - initial room temperature spectrum, b - spectrum after N-transformation.

For sample 1 one spectrum was also recorded at 95 K (Fig. 1c) to compare the results of this work with already published results [1]. The identification of hydroxyl vibration lines was then performed on the basis of this comparison.

The results of application of N-procedure to room temperature spectra are shown in Fig. 1b and 2b. The N-transformation has been done at Δx = 3.86 cm^{-1} and G = 14.15 cm^{-1}. These parameters are favorable to reveal the narrow lines in the spectra under consideration. It is evident that hydroxyl absorption lines in the range 3350 cm^{-1} to 3520 cm $^{-1}$ are narrowed by the N-procedure. Fine structure of strongly overlapped lines which appeared as asymmetric lines with shoulders (e.g., the lines at 3470-3480 cm $^{-1}$ and 3510-3520 cm $^{-1}$) is disclosed after the transformation. Existence of these lines is confirmed by low temperature measurements (Fig. 1c).

There is one important feature of the N-procedure. After the transformation the broader is the component in the experimental spectrum the lower is its relative intensity in N-spectrum if the G parameter is chosen close to the line width of narrow components. As a result in the N-spectra all narrow lines marked by "H", "Li" and "H_2O vapor" became more pronounced and can be definitely included in physically reliable model of the spectra. Hence, the influence of broad line on the accuracy of narrow line parameters is strongly diminished. The whole model of the spectra used for fitting contained three broad components (marked SiO_2) and one broad line at 3513 cm^{-1} and at 3505 cm^{-1} for samples 1 and 2, respectively. There is no proved interpretation of the origin of these broad lines but they definitely exist and should be taken into account.

Fitting has been done for the initial room temperature spectra and for the N-spectra of two samples. For each sample in both cases the model of spectra contained equal number of components and identical relative arrangement. Moreover, the starting values of position and halfwidth of the components in initial spectra were taken from the results of the N-spectra fitting, otherwise additional strong constrains were needed for successful fitting. Calculated line positions, their dispersions and general correlation factors are listed in tab. 1 and 2.

The validity of the decomposition results is characterized (in addition to the traditional dispersions) by the value of general correlation factor. The practice show that for the reliable results this value should not exceed 15-20 otherwise the dispersions are poor estimators of the errors. The general correlation factors for the parameters of overlapped components in initial spectrum are considerably higher in comparison to N-spectrum. They increase strongly for the width and especially for the intensities of overlapped and weak lines (e.g., lines Li (b), Na, K, H (b), Li (d), Li (e)). Hence, the parameters of these lines can be determined with reasonable accuracy only from N-spectrum fitting. For sample 2 the parameters of such lines are ambiguous even in N-spectrum. Interdependence of the parameters of overlapped lines is clearly demonstrated by correlation factor for Li (d) line. Absence of Li (e) line in spectrum of sample 2 has immediately diminished the correlation factor for Li (d) line from 1.8 to 1.0 for N-spectrum and from 85 to very low value for initial spectrum.

Better spectral resolution after the N-transformation permits to calculate the parameters of lines which are separated only by 0.35 of the halfwidth (at approximately equal halfwidths). For nontransformed spectrum about the same correlation factor is achieved for lines with the distance between them equal to about the halfwidth.

Parameters of weak and narrow water vapors lines are not listed in tab. 1 and 2 but the results for both samples coincided within a very small limit of error (about 0.5 cm $^{-1}$ for the position of lines). It may be taken as an indication for the accuracy of the applied procedure.

Table I. Line positions with dispersions (in brackets) and general correlation factors for N-spectrum and for initial spectrum at 300K (sample 1).

N	Line	Position, cm^{-1}	General correlation factor	
			N-spectrum	Init. spectrum
1	SiO_2	3191.1 (5.9)	2.7	1.9
2	SiO_2	3262.9 (7.1)	2.0	17
3	SiO_2	3315.4 (2.4)	3.4	6
4	H (a)	3377.0 (0.5)	1.4	8.8
5	Li (a)	3403.4 (0.7)	2.0	11
6	?	3437.4 (12.6)	54	269
7	Li (b)	3444.1 (0.4)	1.4	89
8	Na	3454.6 (1.2)	1.5	909
9	K	3469.4 (0.9)	52	84
10	H (b)	3471.2 (1.3)	5.3	-
11	Li (c)	3482.3 (0.2)	1.4	9.7
12	Li (d)	3508.1 (0.6)	1.8	85
13	?	3513.4 (3.4)	8.6	887
14	Li (e)	3518.0 (0.9)	2.2	11
15	H (c)	3595.6 (0.3)	1.2	32

Table 2. Line positions with dispersions (in brackets) and general correlation factors for N-spectrum and for initial spectrum at 300K (sample 2).

N	Line	Position, cm^{-1}	General correlation factor	
			N-spectrum	Init. spectrum
1	SiO_2	3197.7 (2.0)	1.2	13
2	SiO_2	3288.5 (14)	12	25
3	SiO_2	3311.9 (4.7)	22	33
4	H (a)	3375.8 (2.0)	6.1	186
5	Li (a)	3403.0 (2.8)	6.2	40
6	?	3432.4 (1.8)	7.0	726
7	Li (b)	3444.0 (1.4)	2.0	28
11	Li (c)	3482.1 (0.1)	1.0	123
12	Li (d)	3508.1 (0.7)	1.0	-
13	?	3505 (11)	10	1.5
15	H (c)	3595.5 (0.4)	1.0	-

In concern to broad lines in the spectra under consideration the N-procedure has advantages and disadvantages. Decrease of the intensity of broad lines permits to reveal and to calculate the parameters of Li (b), Na, K, H (b) lines as well as water vapor lines. On the other hand it hampers the determination of the parameters of broad lines and in some cases it increases the errors of these parameters. To retain the accuracy at the same level for broad lines the parameters of N-transformation (first of all G) should be changed as it was explained above.

CONCLUSIONS

Presented here N-procedure is very useful on the first step of the analysis of spectroscopic data. Numerically improved spectral resolution makes easier visual analysis of the experimental spectrum. It permits to include in theoretical model of the spectrum really existing components which are weak and/or strongly overlapped and, therefore, appear in initial spectrum as shoulders of asymmetric lines.

In general the N-transformation improves the spectral resolution up to about 3 times at the cost of signal-to-noise ratio. At the same time correct fitting results for strongly overlapped lines have been obtained only due to application of the N-transformation. Decrease of the intensity of broad lines in the N-spectrum makes possible to determine the parameters of very weak and narrow lines.

The routine is applicable to all kind of spectroscopic methods like NGR, NMR, EPR, UV and VIS spectroscopy, X-ray diffraction, X-ray fluorescence, etc.

Acknowledgements

This work was partly supported by RFBR (grant N 96-05-65165).

REFERENCES

1. A.Kats. Hydrogen in alfa-quartz. *Philips Research Reports*, 17, 133-279, (1962).
2. I.L.Komov, M.I.Samoilovich. *Natural quartz and its physico-chemical properties*. M., Nedra, 77-93, (1985).
3. G.A.Korn, Th.M.Korn. *Mathematical Handbook for scientists and engineers*. McGraw-Hill Book Company, (1968).
4. G.V.Novikov. Method of analyses of poorly-resolved spectra. Rep. VINITI, N 4112-887, 1-14, (1987).
5. M.S.Paterson. The determination of hydroxyl by infrared absorption in quartz, silicate glasses and similar materials. *Bull. Miner.* 105, 20-29, (1982).

Proc 30th Int'l. Geol. Congr., Vol. 16, pp. 159-175
Huang Yunhui and Cao Yawen (Eds)

Spectroscopic Study of Diatomite in Leizhou Peninsula, China

WANG FUYA, ZHANG HUIFEN, FENG HUANG, CHEN GUOXI, WANG DEQIANG, HE HONGPING

Guangzhou Institute of Geochemistry, Chinese Academy of Sciences, Wushan, Guangzhou, 510640, China

Abstract

In this paper diatomite samples taken from the Leizhou Peninsula have been studied by chemical analysis(24 samples), DTA(23 samples), TG(23 samples), XRD(>100 samples), IR(26 samples), SEM(21 samples) , X-ray Energy Spectroscopy (37 samples, >200 photos), EPR (22 samples), and MAS NMR Techniques (13 samples). The study shows that the diatomaceous genera and their organic contents are variable with buried depth, from *Melosira* to *stephanodiscus* and then to *Cyclotella*. Various impurities in the samples, such as quartz, kaolinite and montmorillonite indicate different sedimentary environments. When heated, the diatom would change in shape due to the phase transformation in which amorphous silica crystallized from disordered opal to ordered cristobalite. The temperatures of phase transformation are different for various diatoms due to the presence of different impurities and constituents of diatomaceous genera and species.There are two existing forms of iron in diatomite: Fe^{3+} in oxide and hydroxide adsorbed on the outer and inner surfaces of diatomite, and Fe^{3+}, which is isomorphously substituted for Al^{3+} in clay minerals such as montomorillonite. The ^{29}Si MAS NMR spectra of diatomites show several signal with different chemical shifts at -110 to -112, -102, -91 and -107 ppm and intensity. These signals belong to ^{29}Si resonances of siliceous sheel of diatoms, kaolinite and quartz.

Key words: diatomite; diatom; chemical composition;MAS NMR; Leizhou Peninsula.

INTRODUCTION

Diatomite is a kind of very important nonmetallic mineral materials and has been widely used in many fields. In fact, diatomite is a kind of siliceous sedimentary rock of biogenesis. It is composed mainly of fossil skeletons of diatoms, which is a kind of single-cell hydrophyte. It is a deposit that has been formed in the stacking process of biomineralization and sedimentary mineralization.

The inner and out siliceous layers constitute the skeleton of a diatom and its main chemical composition is the same as that of opal, $SiO_2 \cdot nH_2O$. The multi- microporous structure was formed as the result of a regular arrangement of pores on the siliceous shell. It is the special chemical components and microporous and skeletal structures that result in many unique physical and chemical characteristics of diatoms, such as lightness, absorbability and non-conductivity. Therefore, diatomite has found wide applications in the fields of filtering materials, insulation, filler, catalyst carrier, ceramic raw materials and so on. It is shown that the study of diatomite is of great significance both in theory and in application.

With the progress in the study of diatomite exploitation and application, a great deal of research work on diatomite has been conducted. Further studies are needed in the field of mineralogy of diatomite.

SAMPLES AND GEOLOGICAL SETTING

It is reported that diatomaceous deposits in the Leizhou Peninsula were formed in the sedimentary layers of the Tianyang Formation during Late Pliocene to Middle Pliocene. The Tianyang Formation consists of clay layers containing various kinds of diatoms, belonging to the accumulations of caldera and volcano-tectonic depression phase. Orebodies occur in the horizontal layers. The footwall is composed of tuffaceous breccia and basalt, and the hanging wall contains sandy clay and peat.

In view of the tectonic background, these diatomaceous deposits lie on the north wing of the Leiqiong down-warping region of the South China fold system. Affected by the Himalayan movement, the bottom of the region broken down and the NW faults were most active with the eruption of a large quantity of basic volcanic rocks. The Jiudouyang, Qintongyang and Tianyang are these craters located along the NW faults . In the inactive period, favorable conditions were created for the formation of diatomite, such as volcanic lake basin, abundant silica, and sufficient nutrient supply.

In Dec. 1992, we carried out geological investigations of these diatomaceous deposits and collected a number of samples. The surface samples were taken from Puchang, Haikan County and Zhitou and jiumu, Xuwen County. And the others were taken from a drill hole at Jiudouyang, Haikan County (ZK-402) and a drill hole at Tianyang, Xuwen County (ZK-001) at different buried depths.

COMPOSITION OF DIATOMITE

Chemical composition

The chemical ompositions of the surface samples of diatomite taken from Puchang, Jiumu and Zhitou and those after water concentrated are given in Table 1. Table 1 shows that the Puchang samples are the best candidates for our study in the Leizhou Peninsula, with the highest content of silica and the lowest content of impurities. There is no obvious difference in the results of chemical analysis between the original samples from Puchang and those treated after water concentration, and so are the samples from Jiumu.

Table 1. Chemical composition of the selected samples

Sample No.	Zhitou 91-2	Puchang 91-18	Jiumu 91-19	Puchang* 91-20	Jiumu* 91-14	Jiumu* 91-15	Jiumu* 91-16	Jiumu* 91-17
SiO_2	58.6	82.94	64.91	82.97	64.76	67.94	68.01	64.55
TiO_2	1.21	0.43	1.03	0.39	1.52	0.90	0.69	0.78
Al_2O_3	14.19	2.89	11.96	3.59	11.89	10.61	10.54	12.66
Fe_2O_3	7.63	0.97	2.88	1.13	5.02	3.82	3.66	4.16
FeO	0.12	0.17	0.12	0.16	0.24	0.12	0.03	0.04
MnO	0.05	/	0.02	/	0.02	0.05	0.03	/
MgO	1.89	0.34	0.35	0.13	1.39	0.71	1.33	0.90
CaO	0.57	1.17	0.98	0.48	0.29	1.61	0.39	0.78
Na_2O	0.03	0.06	0.30	0.04	0.12	0.03	0.15	0.15
K_2O	0.69	0.09	0.35	0.06	0.74	0.43	0.37	0.39
H_2O^+	9.28	5.86	11.90	4.78	6.71	6.94	8.56	7.33
H_2O^-	5.53	4.84	5.51	5.94	7.04	6.54	6.13	7.92
P_2O_5	0.01	/	/	/	/	/	/	/
Total	99.97	99.76	100.31	99.67	99.74	99.75	99.89	99.66

*Sample No. Puchang-91-20 was treated by water concentration: Sample No. Jiumu-91-14 is coarse, Sample No.Jiumu -91-15 medium, Sample No. Jiumu-91-16 fine, and Sample No. Jiumu-91-17 finest in grain size.

The results of microscopic, XRD, DTA, and IR analyses show that the diatomite is composed mainly of diatomaceous skeletons with minor amounts of clay minerals, including montmorillonite, kaolinite, etc. and some detrital minerals such as quartz, feldspar and so on. The impurities in the samples from Puchang are K-feldspar, quartz and muscovite. The samples taken from Jiumu contain some quartz, kaolinite and montmorillonite. In addition ,each of the samples does contain certain amounts of water and organic material.

Diatomaceous genera and species in diatomite

Differences in diatom kinds of the 21 diatomite samples can be found under the microscope and JEOL JSM-35C electron microscope. In the surface samples taken from Puchang, *Pennales* are concentrated, including *Navicula radiosa, Navicula hasta, Achnanthes inflata, Cymbella turgida, Fragilaria construens, Caloneis silicula, Synedra ulna, Eunotia pectinalis*, etc; the Jiumu and Zhitou samples are composed mainly of *Melosira granulata* and their aggregates, belonging to *Melosira Agardh, Centralies.* Also, there is a small amount of *Stephanodiscus* in the samples, for example, *Stephanodiscus astraea* and so on (see Photos 1-2)

In the study of the drill hole samples from Jiudoyang, it is found that sample Jiudou-12 (see table 2) consists mainly of *Melosira agardh* varying in diameter from 15-20µ to 5µ. Besides *Melosira, Navicula hasta, Eunotia pectinalis*, etc. can also be found. Sample Jiudou-8 is composed mainly of *Stephanodiscus* with small amounts of *Melosira Agardh* and *Cyclotella*. The composition of diatom species in sample Jiudou-6 is similar to that of sample Jiudou-8. Sample Jiudou-3 contains a large amount of diatoms, with *Navicula* being dominant, and a small amount of *Melosira* which takes the shape of bamboo. On the contrary, *Cyclotella* is the dominant component in sample Jiudou-2, for example, *Cyclotella stelligera, Cyclotella kutzingiana* and so on with small bodies and their shell surface as large as to be no more than 10μ in diameter. Also, a small amount of *Eunotia pectinalis, Epithemia* and *Cymbella Agardh* is contained. All of this goes to show that the genera and species of diatoms vary with the buried depth, i.e., the dominant genera and species of diatoms are *Melosira, Stephanodiscus, and Cyclotella*, each varying with the buried depth from 20-50m.

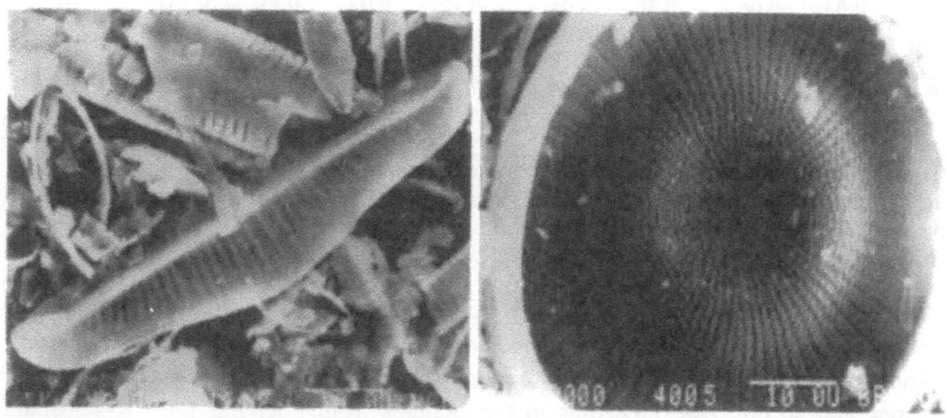

Photo 1 *Caloneis silicula* Photo 2 *Stephanodiscus*

The similar phenomenon that the genera and species of diatoms vary with the buried depth also can be found in the 4 selected Tianyang samples. With the buried depth from 90-240m, the genera and species of diatoms in the diatomite's vary from the group of *Melosira* to that of *Epithemia triceratia, Diploneis ovalis var. oblongella* coexisting with *Stephanodiscus* and *Cyclotella*, and to the group with *Cyclotella rhomboideo-elliptica var. rounda* being the dominant component. These results are shown in Table 2. Obviously, the genera and species of diatoms varying with buried

Table 2. Genera and species of diatoms and buried depth

Location	Sample No.	Depth(m)	Genera & species of diatoms
	Jiudou-12	20–24	Melosira Agardh
Jiudouyang	Jiudou-8	28–30	Stephanodiscus
ZK–001	Jiudou-6	35–37	Stephanodiscus
	Jiudou-3	43–45	Navicula
	Jiudou-2	49–50.7	Cyclotella
	Tianyang-8	94.2–99.65	Melosira
Tianyang	Tianyang-9	182	Epithemia triceratia
			Diploneis ovalis var. oblongella
ZK–001	Tianyang-12	190.88–197.4	Cyclotella
	Tianyang-1	230–240	Cyclotella

depth show the process that some diatoms became extinct while some others just emergencing became thriving. Also, it is shown that there had taken place changes in paleoclimate and sedimentary environment during the rock-forming and mineralization processes in a certain historical period of time in these areas. As viewed from changes in the genera and species of diatoms in the sedimentarylayers of the Tianyang area, it can be seen that the sedimentary environment would change from deep-to shallow-water, and finally to a swamp environment [5].

Chemical composition of diatoms

Diatoms are the main source of SiO_2 in the diatomite. It is reported [4,6] that the main components of a diatom shell are SiO_2 and H_2O, just like the composition of opals, $SiO_2 \cdot nH_2O$, formed in the inner cell walls of living diatoms in the earliest period of time. It can be considered a special variety of silica. By using SEM , the compositional analysis of the diatom shell was conducted on 37 points in various parts of different diatoms. The X-ray energy spectroscopy data and patterns are corresponding with the SEM images. As some points examined are composed chiefly of SiO_2, no other than the peaks of Si can be found in the patterns. Therefore, only the X-ray energy spectroscopic patterns are presented for these examined points. The chemical compositions of the 28 points chosen from being examined are given in Table 3

Table 3 shows that SiO_2 is the main chemical components in the hard shell of diatoms, in most of them, SiO_2 accounts for over 90%. Besides silica, there are contained minor Fe_2O_3, MnO, K_2O, Al_2O_3, TiO_2, etc.

Table 3. The results of X-ray energy spectroscopic analysis

Diatom	Sample No.	Al_2O_3	SiO_2	MoO	K_2O	CaO	TiO_2	Fe_2O_3	Location
	Jiudou-3	0.04	87.15	4.72	0.29	0.42		7.390	Extension of the column
	Jiudou-12	0.0	95.71		0.37	0.55		3.37	Bottom of the column
Melosira	Jiudou-12	1.92	89.30				2.00	3.37	Top membrane
	Tianyang-8	0.00	90.79			0.93		8.82	Outer wall
	Tianyang-8		99.84					0.16	Neck
	Jiudou-6	0.00	93.33	2.09				4.58	Central pore
Stephanodiscus	Jiudou-6	0.00	94.75		0.28	0.56		4.41	Central pore
	Jiudou-6	0.00	89.48	4.94	0.31	0.71	0.80	3.76	Outer shell
	Jiudou-8	0.00	97.56			0.40		1.98	Rib
	Tianyang-8	0.00	98.57					1.43	Membrane
	Jiudou-2	0.00	97.32		0.49	0.42		1.76	Membrane
	Jiudou-3	0.00	97.86		0.18			1.69	Shell surface
	Jiudou-8	0.00	98.61					1.39	Shell surface
Cyclotella	Tianyang-9	0.00	98.67					1.33	Shell surface
	Tianyang-12	0.25	94.38				1.39	3.98	Outer shell
	Tianyang-12	0.00	94.27			1.19		4.55	Inner edge
	Tianyang-12	0.00	97.17					2.83	Center

After careful comparisons were made of the genera and species of diatoms, the studied locations of the diatom bodies and their chemical composition, the following points could be presented.

1. The contents of silica and some inorganic oxides vary with changing genera and species of diatoms. The SiO_2 content in the skeletons of the diatoms tends to decrease from *Cyclotella* (>96%) through *Stephanodiscus* (94%) to *Melosira* (90%), but other impurities, for example Fe_2O_3, show an opposite trend.

2. The SiO_2 content varies from part to part in a diatom body. Generally, the parts with the highest content of SiO_2 are the shell surface of *Cyclotella*, the membrane developed on *Stephanodiscus*, the ribs or smooth zones on the body surface, the transverse ribs of *Epithemia*, the outer shell of *Navicula hasta* and so on. The elemental contents are listed in Table 3. As an example, in the body of *Epithemia*, the content of SiO_2 is about 83-84% in the inner and middle layers, and 95.31% in the outer layer (see photo 3 and 4).

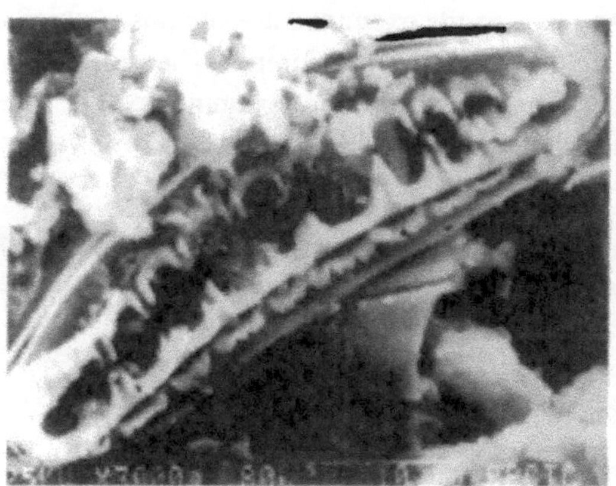

Photo 3 SiO_2-contents in different part of diatom body

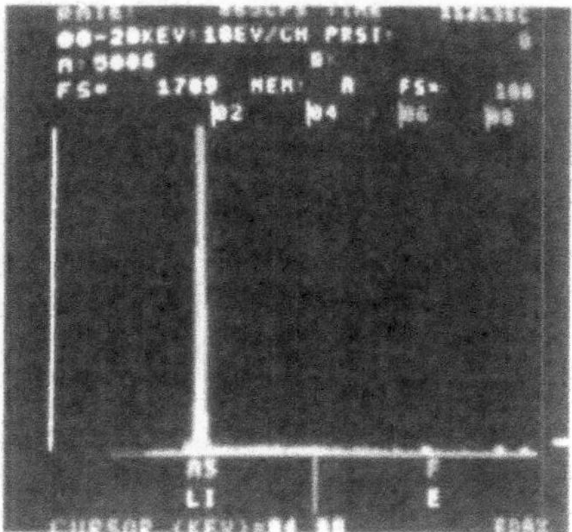

Photo 4 X-ray energy spectroscopy

It could be assumed that the silica contents of the shells of diatoms vary with different genera and species, showing different living environments and different abundances of SiO_2 in the water areas where diatoms live. And, the difference in silica content in the various parts of a diatom body is the reflection of the instinct of life or the physiological process. Further studies are needed in this respect.

It is worthy of note that as the content of water could not be presented only by using X-ray energy spectroscopy and the shells of diatoms do contain a certain amount of water, all the percentage contents of oxides shown in Table 3 only are the relative values which must be larger than the original ones.

THE THERMAL SPECTRUM CHARACTERISTICS OF DIATOMITE IN THE LEIZHOU PENINSULA

The DTA and TG analyses have been performed on 23 diatomite samples and conducted on an LCT-2 differential thermobalance with a sample weight of 50 mg, a temperature rate of 20°C/ min, a paper rate of 4mm/min, DTA measuring scale of 50μV and TG measuring scale of 20 mg. The DTA and TG curves are shown in Fig.1

The DTA curve of the diatomite standard is the same as that of opal, i.e., there is an endothermic peak at 100-250°C due to the loss of absorbed water. The related TG curve shows that the range of weight losses is about 1-10%. The exothermic peak at 1200-1300°C related to the crystallization of α-cristobalite is a piece of important evidence for identification of diatomite. The DTA curve of Puchang diatomite is a standard curve of opal, i.e., there is an exothermic peak at 1239°C. Also, there are a few weak peaks of clay minerals at 450-550°C. This indicates that the diatomite in this area is very pure

In the DTA curves of some samples, the appearance of the exothermic peak around 300°C shows the existence of organic material, though in small amounts. In the DTA curves of the drill hole samples taken from Jiudouyang and Tianyang, there are two connected exothermic peaks near 300°C and 370°C, the former being strong and the latter weak. The TG curves show that caused by the burning of organic material. It is shown that these samples contain a certain amount of organic material. The contents of organic material vary regularly with the buried depth, as can be seen in Table 4.

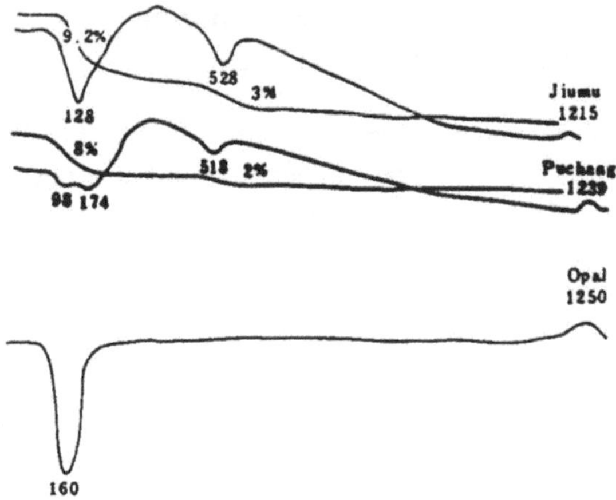

Fig. 1.DTA and TG curves of diatomites from different location

Table 4. The contents of organic material in the samples at different buried depths

Locatin	Sample No.	Depth (m)	Content of organic material(%)
	Jiudou-12	22-24	5.7
	Jiudou-8	28-30	6.0
ZK-402	Jiudou-6	35-37	9.0
	Jiudou-3	43-45	14.7
	Jiudou-2	49-50.7	12.0
	Tianyang-8	94.2-99.65	10.0
ZK-001	Tianyang-9	182	6.5
	Tianyang-2	190.83-197.47	5.5
	Tianyang	230-240	4.5

Table 4 shows that the contents of organic material in the Tianyang samples decrease with the buried depth, but the Jiudouyang samples show an opposite trend. It should be noticed that the Tianyang samples were taken at the depth from about 100m to 240m under the earth surface and the Jiudouyang samples from 20m to 50.7m. At the same time, it should be proved that whether, from the earth surface downward to -50m under the surface, the content of organic material in the samples increases with increasing buried depth and decreases from 100m to 240m under the earth surface. If this relation between the content and the buried depth is not occasional, it might be indicative of a sedimentary environment during diatomaceous mineralization.

STRUCTURAL CHANGES WHEN DIATOMITE IS HEATED

The physical and chemical properties of diatomite are fundamental with respect to its processing and application. As for the application of diatomite. there are two key techniques: acid leaching and calcination with and without flux. The purpose of sintering diatomite is to lump fine particles by melting part of the diatomite and to adjust the distribution of particle size and pore structure. Up to now, no ideal model for sintering diatomite has been established [1,3].

Obviously, an ideal sintering model and processing of diatomite should be established on the basis of the detailed studies of its thermal physical properties and structure. The structural changes of the sintered diatomites taken from the Leizhou Peninsula have been studied with XRD, IR and SEM in order to establish an ideal sintering model. The earth surface diatomites from Puchang and Jiumu are chosen for investigation. Two samples were heated on the LCT-2 differential thermobalance at 350°C, 600°C, 900°C and 10°C intervals from 1000° to 1200°C for 2h, respectively.

XRD analysis

The XRD results (more than 100 samples) are shown in Fig. 2. The close similarity between the XRD patterns of the samples studied and those of opal indicates that the main components of the samples are diatoms. According to the structural difference, the opal can be divided into three types: opal-A, opal-cT and opal-C. The structure of all the samples is similar to that of opal-A which is highly disordered. Both diatomites from Puchang and Jiumu have shown no obvious variations in XRD patterns until 1200°C at which distinct changes could be found. In the XRD patterns, besides the peak of quartz, many new sharp and intense peaks due to cristobalite appeared, whose *d* values are 3.066, 2.493, 2.655 and 3.146, respectively. This shows that the diatom with opal structure has been transformed into cristobalite.

Results of IR investigation

All the samples characterized by the XRD method were measured on a PE580B infrared spectrophotometer. In agreement with the results of XRD, the IR. spectra of all the 26 samples show that the main mineral components of the samples are diatoms with opal structure, as indicated by two strong bands at 1100 cm^{-1} and 470cm^{-1} and a weak band at 790 cm^{-1}. The bands at 3697 cm^{-1}, 3620 cm^{-1}, 1040 cm^{-1}, 910 cm^{-1}, 750cm^{-1}, 690 cm^{-1}, 540 cm^{-1}, 430 cm^{-1} and 345 cm^{-1} indicate the existence of kaolinite, montmorillonite and muscovite. However, the bands of quartz were not identified, probably due to the overlap of other bands. The numbers and intensities of the bands of the samples taken from different locations and depths are not the same, suggesting that impurity minerals in the samples are different in category and content

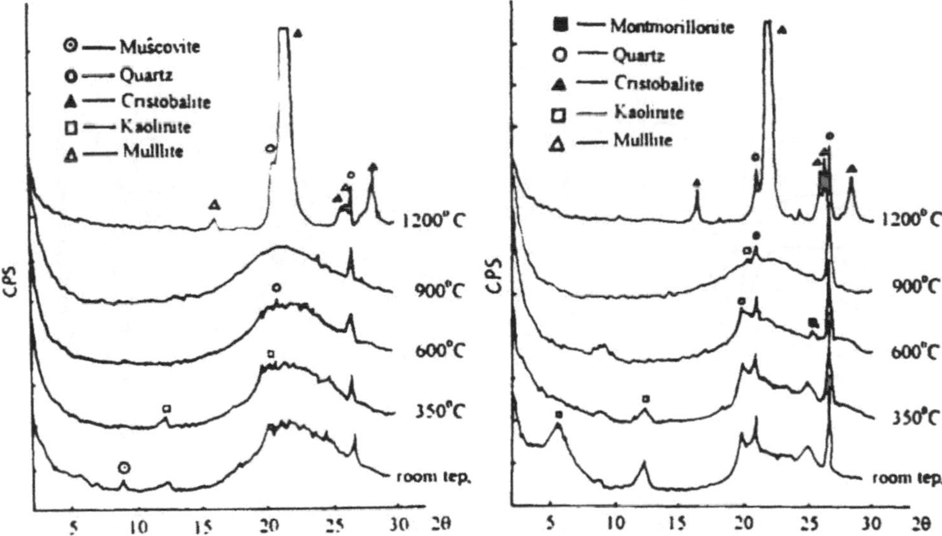

Fig. 2. XRD patters of Puchang (left) and Jiumu (right) diatomite at different temperatures

The IR. spectra of the Puchang samples heated at different temperatures are shown in

Fig. 3 and gradually increasing changes with temperature could be observed. Up to 1050°C, there is no obvious change in the structure of diatoms. With the rise of heating temperature to 1100°C, a weak band at 620cm^{-1} appears and at 1150°C, the peaks at 620cm^{-1} and 300cm^{-1} become clearer while the peaks at 385cm^{-1} and 1200cm^{-1} appear indistinct. Up to 1180°C, the three bands at 300cm^{-1}, 385cm^{-1} and 620cm^{-1} are enhanced strikingly and become still sharper. Also, the peak at 1200cm^{-1} becomes clearer. The spectrum of the sample heated at 1200°C is the same as that at 1180°C. It could be concluded that the violent phase transformation of diatoms could occur at about 1150°C.

The IR spectra of the Jiumu samples heated at different temperatures are similar to those of the Puchang samples. But the appearing temperatures of those new bands are different from those of the Puchang samples. At 1100°C, the two bands at 620cm^{-1} and 385cm^{-1} are evident and the other two at 300cm^{-1} and 1200cm^{-1} indistinct. These bands become stronger and sharper at 1150°C. The spectra at 1150°C, 1180°C (not shown) and 1200°C are coincident. By comparing with the Puchang samples, the temperature of new band appearance is 50°C lower, i.e., the phase transformation of diatom is considered to take place at about 1100°C.

Obviously, it can be seen that the new bands at 300cm^{-1}, 385cm^{-1}, 620cm^{-1} and 1200cm^{-1} should be attributed to cristobalite. Compared with the XRD results, the IR spectra could show the procedure of the structural change of diatom more obviously

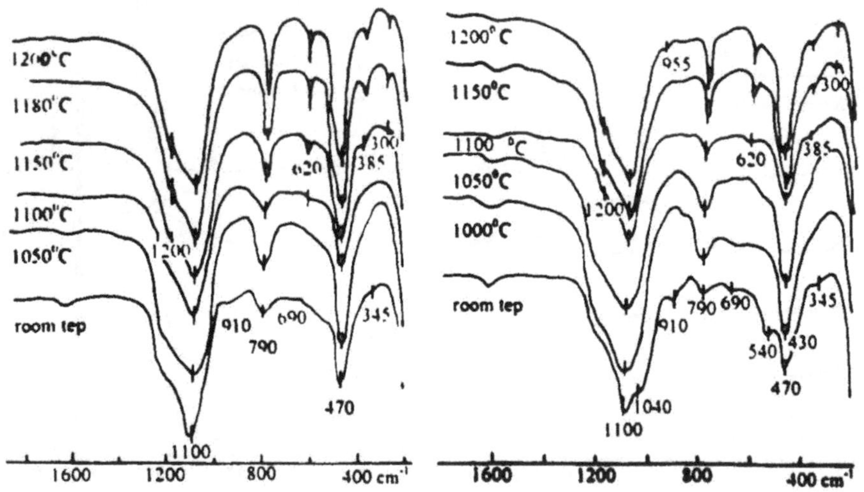

Fig. 3. The IR spectra of Puchang (left) and Jiumu (right) diatomites heated at different temp..

and more exactly when the samples studied are heated. However, at higher temperature, the formation of mullite as a new phase and the existence of the original-quartz could be shown clearly.

SEM EXAMINATION

The samples calcined at 1000-1200°C for various duration have been examined carefully udder the SEM and some of the samples are shown that Variation in shape of various diatoms appear significantly different in the heating process. When heating temperature reaches 1000°C, *Melosira* will change in shape: its body shrinks and becomes thinner, and its surface becomes blurred. When temperature is further increased, its body shrinks, becomes bending and breaks down,but the micropores of most bodies still exist. At 1180-1200°C, its body breaks down severely, with more and more fragments, and the micropores became indistinct.

The surface of *Cyclotella* and *Stephanodiscus* becomes blurred and the shape is broken down severely from 1050-1180°C. The deformation of *Navicula hasta*, *Cymbella turgida* and *Synedra ulna* takes place at higher temperatures. Up to 1200°C, the microstructure of these diatoms becomes blurred but the outline is still retained (Photos 5-6).

RESULTS OF EPR MEASUREMENTS AND DISCUSSION

In order to investigate the valent state and existing forms of iron in diatomite the electron paramagnetic resonance measurements of diatomite samples were conducted with the ECS-106 EPR spectroscope of Bruker Company, Germany. The work frequency was 9.76 Ghz, the sweep width 4000×10^{-4} T, the samples analyzed were in identical weight, and at room temperature. The results obtained are shown in Table 5, and in the figures 4,5 and 6.

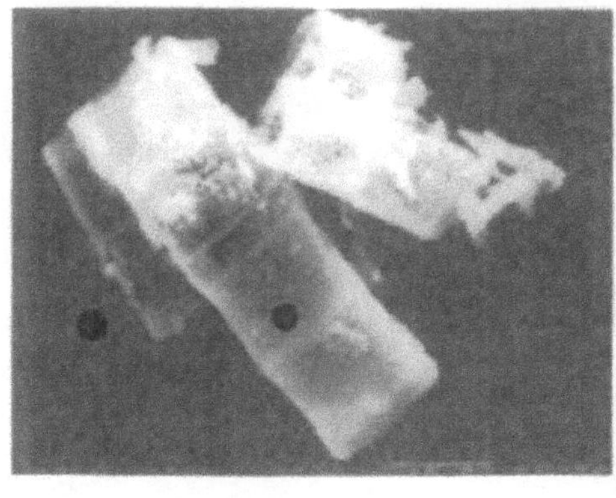

Melosira

Photo 5 X2000

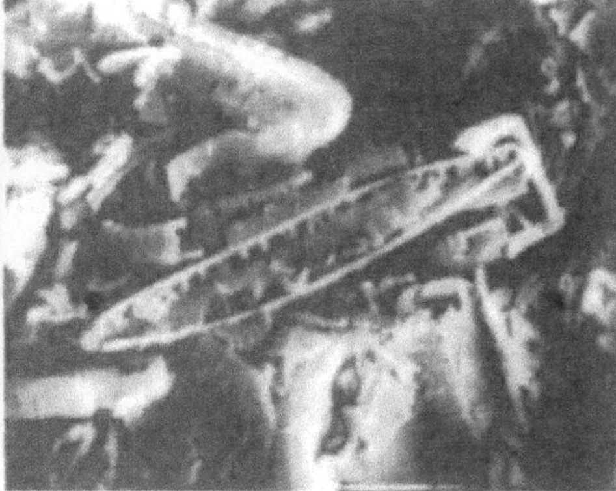

Navicula

Photo 6 X3000

Table 5 EPR parameters of absorption band A in EPR spectra as well as the iron contents in partial diatomite samples

No.	peak high (h)	peak width (ΔH)	peak area (S)	g-factor	FeO(%)	Fe_2O_3(%)	ΣFe_2O_3(%)
1.puchang	**2.5**	**4.0**	**40**	**2.0795**	**0.24**	**0.74**	**1.01**
2.Jiumu	6.0	6.0	216	2 0655	0.20	4 12	4.34
3.Tian-9	2.0	3.9	33 46	2.0076	3.14	3.90	7.39
4.Tian-8	4.0	3.2	40.96	2.0076	2.52	3.71	6.51
5 Tian-1	5.0	5.2	135.2	2.0725	1.23	1 41	2 78
6.Ten-1	6.4	5.0	160.0	2.0619	0.19	2.07	2.28
7 Ten-2	2.2	4.3	40.67		0.13	0.46	0 62
8 Pu-6	9.8	6.0	352.8	2.2359			
9.Pu-9	11.5	4.8	264.9	2.0243			
10.Pu-12	9.9	3.0	89.1	2.0689			
11.Jiu-6	6.4	4.0	102.4	1.9668			
12.Jiu-9	6.2	1.2	89 2	1.9980			
13.Jiu-12	8.8	4.0	140.8	2.0550			

Tian--Tianyang, Ten--Tengchong, Pu-6--Puchang 600 °C, Pu-9--Puchang 900 °C, Pu-12--Puchang 1200 °C, Jiu -6 -- Jiumu 600 °C, Jiu -9 -- Jiumu 900 °C, Jiu -12 -- Jiumu 1200 °C

Description and assignation of EPR spectra

It is revealed from the EPR spectra obtained from 22 diatomite samples that the common feature of these spectra patterns is the presence of a broad, dispersed, but fairly gently absorption band with g-factor in the range of 2.00 to 3.00 (Fig. 1). We call it absorption band A. It is clear that band A is the characteristic spectrum of diatomite, because there exists only this one spectrum for a relatively pure diatomite sample.

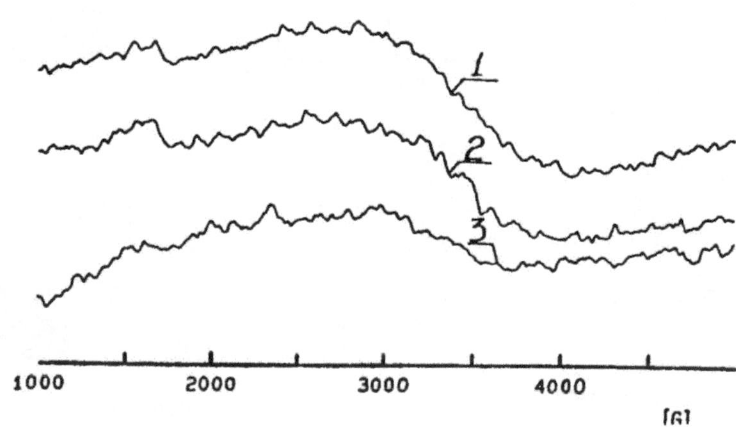

Fig. 4. EPR secpctra of diatomites 1--Jiumu, 2-- Puchang, 3--Guanintang.

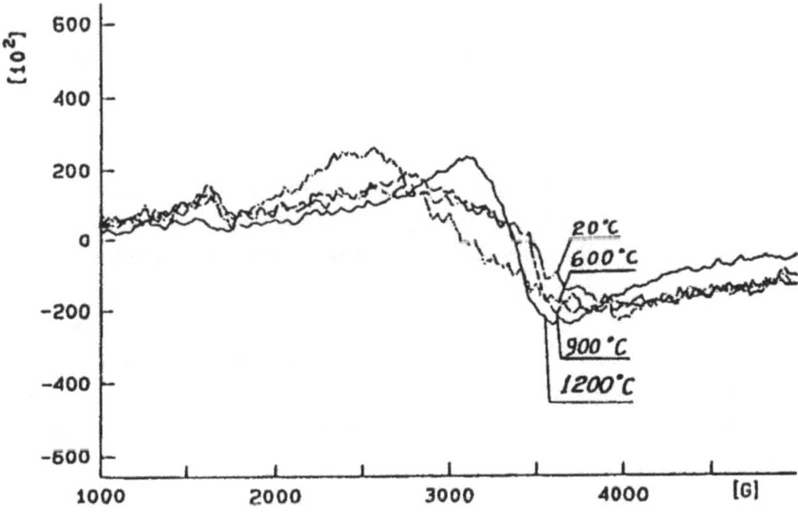

Fig. 5. EPR spectra of the Puchang and Jiumu diatomites at different temperatures.

On the basis of the position, curve shape and g-factor value of absorption spectrum, we deduce that it actually represents the spectrum of Fe^{3+} ion of iron oxides and hydroxides in diatomite. The most probable existing form of these iron oxide and iron hydroxide minerals is in adsorption state on the outer and inner surfaces of diatomite.

For some of the samples studied there are one or two absorption bands of very low intensity with g-factor around 4.00 in the low field region, we call it absorption

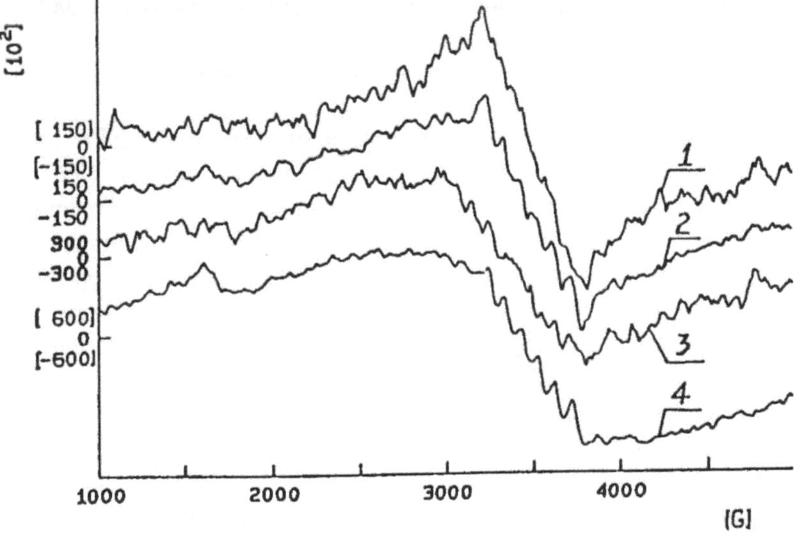

Fig. 6. Mn^{2+} EPR spectra in diatomites. 1--Tianyang-8, 2--Tianyang-9, 3--Tianyang-1,4--Tuantian.

band B. For the sample of high purity, such as that in Fig. 1-C, the band B is so weak that we could hardly reveal it. This band B is believed to be the spectrum of Fe^{3+} ion in the position of low symmetry and exists in the form of isomorphic substitution.

In comparison with spectra of main impurity minerals in diatomite , such as kaolinite and montmorillonite, we can see that the spectra of diatomite are quite similar to those of montmorillonite and completely differ from those of kaolinite. The EPR spectra of diatomite heated at different temperature further indicate that the absorption band B is not due to Fe^{3+} ions contained in kaolinite, because the absorption band with g-factor value around 4 in kaolinite spectra enhances its intensity gradually with temperature and reaches maximum at 800-900°C, while the intensity of band B for diatomite reaches its maximum at 600°C, and it decreases rapidly at 900°C (Fig.5).

It is also necessary to point out that the iron content in montmorillonite is much higher than that in kaolinite, especially in the iron-montmorillonite from Leizhou Peninsula, which contain 6.66-12.08 wt% of Fe_2O_3. Therefore in the EPR spectra of diatomite the resonance absorption of Fe^{3+} ion contained in montmorillorite is more distinct than that in kaolinite.

The EPR spectra of some samples show sixfold absorption band in the vicinity of center field, the g-factor value of which is about 2.00 and it has the characteristic of superfine structure (Fig. 6). This band is assigned to Mn^{2+} ion.

The spectra of a few heated samples we studied show a weak Mo^{5+} absorption band, overlapped with wide Fe^{3+} signals, the g-factor value of which is about 1.9880.

Relationship between EPR spectra of diatomite and iron content

It is revealed that the curve shape of absorption band A in the EPR spectra of diatomite is closely related with the Fe_2O_3 content in the samples, while that of absorption band B is dependent on the content of impurity mineral montmorillonite and that of Fe^{3+} ion ispmorphously substituted for Al^{3+} ion in montmorillonite. After EPR spectroscopic studies, the sample were analyzed for obtaining these FeO and Fe_2O_3 contents. The results are listed in Table 6. Comparing the curve shape of EPR spectra with the results in Table 6, we can easily see the correlation of curve shape with Fe_2O_3 content, e.g. the higher Fe_2O_3 content is, the broader the band A and the higher its intensity, and therefore, larger its enclosed area. For exsample, for two surface samples from Guangdong province, The Puchang sample contains 0.74 % Fe_2O_3, and its area is 40; In the Juimu sample, the Fe_2O_3 content is 4.12%,and its EPR area is 216, which is about 5.4 times as larger as former one. The EPR spectra of two samples from Guanintang and Tuantian of Yunnan province exhibit the similar relationship between EPR parameters and the iron content.

^{29}Si MAS NMR SPECTRA

The ^{29}Si MAS NMR spectra of 13 powered samples of diatomite were recorded at 59.6 MHz on Bruker MSL-300 NMR spectrometer with 5s recycle delay using solid echo sequence. Rotors were spun at 3.5 or 4 MHz. The measurements of spectra were carried out at room temperature with tetramethylilane (TMS) as external standard reference

The ^{29}Si MAS NMR spectra of 6 initial diatomite samples show one broad peak with FWHM (full width at half maximum)=ca. 12 ppm in the range of chemical shift from -110.1 to -112.1 ppm with a mean of -111.2 ppm (Fig.7.a), which belongs to the silicon in Q^4 configuration, indicating that the silicious sell of diatom is amorphous [8] The resonance peaks near -102.5 ppm in samples GIT-1, HH-1 and CHM-1 belong to the silicon in Q^3 configuration. The weak peaks from -106.2 to -107.1 ppm in samples XH-1, CHE-1 and ST-1 were assigned to the alpha-quartz.

The resonance signals ranging from -91.1 to -91.6 ppm (with a mean of -91.4 ppm) are induced by kaolinite in diatomite. As seen in Fig.7.a, this signal gets more strong along with the increasing of kaolinite content in samples. Although the signal induced from kaolinite in samples XH-1, CHE-1 and ST-1 is stronger than that from silicious shell of diatom, it should be indicated that the spectrum features can not be used to estimate the relative contents between kaolinite and diatom in samples, due to the fact that the relaxation time of SiO_2 is very long, and the recycle time (5s) employed to measure the ^{29}Si MAS NMR in the samples is much better for obtaining the resonance signals from kaolinite.

It can be seen from Fig.7.b that an original diatomite sample (HH-1) contains two phases: (1) a broad peak at -112.1 ppm produced from ^{29}Si Q^4 in the amorphous silicon of diatom shell; and (2) a week resonance signal at -101.2 ppm produced from ^{29}Si Q^3 in the diatom shell. ^{29}Si MAS spectra of the sample were measured after heating the sample at 600, 900, 1060, 1100 and 1200 °C for 2 hours respectively. At 900, 1060, 1100 °C, there are little changes in spectrum shapes. At 1200 °C, ^{29}Si spectrum show a sharp resonance signal with FWHM = 2.5 ppm at -110.3 ppm (Fig.7 b.HH-8), indicating the crystallization of cristobalite from siliceous shell of diatom

As stated above , the resonance signal at -91.2 ppm is sharp and was attributed to kaolinite in diatomite. When heated to 600 °C, the peak at -91.2 ppm disappears and a shoulder peak at -102.1 ppm corresponding to Q^3 configuration was reinforced, hence showing that kaolinite was transformed to metakaolinite (Fig.7.b. HH-2). At 900 °C, the shoulder peak at 102.1 ppm disappears(Fig.7.b. HH-3), indicating that all the siliceous shell of diatom consists of Q^4 Si configuration. The amorphous siliceous shell could be maintained to 1100 °C. At 1200 °C, the peak at -110.3 ppm gets very sharp (FWHM=2.5 ppm) showing that the amorphous siliceous shell of

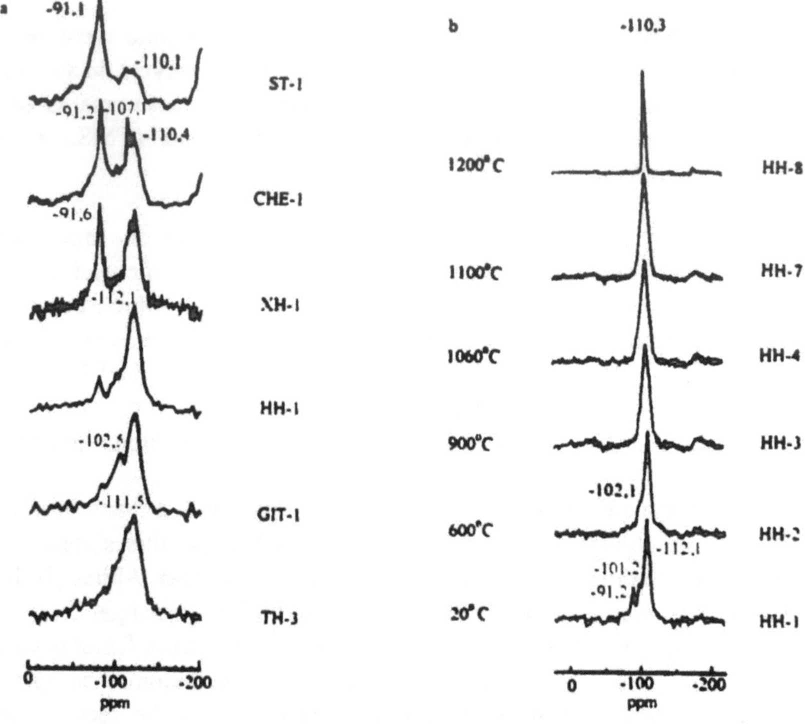

Fig. 7. The MAS NUR characteristics of diatomite in Leizhou peninsula

diatom was transformed to cristobalite. which is consistent with the X-ray data.

CONCLUSIONS

The contents of silica in diatom bodies vary with diatomaceous genera and species while the genera and species vary with buried depth. Also, the contents of silica vary from one part to another of a diatomaceous body

Structural changes of the diatoms sintered are due to their inner microstructure transformation, from amorphous or aphanitic opal structure to the high-ordered structure of cristobalite. The results of XRD, IR , SEM, EPR and ^{29}Si MAS NMR are coincident with one another.

Diatoms taken from different locations show changes in shape at different temperatures, for example, the Jiumu samples show changes in shape at lower temperature than the Puchang ones. That is because the diatomaceous genera and species of Jiumu are more simple in composition than those of Puchang, the former mainly consisting of *Melosira* and the latter of various species of *Pennales*. *Melosira* is more fragile than *Pennales*. Also, the presence of impurity minerals and iron elements in large amounts could easily cause the deformation of diatoms.

Iron in diatomite exites mainly in the form of Fe^{3+} in iron oxide and hydroxide minerals, which are in adsorption state on the outer and inner surfaces of diatomite. the another existing form of iron is Fe^{3+} ion isomorphic substituteed Al^{+3} in impurity clay minerals, especially montmorillonite.

The ^{29}Si MAS NMR spectra of powered diatomite samples consist of mainly a Q^4 Si and a little amount of Q^3 Si enviroment, The Q^3 Si configuration stabilizes under 900°C, and above this temperature transforms into Q^4 Si configuration.

Acknowledgements:

This project was supported by National Natural Science Foundation of China and Science Foundation of Guangdong Province.

REFERENCES

1. E. L. Calacal, and O. J. Whittemore. The sintering of diatomite. *Am. Ceram. Soc. Bull.*, **66**, 790-793 (1987).
2. Zhiqiang He, Evaluation of diatomite and its exploitation. *Journal of Geological Science and Technology in Guangdong Province.* **4**, 19-21(in Chinese)(1988).
3. Yanchen Huang, et al.. Diatomite in China and Its Application. Beijing, Science Press, pp.1-44, 87-91 (in Chinese) (1993).
4. Baolin Shen , Yeshen, and Jinman Fang. Diatomite, Industrial Minerals and Rocks of China, Chapter 17, pp 406-419(in Chinese) (1987).
5. The Second Oceanic-Geologic Team of the ministry of Geology and mineral Resources. A research report on Quaternary sediments and Paleoclimate of Tianyang Lake Basin, pp. 3-7, 17-23, 99-104 (in Chinese) (1986)
6. Kaichang Zhou. The Geologic characteristics of diatomite deposits in China. *Geology of Building Materials*, **2**, 28-33 (in Chinese) (1988).
7. Yongqin Zhou . The Geologic characteristics and the exploitation foreground of diatomites in the Leizhou Peninsula. *Guangdong Geology*. **2**, 22-27 (in Chinese) (1990).
8. Adams S.J., Hawkes G.E., Curzon EH, Asolid state 29Si nuclear magnetic resonance study of opal and other hydrous silicas, *American Mineralogist*, **76**: 1863-1871 (1981).
9. Graetsch H., Gies H., Topalovic I., NMR, XRD and IR study on microcrystalline opals, *Phys Chem Minerals*, **21**: 166-175 (1994).

Proc. 30th Int'l. Geol. Congr., Vol. 16, pp. 177-190
Huang Yunhui and Cao Yawen (Eds)

Physical Properties of Silicate Glasses

DOROTHEE J.M. BURKHARD
Scientific Center for Materials Research, and Institute for Mineralogy, University of Marburg, , D-35032 Marburg, Germany, and Hawai'i Institute of Geophysics, University of Hawai'i, Honolulu, HI 96822, USA

Abstract

Physical properties of silicate glasses are of specific interest in mineral physics with respect to the relation to structure. However, this relation is cryptic because of the difficulty to access the glass structure per se. This paper discusses elastic moduli of binary alkali silicate glasses from the literature and new experimental data of glasses in the systems SiO_2-Fe_2O_3-Li_2O and SiO_2-Fe_2O_3-Na_2O. With the focus on how these properties relate to glass structure, it is shown that a comparison to calculation (method of Makishima and Mackenzie) and to data from Mössbauer spectroscopy provide useful constraints.

Keywords: structure of silicate glasses, elasticity, ultrasonic methods

INTRODUCTION

A principal request in mineral physics is the investigation of physical properties of minerals, with two closely related goals, to understand and model physical properties of the solid earth, and to relate physical properties of a specific crystal to its crystal structure. In a similar fashion, silicate glasses and melts have moved into the focus of interest to understand and model the liquid earth, i.e. magmas and magmatogene processes, and to develop concepts of how physical properties and structure may be correlated. If one seeks a relation between property and structure on the atomic scale one is often forced to start with glasses and to consider glasses as a model for melts in a first approach. It is attempted here to consider physical properties of glasses with the question to what extent a relation to structure may be possible. Of the many physical properties, elasticity is fundamental in the light of the desire to model rheologic properties of magmas, and for that reason, elasticity will be given the main emphasis here.

The given goal confronts one with the need to consider simple systems. Most research has therefore been devoted to binary silicate glasses. Here, we shall progress to ternary systems, alkali silicate with iron oxides to investigate how iron affects elastic properties of alkali silicate glasses, in four parts:

- Structure of silicate glass
- Methods to determine elastic properties
- Binary alkali silicate glasses
- Alkali silicate glasses with iron:
  experimental, results; suggestion for a structure-property relation

STRUCTURE OF SILICATE GLASS

In contrast to crystals which are characterized by a periodicity of the atomic arrangement giving rise to a long range order, in glasses, atoms are arranged in short range order. The fact that an order can be observed is an outcome of high energy X-ray diffraction and spectroscopic investigations of the last decades, and these discoveries changed the previous concept of a random network, the Zernike-Prins-Bernal-Fowler-Warren hypothesis (in: [1]). New spectroscopic techniques also confirmed postulates by Zachariasen [2] that the glass forming tendency is a result of the flexibility of linkages. The flexibility was postulated in that oxygen polyhedra share edges, but no corners nor faces. For example, the flexibility of the intertetrahedral Si-O-Si angle is the reason why silicates may form glasses. The postulated flexibility had been ascertained by X-ray diffraction Mozzi and Warren [3] or by Silva et al. [4] and in the eighties, using NMR spectroscopy, by Gladden et al. [5] and DeJong [6]. The principal result is that in silica glasses the Si-O-Si angle varies between ca. 120 and 170°, showing a maximum at 151° [5] or 147° [6]. Note, that this angle is larger than the intertetrahedral angle of any crystalline SiO_2 polymorph.

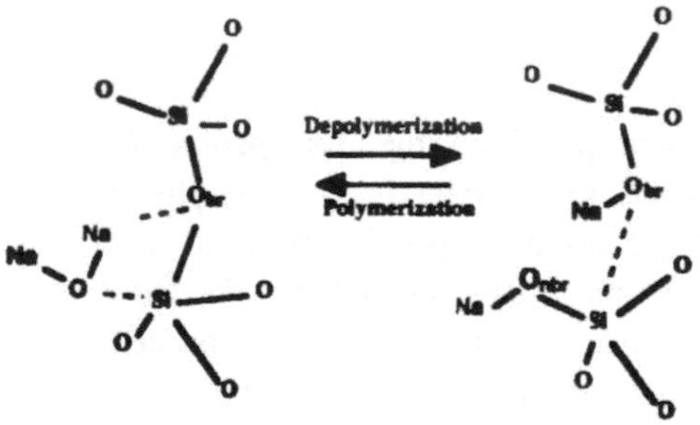

Figure 1. A network modifier, here Na, breaks up Si-O_{br}-Si bonds and thus depolymerizes the network.

In part based on Goldschmidt [7] Zachariasen introduced the terminology of network formers and network modifiers [2]. Network formers are, for example, Si, B, P, Ge, As, Be, etc., with a coordination number of 3 or 4. Network modifiers have a coordination number of 6 (Na, K, Ca, Ba, etc.) or between 4 and 6 (Li, Al, Mg, Zn, Pb, Nb, Ta, etc.). Without network modifiers, a glass network consists of network formers, only, and is completely polymerized, e.g. SiO_2 glass. With the addition of network modifiers, the network breaks up in that Si-O-Si linkages are broken up, a process called depolymerization (Fig. 1). Hence, as for crystals, the chemical composition determines the degree of polymerization. For example, a ratio of alkali oxide to silica of 1:1 results in two bridging oxygens per tetrahedron, expressed as connectivity, Q^2 (notation from Engelhardt et al. [8], and used in NMR spectroscopy), giving rise to the pyroxene structure. The decisive difference between crystal and glass is, however, that in a glass not only structural elements of a pyroxene occur, but also, of a sheet silicate, Q^3 , three bridging oxygens per tetrahedron), or of disilicates, Q^1 (one bridging oxygen per tetrahedron).

Comparing crystal and glass, one finds that bondlengths are of the same order of magnitude, and coordination numbers are the same in crystal and glass for a certain composition (review in [9]). Angles show a distribution with a maximum and average value in the range of the chemically corresponding crystal. Recent reviews on glass structure are given based on X-ray scattering and spectroscopy, in [9], based on NMR techniques, in [10], and using Raman spectroscopy, in [11]. While the local environment of glasses may be readily accessed using spectroscopic techniques, it is the extended environment that is difficult to approach. This uncertainty, unfortunately, also dictates the limits for a correlation of bulk property to structure.

METHODS TO DETERMINE ELASTIC PROPERTIES

In principle one may distinguish two different categories of techniques for the experimental determination of elastic properties, besides the theoretical approach of calculation:

(a): Static methods
(b): Dynamic methods.
(c): Calculation

(a) Static methods follow the definition of moduli, such as the elongation or bending of a body, or its hydrostatic compression:

- Elongation : Young's Modulus E, $E = \frac{\Delta l}{l}$
- Bending : Shear Modulus G, $G = \frac{\tau}{\gamma}$
- Compression : Bulk Modulus K, $K_T = -V(\frac{\partial P}{\partial V})_T$, $K_S = -V(\frac{\partial P}{\partial V})_S$

where l is the length of the sample rod, τ is the shear stress, γ is the tangents of the bending angle, K_T is the isothermal, and K_S is the adiabatic bulk modulus (inverse compressibility); V: the volume, and P, the hydrostatic pressure.

(b) Dynamic methods determine elastic moduli from the velocity of a sound wave, either by ultrasonic methods in the time or frequency domain, or by Brillouin scattering. The advantage of the dynamic approach is that one obtains not only one modulus, but all elastic moduli with one technique. Using ultrasonic techniques, one needs to determine the density ρ of the specimen, and the velocity of P-waves, V_p, (compressional, or longitudinal) and of S-waves, V_s (shear or transversal) in this sample. Elastic moduli may be determined from the following equations:

from V_l : $$V_l = \sqrt{\frac{L}{\rho}}, \qquad V_l = \sqrt{\frac{K - \frac{4G}{3}}{\rho}} \tag{I.a,b}$$

from V_s : $$V_s = \sqrt{\frac{G}{\rho}} \tag{II}$$

From these, all elastic moduli may be calculated,: L, longitudinal modulus; G, shear modulus; E, Young's modulus; K, bulk modulus, and ν the Poisson's ratio. An introduction to various ultrasonic methods is given, for example, in [12].

A specific ultrasonic technique is the phase comparison method. This method was applied in the present study and is therefore introduced in the following.

Phase comparison method is a high precision interferometric technique that permits an investigation of specimen very small in size, [13-15]. A rectangular pulsed frequency from a radio frequency generator is transferred to a pulsed train of mechanical, *i.e.* acoustic waves, via a transducer (piezoelectric effect). This wave travels through a buffer rod (quartz). While part of the wave is reflected at the end of the rod, the other part travels through the sample specimen placed on top of the rod (Fig.2). Again, part of this wave is reflected at the boundary of the specimen and interferes with the incoming wave. Depending on the elastic properties of the specimen and its thickness d, the applied frequency f can be adjusted to an f_o such that incoming and reflected waves are out of phase, resulting in a cancellation. One can determine an integral number n of frequencies, or an integral number of half waves (λ/2) or λ in a round trip. Because the sound velocity is wavelength times frequency, V is:

$$V = \frac{2df_o}{n' + \beta/360} \tag{III}$$

Here, n' is the average over all n. The effect of the bond, for example honey between sample and buffer rod, is expressed in $\beta/360$. If one carefully tries to minimize the influence of the bond, β may be neglected. A lateral interaction of the waves with the sample boundaries results in dispersion. It is therefore necessary to determine the free space velocity. This may be done by extrapolation towards infinite frequencies [14]. The uncertainty of this technique is 0.125% [14].

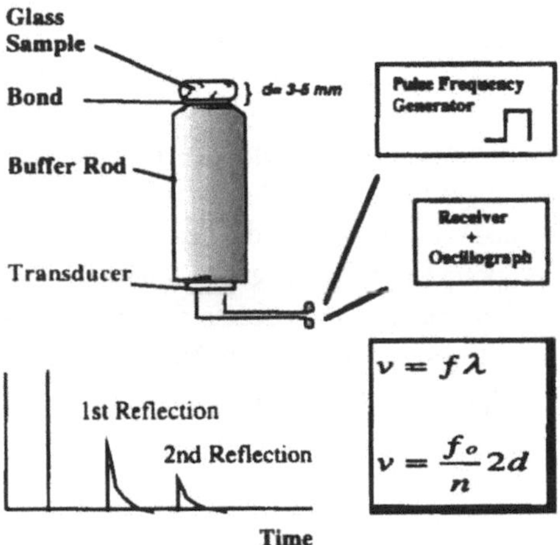

Figure 2. Principle of the phase comparison method; the frequency of a pulsed wave train passing through the sample is changed such that the it cancels the wave reflected at the sample boundary. Under this condition, the wave length λ is twice the sample thickness d, and the velocity v may be calculated.

(c) Calculation: One approach to model the structure in relation to physical properties is the calculation of these properties based on structural parameters. In crystallography such calculations relate the Young's modulus E to the fourth power of atomic spacing r_o:

$$E = 2a\frac{e^2}{r_o^4} = 2a\frac{U}{r_o^3} = 2\frac{U_m}{r_o^3} \tag{IV}$$

where α is the Madelung constant, e is the charge of a monovalent ion, and U is the electrostatic energy of attraction, $U=e^2/r_o$. The binding energy that includes all interactions between ions within a crystal is $U_m = \alpha U$.

For glass it is difficult to find a useful estimate of the Madelung constant. However, equation (IV) illustrates, that the Young's modulus is given by twice the binding energy. This relationship is used in one of the often applied model for glasses (Makishima and Mackenzie Model [16]) to estimate the binding energy from the atomization enthalpy per unit volume, G_{Gl}. To take account of the distorted arrangement of units in a glass, and the usually lower density, compared to crystals, one considers the packing density of oxides, V_{Gl}, to derive the Young's modulus E:

$$E = 2\,G_{Gl}V_{Gl} \tag{V}$$

In a multi-component glass, G_{Gl} and V_{gl} have to be calculated as a summation of the contribution of the individual oxides:

$$G_{Gl} = \sum_i (X_i \cdot \frac{U_i}{M_i}), \tag{VI}$$

$$V_{Gl} = \frac{1}{M_{GL}} \cdot \sum_i (X_i V_i), \text{ whereby } V_i = (a r_m^3 + b r_O^3) \cdot N \frac{4}{3} \pi \qquad \text{(VII.a,b)}$$

with X_i and M_i, the molar fraction and molar volume of oxide i, respectively. U_i is the energy of atomization of oxide i, *and* M_{GL} is the molar volume of the glass. V_i is calculated from the ionic radii of the metal ion r_m and oxygen, r_o. a and b are the number of cations and oxygens, respectively, in the oxide compound i. N is Avogadro's number. Ionic radii may be taken from Shannon [17]. Thermodynamic data to calculate the enthalpy of atomization and partial molar volumina may be taken from tables of the NBS [18]. The enthalpy of atomization calculates as the sum of enthalpies involved in the following processes (example of SiO_2): SiO_2 ->$Si_{(sol)}$ + $O_{2\,(gas)}$ -> $Si_{(gas)}$ + 2 $O_{(gas)}$.

BINARY ALKALI SILICATE GLASSES

Some early investigations of elastic moduli of binary silicate glasses are mentioned, for example in [19]. In Fig.3, elastic moduli of binary alkali silicate glasses are plotted versus the alkali content (data from [20]). Young´s and shear moduli decrease and the Poisson's ratio increases with the amount of network modifier. Qualitatively, these data may be understood in the light that each alkali creates a non-bridging oxygen and thus causes depolymerization of the network (Fig. 1). Tracing the effect of a monovalent ion, such as Na, similar results are observed in multi-component systems, vor example, mixed alkalis, *e.g.*, [21,22]. In the case of divalent cations such as the earth alkalis, the situation is different. Here, an increasing amount of cations results in a strengthening of the structure [23,24].

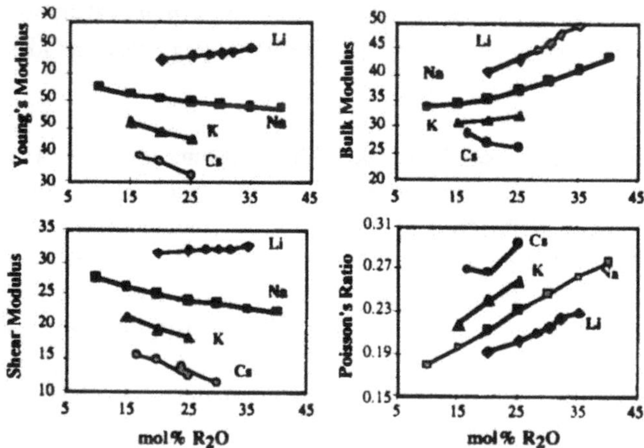

Figure 3. Elastic moduli of binary alkali silicate glasses, as a function of composition alkali content R_2O, R: Li, Na, K, Cs. Data are from [20].

Young´s moduli of the binary silicates in comparison to calculated moduli (according to equations VI to VII), are shown in Fig.4. Only in the case of sodium silicates, experimental and calculated moduli overlap to some degree of satisfaction. For lithium silicates, experimental moduli are significant higher than calculated and for potassium and cesium silicates, experimental moduli become increasingly smaller than calculated. The reason for these observations could be in the input parameters of equation (V), *i.e.*, either the partial molar volumina of the oxides M_i, in equation (VI), or the packing density V_{Gi}, in equation (VII.a,b). If one compares experimentally derived molar volumina with theoretical ones of the these glasses one observes for all these glasses systematically higher experimental value. Therefore, the 'static parameter' packing density cannot be the reason for the observed difference, experiment-calculation, of elastic moduli. Yet, what is not taken into account in Makishima-Mackenzie's calculation are 'dynamic parameters' such as the compressibility of structural units. In the case of potassium and cesium it is an increasing compressibility of the electronic shells of these ions which is likely to provide an overall lower force constant and hence smaller elastic moduli than calculated. For glasses with lithium a different mechanism will be suggested below.

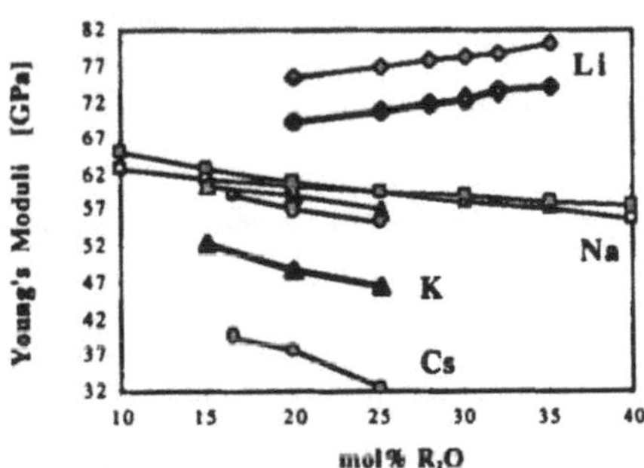

Figure 4. Comparison of experimental data of Young's moduli, as shown in Fig. 3 (filled symbols) with calculation after [20] (corresponding open symbols). Only for sodium silicate glasses experiment and calculation match well.

ALKALI SILICATE GLASSES WITH IRON

Experimental

Glass samples were prepared in the systems SiO_2-Fe_2O_3-Li_2O (SFL) and SiO_2-Fe_2O_3-Na_2O (SFN) with varying silica to alkali ratios, at nominal 2, 5, and 10 mol% Fe_2O_3 (Fig.5). The synthesis used regent grade quartz, Fe_2O_3 powder and Li_2CO_3 and Na_2CO_3, respectively. After decarbonatization at temperatures below the carbonate liquidus, the mixtures were molten between 1300 to 1400°C for ca. 1h, quenched on a metal plate, and stored in a desiccator. XRD ascertained the glassy state and an investigation under transmitted and reflected light showed that crystal exsolution, if present at all, is less than 0.1 vol.%. Microprobe analyses showed homogeneity of the samples by ±2%. In SFL glasses with about 42 mol% Li_2O, several vol.% of crystals are observed showing that the glass forming region of glasses with Li is smaller than for glasses with sodium.

Density of all glasses was determined by the Archimedean method, using destilled water for glasses with Li and ethanol for the hygroscopic glasses with Na. Data are plotted in Fig.6. Larger glass pills of a thickness between 3 and 5 mm were polished coplanar to be used for the phase comparison method described above. Longitudinal and transverse velocities were determined using a transducer of frequency 20 MHz and 10 MHz, receptively. The frequency range for the measurements was in-between ca 18 to 33 MHz for longitudinal and in-between 15 to 23 MHz for shear waves.

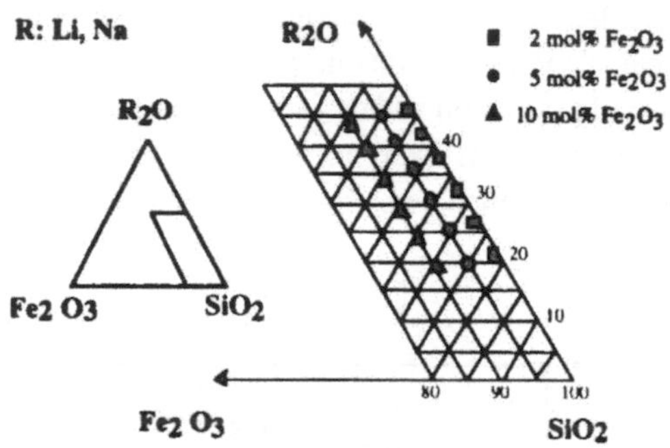

Figure 5. Compositions of glass samples synthesized in the systems Li_2O-SiO_2-Fe_2O_3 (SFL) and Na_2O-SiO_2-Fe_2O_3 (SFN).

Results

For SFL glasses, longitudinal and transversal velocities are between 6. 086 to 6. 469 km/s, and 3.714 to 3.766 km/s, respectively. For SFN glasses, longitudinal and transversal wave velocities ranged between 5.2940 to 5.4412 km/s and 2.9794 to 3.2419 km/s, respectively. From these velocities and the density, elastic moduli were derived according to equations (I.a,b) and (II). The results are plotted in Fig. 7, in comparison to moduli of the binary systems. Elastic moduli of SFL glasses are significantly higher than those of SFN glasses. In SFL glasses both Fe_2O_3 and Li_2O effect elastic moduli significantly, much in contrast to Fe_2O_3 and Na_2O in SFN glasses. It should be noted that the slope of each line in Fig.7 is the same as for the corresponding binary system.

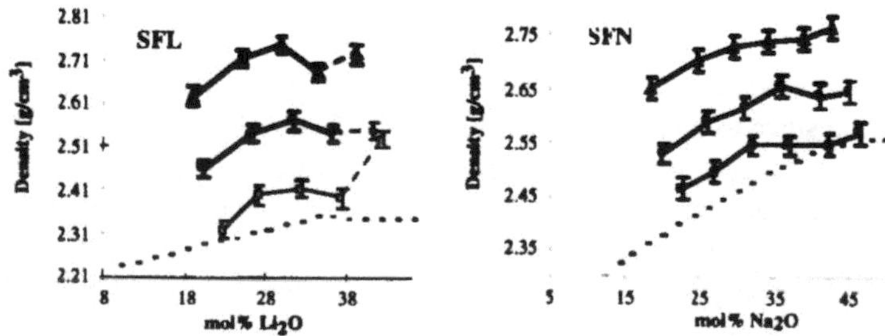

Figure 6. Density of SFL and SFN glasses, determined by the Archimedean method. Signatures refer to the nominal mol.% of Fe_2O_3: squares, 2, circles 5, and triangles 10 mol.%. The dashed lines show the density of glasses in the corresponding binary system. Samples high in Li_2O are in part crystalline.

Discussion

The glass strengthening effect of Fe_2O_3 is known already since Gehlhoff and Thomas (1927). More recent studies in three component systems showed that for Fe_2O_3 in sodium germanate glasses elastic moduli depend on whether Fe^{3+} acts as network former or network modifier [26]. For the case that the concentration of Fe_2O_3 is less than that of Na_2O, the Young´s modulus decreases if Fe^{3+} acts as a network former, and it increases if Fe^{3+} acts as a network modifier. In sodium silicate glasses Fe^{3+} is assumed to act as network former. This is suggested by the overall tetrahedral coordination of Fe^{3+} in such glasses [27-30] and by a generally increasing viscosity of corresponding melts with the addition of Fe^{3+} [31]. In the light of the work by Livshits *et al.* [26], decreasing Young´s moduli of SFN glasses confirm a network forming role of Fe^{3+}. According to Mössbauer spectroscopy, Fe^{3+} is for the most part tetrahedrally coordinated also in SFL glasses [29,30]. It is therefore likely that Fe^{3+} acts as network former in these glasses as well. This is confirmed by an extended glass forming region; specimen SFL 25 with 10 mol% nominal Fe_2O_3, contains more glass than other samples with less Fe_2O_3 but with a comparable content of Li_2O.

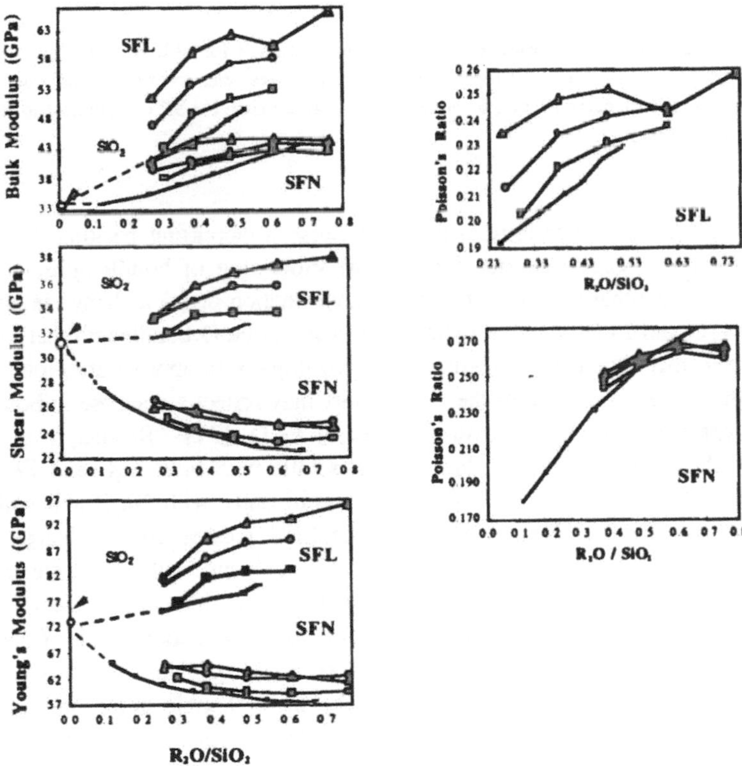

Figure 7. Bulk, shear and Young's moduli, and the Poisson's ratio of SFL and SFN glasses. Signatures refer to the nominal mol.% of Fe_2O_3: squares, 2, circles 5, and triangles 10 mol.%. Solid lines show the moduli of glasses in the corresponding binary systems, extrapolated to silica glass (dashed lines). The open triangle refers to an SFL sample high in Li_2O and partially crystalline. The dashed line for the Poisson's ratio marks the Cauchy relation.

Suggestion for a structure-property relation

Experimental Young's moduli of SFL and SFN glasses are compared to calculation in Fig.8. For SFN glasses, experiment and calculation agree reasonably well. For SFL glasses, however, experimental moduli are by some orders of magnitude larger, and the difference to calculation increases with the iron content. To evaluate a possible reason it is necessary to first consider the influence of the packing density.

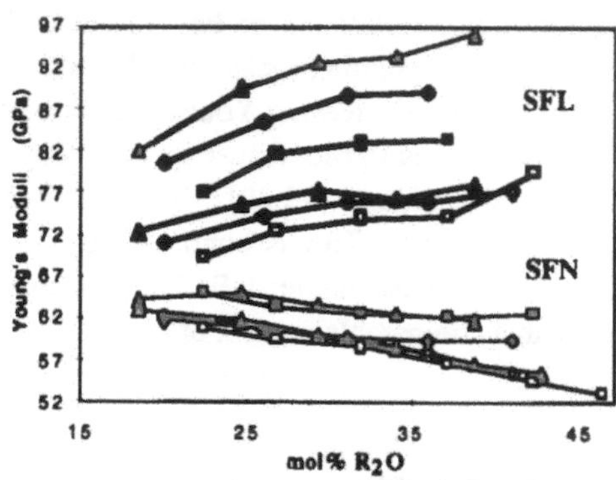

Figure 8. Comparison of experimental data of the Young's moduli for SFL and SFN as in Fig. 7 (filled symbols) with calculated data after [20] (open symbols). Signatures refer to the nominal mol.% of Fe_2O_3: squares, 2, circles 5, and triangles 10 mol.%. Note the reasonable match for SFN in contrast to the significant differences for SFL glasses.

Classically, there are three possibilities that may cause a increasing packing density, either, interatomic distances become shorter due to a (a) shortening of bondlengths, or the metric changes, which may either be (b) an increase in coordination or (c) a change in bond angles. Because the Si-O and Li-O bond are rather rigid, it is the Fe-O bondlength that may change. The ^{57}Fe isomer shift obtained by Mössbauer spectroscopy is an appropriate probe for the Fe-O bondlength, in that a decrease of the isomer shift may reflect a decrease in bondlength, as evident, for example from high pressure studies of Fe_2O_3 [32]. Results from Mössbauer spectroscopy show that the isomer shift decreases little with both Li_2O and Fe_2O_3 in SFL glasses. In contrast, the isomer shift decreases significantly with Na_2O and with Fe_2O_3 in SFN glasses [30] (Fig.9). Yet, elastic moduli of SFN glasses do not change much as a function of composition. It is therefore unlikely that Fe-O bondlength have a significant effect on elastic moduli. There is also no indication that Fe may increase its coordination as a function of Li_2O or Fe_2O_3, and it is unlikely that Li does. A coordination of four is rather stable, as shown by ^7Li NMR, even at elevated pressures [33]. This leaves us with the third option (c), that is a decrease in bond angles. The most flexible angle one encounters in silicates is the intertetrahedral Si-O-Si angle, known from numerous studies, *e.g.*, [34-37]. We therefore suggest here that it is this Si-O-Si angle that is bended such that the average Si-O-Si angles narrows with increasing amount of Li_2O and Fe_2O_3. For the Li_2O-SiO_2 system much evidence has been given in the literature that the average Si-O-Si angle is around 138° or less [38-39], hence, significantly smaller than in SiO_2 glass [5,6], but similar to crystalline phases [41,42]. A visualization of the mechanism of angle bending was provided by *ab initio* calculations of the model $H_6Si_2O_7$. There, the attraction of a hydrogen by an oxygen of the adjacent tetrahedron resulted in a H-bridge and concomitant contraction of the molecule [43]. Since it is likely that cations with ahigh cationic field strength induce the Si-O-Si bending, both Li and Fe are favorite candidates [39]. For geometric reasons, silicate structures with narrow Si-O-Si angles have a higher packing density and a higher bulk density which directly triggers the wave velocity and hence elastic moduli (equations, I, II).

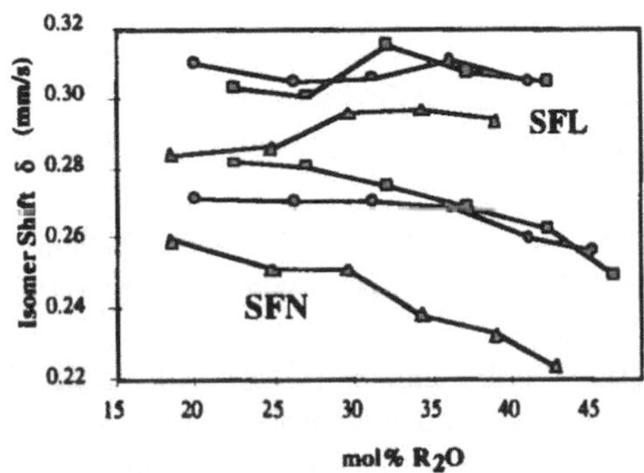

Figure 9. ^{57}Fe Mössbauer isomer shift of Fe^{3+} in SFL (upper three lines) and SFN glasses (lower three lines), from [30]. Signatures refer to the nominal mol.% of Fe_2O_3: squares, 2, circles 5, and triangles 10 mol.%. For SFL glasses the δ, an indicator of the Fe-O bond length, is relatively constant for a given Li_2O content and increases slightly with Fe_2O_3. Corresponding effects are much better developed in SFN glasses.

However, we feel that it is not so much this static characteristic that is responsible for the observed high elastic moduli of SFL glasses. As in the case of binary alkali silicates, dynamic parameters, *i.e.* the compressibility of structural units is likely to be much more relevant. Steric arguments [44] and *ab initio* molecular calculations [35] suggest that bending of the Si-O-Si angle is limited to 120°. The reason for this limit are oxygen-oxygen repulsive forces. In correspondence to these results, one finds that in crystals the distance between oxygens is limited [45]. Hence, if the bending of Si-O-Si angle is limited, structures with already narrow Si-O-Si angles cannot be bended much further. Force constants will be large, and structures are likely to react more rigid, giving rise to high elastic moduli.

It is interesting to note that also for melts and mulit-component systems bulk properties show similar trends as a function of chemistry as described here for glasses. For example, bulk moduli of binary alkali melts show trends similar to those shown in Fig.3 [46]; and the molar volume of haplogranitic melts decreases with increasing Li_2O added [47], in correspondence to Fig.6. If in such melts the effect of specific elements on the bulk property is comparable to what we find in simple glass systems, it is likely that also affiliated structural mechanisms in the melts are comparable to those suggested for glasses. Hence, glasses serve as suitable model to explore concepts relevant for geodynamic processes.

SUMMARY AND CONCLUSION

Elasticity, one of the most important physical properties of glasses, was considered in the light of possible relations to characteristics of the glass structure. Therefore, simple systems were investigated, binary alkali silicates (literature data) and new data sets of alkali silicates with iron. Elastic data were obtained with a special ultrasonic interferometric technique, the phase comparison method. Experimental values were compared with theoretical Young's moduli calculated after the method of Makishima and Mackenzie. The following characteristics are found:

- Elastic moduli increase with the cationic field strength of the alkali metal.
- Experiment and calculation agree well for all glasses with sodium, in the binary and ternary system.
- Experimental values are increasingly smaller than calculated for glasses with potassium and cesium. This effect is related to an increasing compressibility of the electron shells with decreasing cationic field strength.
- Experimental values of glasses with lithium are significantly larger than calculated. And this difference increases with increasing amount of iron. Based on Mössbauer isomer shift data for Fe^{3+}, decreasing Fe-O bond lengths with increasing amount of iron provides an only minor contribution, if at all. It is suggested that with increasing amount of lithium, and specifically with iron in the glass, the average Si-O-Si angle is narrowed. Because in any silicate the Si-O-Si angle is limited to 120°, an already narrow angle would make the material incompressible. Hence, on average narrow Si-O-Si angles could be the reason for the observed high elastic moduli, and specifically for the high bulk moduli.

ACKNOWLEDGMENTS

I thank H. Pentinghaus for the access to his laboratory to synthesize most of the glass specimen discussed here. M.H. Manghnani kindly invited me to the Hawai'i Institute of Geophysics and enabled the use of facilities such as the phase comparison method. S.S. Hafner's open lab and hospitality at Marburg is warmly acknowledged. The work was carried out with the Heisenberg Program of the Deutsche Forschungsgemeinschaft (DFG). This, and travel grants from the DFG to participate in the IGC meeting are acknowledged.

REFERENCES

1. J.T. Randall. The diffraction of X-rays and electrons by amorphous solids, liquids, and glasses. *Wiley, New York*, 290p. (1934)
2. W.H. Zachariasen. The atomic arrangement in glass. *J. Am. Chem. Soc.* **54**, 3841-3851 (1932).
3. R.L. Mozzi and B.E. Warren. The structure of vitreous silica. *Scientific Am.* **257**, 78-85 (1969).

4. J.R. Da Silva, D.G. Pinatti, G.E. Anderson and M.L. Rudee. A refinement of the structure of vitreous silica. *Philos. Mag.* **42**, 713-717 (1975).
5. Gladden L.F., T.A. Carpenter, S.R. Elliot. ^{29}Si MAS NMR studies of the spin-lattice relaxation time and bond angle distribution of vitrous silica. *Philos. Mag.* **53**, L81-L87 (1986).
6. B.H.W.S. DeJong. Glass. Ullmann's Encyclopedia of Industrial Chemistry. *VCH Verlagsgesell. Weinheim*. vol.A **12**, 365-432 (1989).
7. V.M. Goldschmidt. Geochemical distribution laws of the elements. VIII. Reseraches on the structure and properties of crystals. *Skr. Nor. Vidensk.-Akad. (Kl). 1: Oslo I. Matemat-Naturv. Kl.* **8**, 7-156 (1926).
8. G. Engelhardt, D. Zeigan, H. Jancke, D. Hobbel and W. Wieker. Zur Abhängigkeit der Struktur der Silicatanionen in wäßrigen Natriumsilikatlösungen vom Na : Si Verhältnis. *Z. Anorg. Allg. Chem.*, **418**, 17-28 (1975).
9. G.E. Brown Jr., F. Farges and G. Calas. X-ray scattering and X-ray spectroscopy studies of silicate melts. in: Structure, dynamics and properties of melts, Mineral. Soc. Am., Rev. Mineral. **32**, 317-410 (1995).
10. J. Stebbins. Dynamics and structure of silicate and oxide melts: nuclear magnetic resonance studies. in: Structure, dynamics and properties of melts. *Mineral. Soc. Am., Rev. Mineral.* **32**, 191-246 (1995).
11. P.F. McMillan and G.H. Wolf. Vibrational spectroscopy of silicate liquids. in: Structure, dynamics and properties of melts. *Mineral. Soc. Am., Rev. Mineral.* **32**, 245-315 (1995).
12. E. Schreiber, O.L. Anderson and N. Soga. Elastic constants and their measurements. *MacGraw-Hill*, 196p. (1973).
13. H.J. McSkimin. Ultrasonic measurement techniques applicable to small specimens. *J. Acoust. Soc. Am.* **22**, 413-418 (1950).
14. E.S. Fisher. The adiabatic elastic moduli of single crystal alpha uranium at 25°. *Argonne Nat. Lab.* (ANL-6096), 91p. (1960).
15. E.S. Fisher and C.J. Renken. Single crystal elastic moduli and the hcp-bcc transformation in Ti, Zr, and Hf. *Phys. Rev.* **135**, A482-A494 (1964)
16. A. Makishima and J.D. Mackenzie. Direct calculation of Young's modulus of glass. *J. Non-crystal. Solids* **12**, 35-45 (1973).
17. R.D. Shannon Revised effective ionic radii and systematic studies of interatomic distances in halides and chalcogenides. *Acta. Cryst.* **32**, 751-767 (1976).
18. National Buro of Standards, Tables of chemical thermodynamic properties. *J. Phys. Chem. Ref. Data* **11**, Supplement 2 (1982).
19. G.W. Morey. The properties of glass, 2.nd ed. *Am. Chem. Soc. Monograph* **77**. Reinhold Publ. Corp. New York. (1954)
20. N.P. Bansal and R.H. Doremus. Handbook of glass properties. *Academ. Press* London 680p. (1986).
21. K. Matusita, S. Sakka, A. Osaka, N. Soga and M. Kunugi. Elastic modulus of mixed alkali glass. *J. Non-cryst. Solids* **16**, 308-312 (1974).
22. G.B. Rouse, E.I. Kamitsos and W.M. Risen Jr. Brillouin spectra of mixed alkali glasses $xCs_2O(1-x)Na_2O5SiO_2$. *J. Non-cryst. Solids* **45**, 257-269 (1981).
23. Y. Vaills, Y. Luspin, G. Hauret, B. Coté and F. Gervais. Elastic properties of sodium calcium silica glasses by Brillouin scattering. *Solid State Com.* **82**, 221-224 (1992).
24. Y. Vaills, Y. Luspin, G. Hauret and B. Coté. Elastic properties of sodium magnesium silica and sodium calcium silica glasses by Brillouin scattering. *Solid. State. Com.* **87**, 1097-1100 (1993).
25. G. Gehlhoff and M. Thomas, in G.W. Morey. The properties of glass, 2.nd ed. *Am. Chem. Soc. Monograph* **77**. Reinhold Publ. Corp. New York. (1954).
26. V.Y. Livshits, M.G.Potalitsyn and R.A.Nakhapetyan. Elastic properties and energetics of sodium germanate glasses containing oxides of group II - VI elements. *Sov. J. Glass Phys. Chem.*, **18**, 159 (russ., engl. abstr.) (1992).
27. K. Hirao, T. Komatsu and N. Soga. Mössbauer studies of some glasses and crystals in the Na_2O-Fe_2O_3-SiO_2 system. *J. Non-cryst. Solids* **40**, 315-323 (1980).

28. B.O. Mysen, F. Seifert and D. Virgo. Structure and redox equilibria of iron-bearing silicate melts. *Am. Mineral.* **65**, 867-884 (1980).

29. D.J.M. Burkhard. ^{57}Fe Mössbauer-Spectroscopy and magnetic susceptiility of iron-alkali silicate glasses (subm.)

30. D.J.M. Burkhard. Structure of alkali-silicate glasses with iron (in prep.).

31. D.B. Dingwell and D. Virgo. Viscosities of melts in the Na_2O-FeO-Fe_2O_3-SiO_2 system and factors controlling relative viscosities of fully polymerized silicate melts. *Geoch. Cosmochimica Acta* **52**, 395- (1988).

32. Y. Syono, A. Ito, S. Morimoto, T. Suzuki, T. Yagi and A. Akimoto. Mössbauer study of the high pressure phase of Fe_2O_3. *Solid State Com.* **50**, 97-10 (1984).

33. D.J.M. Burkhard and G. Nachtegaal. Nucleation and growth behavior of lithium disilicate at pressures to 0.5 GPa investigated by NMR spectroscopy. *J. Non-cryst. Solids* (in press) (1996)

34. R.J. Hemley, H.K. Mao, P.M. Bell and B.O. Mysen. Raman spectrsocopy of SiO_2 glass at high pressure. *Phys. Rev. Lett.* **57**, 747-750 (1986).

35. R.M. Hazen L.W. Finger, R.J. Hemley and H.K. Mao. High-pressure crystal chemistry and amorphization of alpha quartz. *Phys. Rev.* **B72**, 507-511 (1989).

36. J.R. Chelikowski, H.E. King, N.Troullier, J.L. Martins and J. Glinnemann. Structural properties of alpha quartz near the amorphous transition. *Phys. Rev. Lett.* **65**, 3309-3312 (1990).

37. J.R. Chelikowski, N. Troullier, J.L. Martins and H. King. Pressure dependence of the structural properties of alpha quartz near the amorphous transiton. *Phys. Rev.* **B44**, 489-497 (1991).

38. U. Selvaray, K.J. Rao, C.N.R Rao, J. Klinowski and J.M. Thomas. MAS NMR as a probe for investigating the distribution of th Si-O-Si angles in lithium silicate glasses. *Chem. Phys. Lett.* **114**, 24-27 (1985).

39. D.J.M. Burkhard. Zur metrischen Struktur amorpher Silikatsysteme: H_2O-SiO_2 und Li_2O-SiO_2. Korrelierbare physikalisch-chemische Eigenschaften. *Habilitationsschrift, Univ. Marburg*. (1993).

40. D. Sprenger. Spektroskopische Untersuchungen und Berechnungen zur Struktur anorganischer Gläser. *Dissertation, Univ. Mainz*, (1996).

41. F. Liebau. Untersuchungen and Schichtsilikaten des Formeltyps $A_m(Si_2O_5)_n$. I. Kristallstruktur der Zimmertemperaturform des $Li_2Si_2O_5$. *Acta Cryst.* **14**, 369-395 (1961).

42. K.F. Hesse. Refinement on the crystal structure of lithium polysilicate. *Acta Cryst.* **B33**, 901-902 (1977).

44. D.J.M. Burkhard, B.H.W.S. DeJong., A.J.H Meyer, and J.H. van Lenthe. $H_6Si_2O_7$: *Ab initio* molecular orbital calculations show two geometric conformations. Geoch. Cosmochimica Acta **55**, 3453 (1991).

44. Y.T. Thathachari and W.A. Tiller. Steric origin of the silicon-oxygen-silicon angle distribution in silica. *J. Appl. Phys.* **53**, 8615 (1982)..

45. J. Zemann. The shortest known interpolyhedral O-O distance in a silicate. *Zeitschr. Kristallogr.* **175**, 299-303 (1986).

46. C.T. Herzberg. Magma density at high pressure Part 1: The efect of composition on the elastic properties of silicate liquids. Magmatic Processes, Physicochemical Principles. *The Geoch. Soc. Spec. Publ.* (B.O.Mysen ed.) 25-58 (1987).

47. R. Knoche, D.B. Dingwell, and S.L. Webb. Melt densities for leucogranites and grantic pegmatites: Partial molar volumes for SiO_2, Al_2O_3, Na_2O, K_2O, Li_2O, Rb_2O, Cs_2O, MgO, CaO, SrO, BaO, B_2O_3, P_2O_5, F_2O_{-1}, TiO_2, Nb_2O_5, Ta_2O_5, and WO_3. *Geoch. Cosmochimica Acta* **59**, 4645-4652 (1995).

Proc. 30th Int'l. Geol. Congr., Vol. 16, pp. 191-202
Huang Yunhui and Cao Yawen (Eds)

Plasticity of $YAlO_3$ and YAl_3O_5 at High Temperature: Implication in Searching for Strong Materials

ZICHAO WANG
Department of Geology and Geophysics, University of Minnesota, MPLS, MN 55455, USA

Abstract

High temperature plasticity of YAP and YAG single crystals was experimental investigated to understand the correlation between plasticity and crystal structure. YAG showed a very high resistance to creep and the plastic deformation was attributed largely to the dislocation glide. In the case of YAP, plastic deformation resulted from dislocation climb and the resistance to creep was significant lower than that of YAG at a comparable condition. The difference showed a clear control of creep mechanism over the high temperature plasticity. Crystal plasticity is primarily determined by the Peierls stress when dislocation glide dominant in creep and a well defined plasticity-systematics found among YAG and other oxide garnets. Similar plasticity systematics is failure in the YAP (weakest slip system) and the other perovskites, in which dislocation climb is involved in creep. This observation implies that plasticity systematics works only for the materials undergoing creep through dislocation glide, and gives the constrain to searching for strong materials based on the plasticity of known materials.

Keywords: *plasticity, high-temperature, garnet, perovskite*

INTRODUCTION

Materials with high creep strength ($T>0.5T_m$, T_m-melting temperature) are of many applications. Continuous efforts in studding creep behavior of materials at high temperature have resulted in tremendous progresses in the application of new materials in industry since last two decades. Two of the most important issues in material science are understanding high temperature plasticity and investigating plasticity-systematics for searching new materials with high creep strength. The general knowledge, we have so far, is that the creep strength, resistance to creep at it's steady-state stage, of crystalline materials is determined by the resistance to plastic deformation, which occurs usually by number of mechanisms including twinning and dislocation motion. Plastic deformation due to twinning occurs at relative low temperature ($T<0.5\ T_m$), the strength of materials undergoing twinning is basically determined by lattice resistance to the motion of twinning dislocation. Plastic deformation due to dislocation motion is coupled with either glide and/or climb. The creep solely due to dislocation glide is named as glide-controlled creep [1]. The strength of material undergoing dislocation glide is determined by the Peierls stress (τ_p), which is the intrinsic resistance to dislocation glide on a particular slip plane due to lattice periodicity. Previous works have found a good correlation between Peierls stress τ_p and crystal-geometrical factor so that the creep strength of crystal can be predicted based on the data from known materials with similar crystal structure [2]. However at high temperature, dislocations acquire a new degree of freedom: dislocation can climb as well as glide. Mechanisms which are based on this climb-plus-glide sequence are refereed as climb-controlled creep [3,4]. The average velocity of dislocation motion in climb-controlled creep is determined by the climb step, which is in turn determined by the diffusive motion of single ions or vacancies to or from the climbing dislocations so that no unique correlation between τ_p and crystal-geometrical factor presence. This difference implies an important role of the nature of dislocation motion, i.e. glide or climb, in determining the plasticity systematics of materials. However, only very a few critical discussions were made about the correlation between plasticity systematics and the nature of dislocation motion, or the creep mechanisms, in the past, so that the searching for an plasticity systematics of materials with identical crystal structure is in argument [5,6].

This paper will discuss high temperature plasticity of materials from the crystal structure and the nature of dislocation motion view points. We will focus on the plasticity of materials at high temperature ($T>0.5T_m$) so that twinning isn't an important process in creep. We studied experimentally two yttrium aluminum oxide crystals, YAP ($YAlO_3$) and YAG (YAl_3O_5), as model materials. These two crystals have identical chemical composition, but YAP is of perovskite structure and YAG of garnet structure. Experimental results showed significant difference in creep behavior, which can be attributed to the difference in nature of dislocation motion due to the difference in crystal structures of the materials. These results provide a physical interpretation for observed plasticity-systematics so that the searching for new materials with high creep strength at extreme conditions is practically possible.

CREEP OF YAP AND YAG AT HIGH TEMPERATURE

The chemical composition, physical properties and crystal lattice dimension of YAP and YAG are listed in Table 1. YAP crystallizes into orthorhombic perovskite structure (space group Pbnm) and it is stable at temperature up to it's melting point. Plastic deformation occurs mainly via the activity of the [100](001) slip system when crystal is compressed along [110] direction at $T>0.8T_m$ [7]. YAG grows into garnet structure (space group Ia3d) and it is stable at temperature also up to melting point. Plastic deformation for YAG occurs via the 1/2<111>{110} slip systems when crystal is compressed along <100> direction at $T>0.8T_m$[8]. Both orientations along which YAP and YAG were deformed are assumed to be the weakest orientations (that means the weakest slip systems are active and dominant during creep).

1. Macro-creep behavior

From the creep experiments performed at the conditions of T (temperature)=0.8-0.95T_m, P(total pressure)=0.1 MPa, f_{O_2} (oxygen fugacity) =10^{-20}-10^{-4} MPa and σ (applied stress) =50-300 MPa, we found that plastic deformation occurs only at fairly high temperature for both YAP and YAG (T>0.8 Tm). Steady-state creep was usually reached after initial transit creep (<3% in strain). Creep rate of YAP and YAG in their steady-state stage can be described with a power law, viz.,

$$\dot{\varepsilon} = A\sigma^n f_{O_2}^m \exp\left(-\frac{Q}{RT}\right) \tag{1}$$

where A is a constant, $\dot{\varepsilon}$ the strain rate, n the stress exponent, m the oxygen fugacity exponent, Q the activation energy and R the gas constant. We noticed that measured n is close to 3 for both YAP and YAG (Table 2), indicating that creep of both YAP and YAG results from dislocation motion. However, measured activation energy was not strictly constant in both YAP and YAG suggesting a dependence of activation energy on stress, namely, $Q = Q_0 - B\sigma$. We determined Q_0 = 803 kJ/mol and B = 0.27 (kJ mol MPa) for YAG and Q_0 = 1035 kJ/mol and B = 0.57 (kJ mol MPa) for YAP respectively (Fig. 1). One of the most important observation in this work is the significant difference in creep strength between YAP and YAG (Fig. 2). Creep strength of YAG is about a factor of 3 higher than that of YAP at a comparable condition.

Table 1. Physical Properties of YAP and YAG

	YAP	YAG
Chemical Composition	$YAlO_3$	$Y_3Al_5O_{12}$
Melting Point	1870±10 (°C)	2000±10 (°C)
Symmetry	orthorhombic	cubic
Unit cell dimension	a: 5.180 (Å) b: 5.320 (Å) c: 7.375 (Å)	a: 13.0 (Å)
Density	5.35 (g/cm^3)	4.56 (g/cm^3)
Expansion coefficient	a: 4.2x10^{-6} (1/°C) b: 3.9710^{-6} (1/°C)	7.7x10^{-6} (1/°C
shear modulus	100 (GPa)*	114 (GPa)
bulk modulus	180 (GPa)*	185 (GPa)

*-estimated from elastic systematics.

Table 2 Experimental determined creep parameters for YAP and YAG

	YAG	YAP	
n*	3.1±0.1	2.8±0.2	
Q_0 (kJ/mol)	803±75	1035±120	
B (kJ.mol.MPa)	0.27	0.57	
m	0.0	0.0	(fO_2 >10^{-10}MPa)
	0.0	-1/3	(fO_2 < 10^{-10}MPa)
σ_i/σ	0.6	0.82	

* - average from total 15 and 12 experimental data for YAP and YAG respectively.

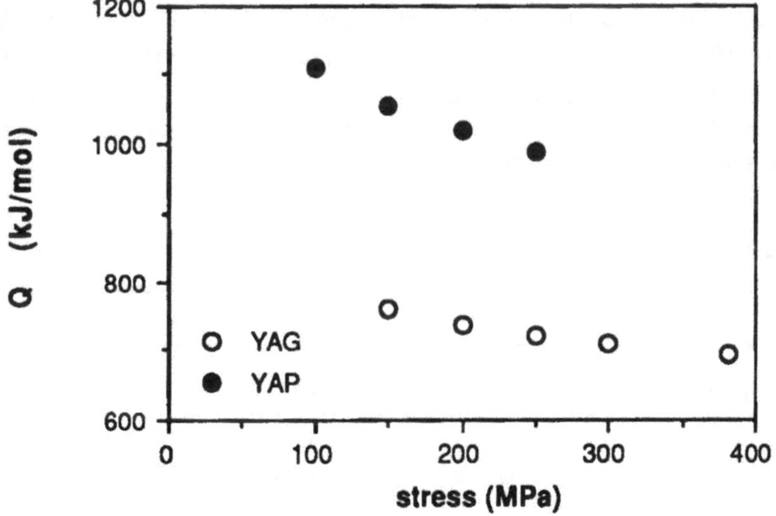

Figure. 1 Variations of Q with applied stress for YAP and YAG

2. Influence of oxygen fugacity (fO_2) on creep

The dependence of creep rate on oxygen fugacity (fO_2) was examined for both YAP and YAG(Fig.3). We did not observed clear effect of fO_2 on creep rate for YAG. However, a substantial change in creep rate was detected for YAP at $fO_2 < 10^{-13}$ MPa. It has been proposed that thermodynamic environment can dramatically affect the creep at high temperature due to the change of point defects concentration arising from the corresponding change in fO_2, which will finally result in a change in the velocity of dislocation motion.

The influence from point defects may become more significant in high temperature creep when it involves in both diffusion and diffusion-assistant dislocation (e.g. dislocation climb). The clear dependence of creep rate on oxygen fugacity may be a good evidence for a dislocation-climb dominant creep in YAP.

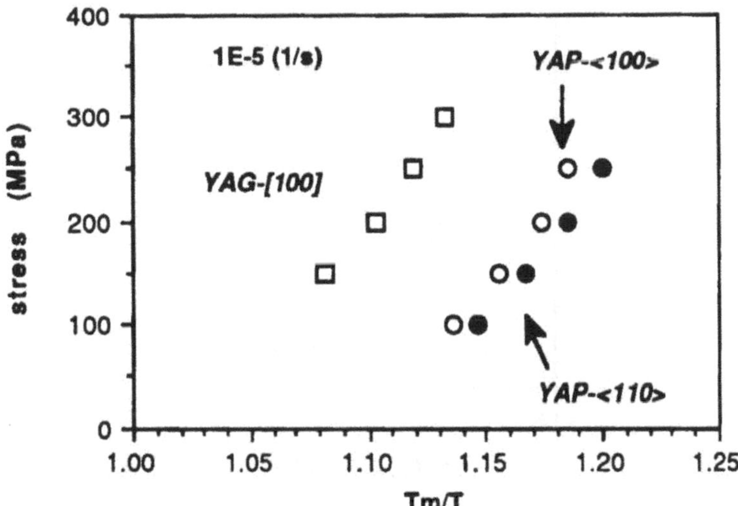

Figure. 2 Creep strength of YAP and YAG at various temperatures. Data for YAP from two orientations are presented here.

3. *Micro-dynamics of dislocation motion*

We generally assume that the flow stress, driving the dislocation motion, may be separated into two component: an effective stress (σ) and an internal stress (σ_i) which are basically different in character from each other[1]. The σ is in the sense of the stress that acts effectively for a dislocation to glide over obstacles by a thermally activated process and it is the dominant component in a glide controlled creep. The σ_i is an athermal resistance which a dislocation can overcome only by increasing flow stress of by reducing the obstacle strength through restoration. Thus, the σ_i is a main part of flow stress in a dislocation climb-controlled creep. The results of stress-dip tests, which provide a measure of the importance of the σ_i during creep, is given in Fig.4.

As illustrated in Fig.4a, a decrease in strain-rate occurs soon after stress-dip of 3% of the total applied stress for YAG, but the absolute magnitude was relatively small. The stress-dip did not cause deformation to halt as seen in NaCl but only led to a small decrease in strain rate. In other words, the creep was continuous for a small instantaneous change of stress [9]. The ratio of σ_i/σ was estimated to be < 0.65 for YAG. Similar test was performed for YAP, a ratio of σ_i/σ > 0.82 was obtained. It can be seen that the ratio of σ_i/σ for YAG is significantly smaller than that for YAP indicating the relative dominance of dislocation glide and dislocation climb in YAG and YAP respectively. Table 3 listed the σ_i/σ for a number of materials demonstrated a clear sequence of reduced σ_i/σ when dislocation glide become more important .

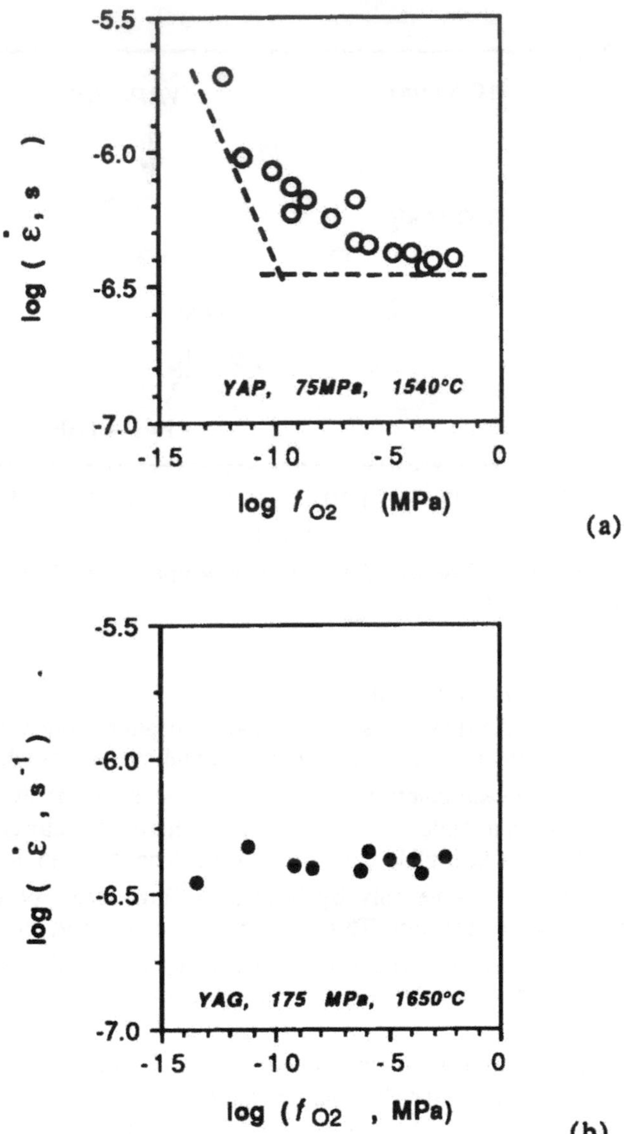

Figure.3 variation of creep rate with f_{O2} for YAP (a) and YAG (b)

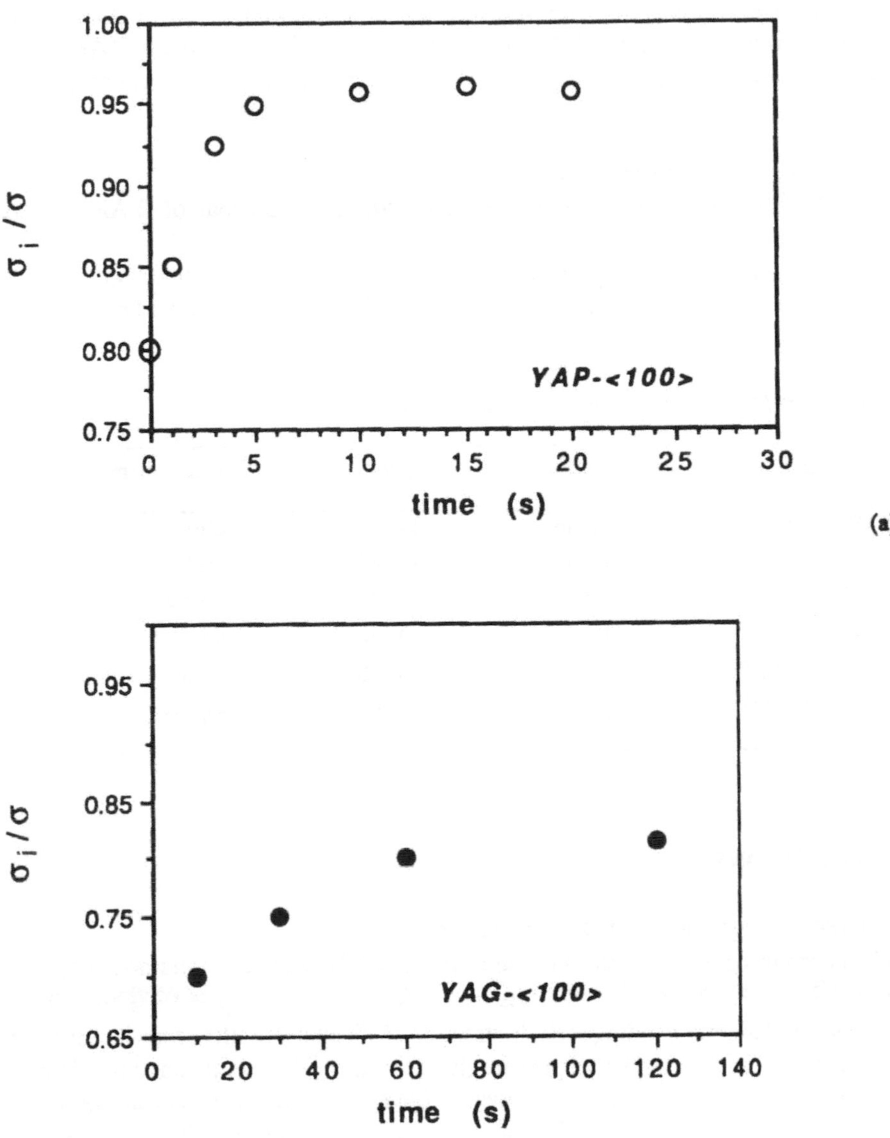

Figure 4. Measured internal stress σ_i for YAP (a) and YAG (b). The σ_i was obtained by extrapolating data to t=0

4. Dislocation structures

Typical dislocation structures observed in deformed YAG are isolated straight dislocation line and glide loop. Numerous triple junctions are common, which are due to reaction of dislocations with different Burgers vector. No subgrain boundaries were observed and the distribution of dislocations is uniform. All these observations suggest dislocation glide as dominant mechanism responsible for the creep. However, the free dislocations observed in YAP are short and curved. Significant dislocation dissociation, dislocation junctions and prism loops were found in YAP. The nature of dislocation observed in YAP indicated that creep was controlled by climb. There is not much evidence of dislocation glide observed in YAP.

We can summarize the present observations as follows:
(1) YAG showed a significant higher creep strength than that of YAP at comparable conditions;
(2) Dislocation glide is responsible for the creep of YAG at high temperature, however dislocation climb contributes more to the creep of YAP at high temperature.

Table 3. σ_i/σ for materials with different crystal structures

Materials	Crystal Structures	σ_i/σ	mechanism*
YAP-[110]	ortho.	>0.82	climb
YAG-[100][8]	b.c.c.	<0.65	glide
NaCl-[100][9]	f.c.c.	~1.0	climb
$SrTiO_3$-[100][6]	f.c.c.	>0.85	climb
$SrTiO_3$-[110][6]	f.c.c.	<0.80	glide
GGG-[100][13]	b.c.c.	<0.65	glide

DISCUSSIONS

1. Creep strength of garnet and perovskite

It has been recognized that the difference in strength between diamond and copper is due to the difference of the Peierls stress (τ_p), the big difference of τ_p is obviously attributed to the nature of atomic bonding. High strength of diamond results from its strong covalent bonding and easy in creep of copper results from its weak metallic bonding. However, this is not the case for YAP and YAG, which have very similar chemical composition and bonding structure. One clear difference between these two materials is the crystal structure. YAP has perovskite structure (Pbnm) and YAG garnet structure (Iad3). There is no, so far, any discussions over how the crystal structure affects creep mechanisms, and further the resistance to creep at high temperature. The fact is that creep of YAP was most likely due to dislocation climb, or recovery-controlled creep and dislocation glide, in contrast, contributed mainly to the creep of YAG. This observation may imply a close correlation among dominant creep mechanism, creep strength and crystal structure. One of the most fundament questions in regarding this question is the relative role of dislocation multiplication (dislocation glide) and migration (climb or cross-slip verse glide). Both dislocation multiplication and migration are necessary to maintain steady-state creep. The slower of the two processes will control the overall rate of steady-state creep. The strength of material , at the other hand, is determined by the stress that acts effectively for a motion of dislocation. Experimental observations supported that the resistance to creep was arise from the dislocation glide over obstacles in the case of YAG. The measured activation energy showed a clear stress dependence giving $Q = Q_0 - \sigma b \Delta A$, where Q_0 is the height of the energy barrier at $\sigma = 0$ and $\sigma b \Delta A$ is the work done by applied stress to overcome the barrier and thus effectively lowering the height of barrier; ΔA is the area swept in the glide plane; b is the magnitude of the Burgers vector or distance between two successive energy troughs. Our data yield $\Delta A \approx 10^{-18}$ m^2, which is consistent with $\Delta A \approx b^2$ with b=1/2<111>, implying that a small scale barrier (namely the Peierls potential) is involved in the dislocation glide process. Fairly high internal stress was observed ($\sigma_i > 0.8$) in the case of YAP, which is thought to arise from both long range interaction between dislocations and the self stress of dislocation. The measured higher internal stress supported the dominance of dislocation climb, which is consistent with observed dislocation structures in YAP. Thus, we interpret that the high creep strength of YAG is due to the high Peierls stress, which in turn is due to the crystal structure of garnet: a large unit cell (a = 13Å) and the b.c.c packing tends to increase the Peierls stress [8]. The creep strength of YAP isn't determined only by the Peierls stress so that the crystal structure periodicity may not be the only critical factors controlling the resistance to creep.

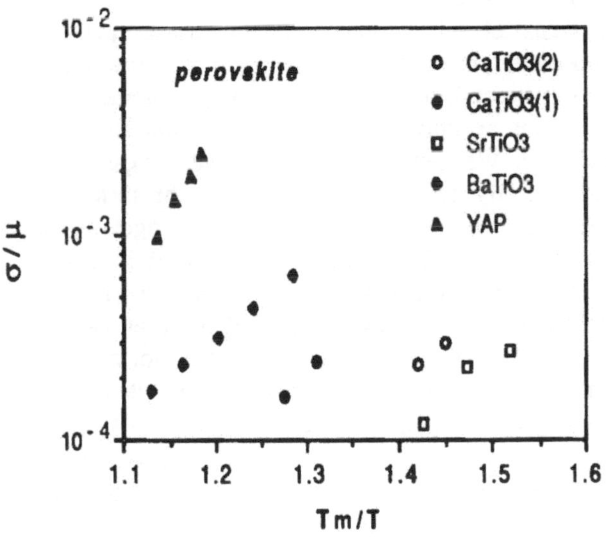

(a)

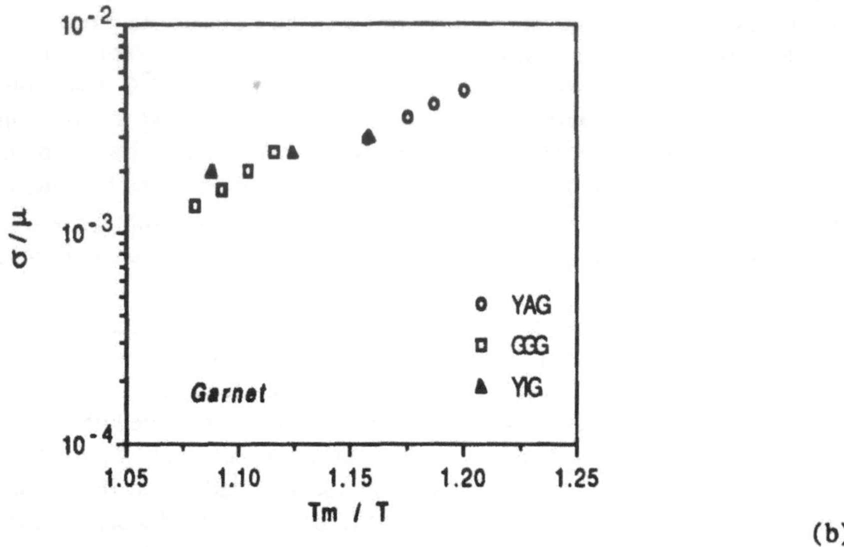

(b)

Figure 5. Normalized creep strength for perovskites (a) and garnets (b)

2. .Plasticity- crystal structure systematics

One of the most important goals of this study is to investigate the systematic relationships between high temperature plasticity (creep strength) and crystal structure to understand the origin of high creep strength and search for strong materials. However, previous works suggested that there is no universal systematic relationship found, but it may depend on crystal structure [9,10]. We compiled the data from oxide garnets and oxide perovskite for a comparison [11,12]. By normalizing the data in the form of s/m and T/Tm, we will see the presence of any correlation between creep strength and the other properties for a given crystal structure. The result was illustrated in Fig.5. The convergence of a large of data sets into a single trend is remarkable for materials with garnet structure and suggests that most of the oxide garnets belong to an iso-mechanical group [12]. On the other hand, the data from oxide perovskite does not converge into a single trend. We should point out that plasticity of oxide perovskites is of strong anisotropy. For the hard slip system in a cubic perovskite ($<110>\{1\bar{1}0\}$ slip system), normalized creep data did converge into a single trend [9]. The study indicated that dislocation glide was the controlling process in this case. We interpret this due to the difference in creep mechanisms: creep of garnets is controlled mainly by dislocation glide owing to their high Peierls stress, which in turn is determined essentially by the crystal structure. This provide a natural explanation for the presence of plasticity-systematics. In contrast, the failure of plasticity-systematics for perovskite reflects the fact that their creep is largely controlled by dislocation climb, which is controlled not only by the crystal structure but also involved in many different factors such as impurity concentration.

It can be summarized that YAG and YAP showed different creep behavior at comparable conditions, we interpret this due to the difference in crystal structures and the resulting dominant creep mechanisms. YAG is of garnet structure and crept through dislocation glide so that the creep strength is primarily determined by the Peierls stress, which is significant high due to it's large garnet unit cell and b.c.c packing. The sole dependence of creep strength on Peierls stress gives a universal systematics for materials crystallizing into garnet structure so that a predication on strength of materials with similar crystal structure is possible. YAP is deformed mainly through dislocation climb, the creep is rate-controlled by the diffusion rate of slower species. Many factors rather than Peierls stress alone will affect the strength so that a universal plasticity systematics does not apply. We conclude based on this study: (1) yttrium aluminum oxide with garnet structure is of higher resistance to creep at high temperature than that with perovskite structure; (2) In order to search a material with high strength based on known materials, it is necessary to know whether the material crystallizes into a similar structure and it's deformation is controlled by a glide or a climb process. Predication may only be successful when creep is controlled by dislocation glide.

REFERENCES

1. H.Kurishta, H.Yoshinaga and H.Nakashima, The high temperature deformation mechanism in pure metals, Acta Metal, 37: 499-505 (1989).
2. T Suzuki and S Takeuchi, Intrinsic strength of crystals and Deformability of ceramics, in lattice Defects in ceramics, ed. by T.Suzuki and S Takeuchi, 9-15 (1989).
3. J. Weetman, Dislocation climb theory of steady-state creep, Am.Soc.Metal, Quart.Trans., 61: 681-694 (1968).
4. J. Weetman, High temperature creep produced by dislocation motion, In: Rate processes in plastic deformation of materials (ed. J.C.M. Li and A.K.Mukherjee), 315-336, Am.Soc.Metal. (1975).
5. S Karato, Plasticity-crystal structure systematics in dense oxides and its implications for the creep strength of the earth's deep interior: a preliminary result. Earth.Planet.Sci,Lett 86: 365-376 (1989).
6. Z.-C Wang, Shun-ichro Karato and Kiyoshi Fujino, High temperature creepof single crystal strontium titanate ($SrTiO_3$): a contribution to creep systematics in perovskite, Phys.Earth.Planet.Int. 79: 299-312 (1993).
7. Z.-C Wang, S Karato and C Dupas, High temperature creep of an orthorhombic perovskite-$YAlO_3$: implication to the dynamics of the lower mantle. EOS. supplement, AGU Fall-meeting (1996).
8. S.Karato, Z-C Wang and K.Fujino, High temperature creep of yttrium-aluminum garnet single crystals, Journal of Materials Science 29: 6458-6462 (1994).
9. S. Karato, Z-C Wang, B Liu and K Fujino, Plastic deformation of garnets: systematics and implications for the Rheology of the mantle transition zone. Earth.Planet Sci.Lett. 130: 13-30 (1995).
10. J.Rabier and H.Garem, Plastic deformation of oxides with garnet structure, In : Tressler RE (ed.) Deformation of Ceramic Materials II, Materials Science Research, 18. Plenum Press (1983).
11. H.J. Frost, and Ashby MF, Deformation - Mechanism Maps. Pergamon Press,

Oxford (1982).

12 Z.-C Wang, S-i Karato and K.Fujino, High temperature creep of single crystal gadolinium gallium garnet, Phys.Chem.Minerals, 23: 73-80 (1996).

Proc. 30th Int'l. Geol. Congr., Vol. 16, pp. 203-218
Huang Yunhui and Cao Yawen (Eds)

Revisiting the Kaolinite-Mullite Reaction Series: A ^{29}Si and ^{27}Al MAS NMR Study*

GUO JIUGAO, HE HONGPING, WANG FUYA, WANG DEQIANG AND ZHANG HUIFEN
Guangzhou Institute of Geochemistry, Academia Sinica, Guangzhou 510640, China)
HU CHENG
National Laboratory of Solid State Microstructural Physics, Nanjing University, Nanjing 210093, China)

Abstract

The kaolinite-mullite reaction series of eight kaolin samples have been studied with XRD, IR, DTA and MAS NMR spectra. The acquired results clearly show that the evolution of ^{29}Si and ^{27}Al spectra of the various samples may differ from each other significantly. In the metakaolinite studied, the Al^V is universally present and it appears as a sign of the variation of the structure and some properties of the metakaolinites. It can be suggested from many experimental results that the high-temperatural phase occurring after the exothermal peak should be γ-Al_2O_3 rather than Al-Si spinel, and the mullite-formation path in the studied samples involve two steps: first, the primary mullite formed directly from the metakaolinite around 850 ℃; Second step, the secondary mullite developped extensively by reaction between segregated SiO_2 and γ-Al_2O_3 at about 1200-1300 ℃ and the composition or/and structure of secondary mullite may be visibly different depending on the content of the impurity mica minerals.

Keywords: Kaolinite, Metakaolinite, γ-Al_2O_3, Mullite, ^{29}Si and ^{27}Al MAS NMR, XRD, IR

INTRODUCTION

The series of reaction by which kaolinite transforms to mullite is perhaps the most important in the entire feild of ceramic technology. It has been extensively studied for more than 100 years, but nevertheless retains many unsolved questions. The main ones are the structure of metakaolinite and the relationship between the various high temperature phases[1,2]. Despite of the application of many experimental techniques to the problems, the products of thermal transformation of kaolinite in a wide range of temperatures are usually microcrystalline or amorphous which makes them difficult to be studied by conventional X-ray diffraction techniques, so the studied

* This project supported by National Natural Science Foundation of China.

areas remain in question. Since 1985, the Magic-Angle Spinning Nuclear Magnetic Resonance (MAS NMR) has been introduced into the study of kaolinite-mullite transformation[3-12]. However, most of previous studies are to pay main attention to observe the structure of metakaolinite, particularly the 5-fold coordination aluminium. Watanabe et al.[5] only detected the signal of aluminium atoms in 4- and 5-fold coordination in metakaolinite at 7T. Later, the others have given the ^{27}Al spectra of metakaolinite with a strong peak of 5-coordination aluminium signal at from 8.4T to 11.7T[7,8,9,11]. Lambert et al.[8] considered that Watanabe et al. were unable to detect the Al^{V} signal is likely to be due to a number of experimental problems, such as the effect of the low magnetic field on the second-order quadrapolar broadening[8]. Although many people demonstrated the existence of five-fold coordinated aluminium in metakaolinite, the aluminium distribution in the different sites in metakaolinite reported are rather different. Massiot et al.[11] suggested that this can be partly due to the different NMR experiment condition.

The problem regarding the nature of the high temperature phase appearing after the exothermal reaction around 980 ℃ has been strongly debated at home and abroad, and it is the point at issue that this phase is Al-Si spinel or γ-Al_2O_3 which is associated with the mechanism of the formation process of mullite, whethere formed by the topotactic formation from metakaolinite through the Al-Si spinel, or by the interaction of segregated free aluminia (γ -Al_2O_3) and silica at higher temperature?[13]

In the past, the people only studied one or at most two initial kaolinite samples respectively. However, kaolinite minerals in nature have various geological origin and different crystallinity, and usually inevitablely coexist with other impurity minerals. Then, what influence will be made to the products of thermal transformation of kaolinite by different starting samples? In this work, we have chosen eight kaolin samples as starting materials and systematically investigated using the ^{29}Si and ^{27}Al MAS NMR spectra of thermal transformation of these samples.

EXPERIMENTAL PROCEDURE

The features of the samples studied are shown in the Table, in which most of kaolin samples are natural, except the sample YKB and SK are purified.

Thermal treatment of natural kaolin samples was carried out in controlled rate LCT-2 differential thermobalance at the set temperature (450-1200 ℃) for 1h at 50 ℃ intervals and the treatment of samples above 1200 ℃ was conducted with a high-temperature electric furnance, followed by quenching to room temperature. The specimens were ground to powders less than 75 μ m..

Table: The Hinkley index, component, occurrence and locality of kaolin.

Code	HI	Component	Occurrence*	Locality
XK	1.5	K. 93%, Q. 5% A. very little	hydrothermal alteration	Shalinzi, Xunhua Hebei Province
Y	0.88	K. 97% Q. very little	tonstein	Hunyuan County, Shanxi Province
YKB	0.6	K. 93%, Q. 5% M. very little	weathered residual	Yangjiang County Guangdong Province
P1	1.1	K. 89%, G. 9% A. 2%	tonstein	Hunyuan County, Shanxi Province
YK	0.6	K.57%, M.24%, Q.17%, F. very little	weathered residual	Yangjiang County, Guangdong Province
KK	<0.1	k. 72%, I. 17% Q.11%	sedimentary	Kanglong, Nanan, Fujian Province
P2	amorphous	Metak. 92% All. 6% A.very little	natural metakaolinite	Hunyuan County, Shanxi Province

* The component of kaolin samples is calculated from the intensity of X-ray diffraction lines of corresponding minerals in addition to P2, in which the contents of K. and All. are estimated by the area of ^{29}Si peak of minerals in NMR spectra.
K.-kaolinite, M.-muscovite, Q.-quartz, I.-illite, A.-anataze, All.-allophane, G.-gibbsite, F.-feldspar. Metak -metakaolinite

The initial and fired samples were examined using XRD, IR, DTA and ^{29}Si and ^{27}Al MAS NMR. Powder XRD patterns were recorded on a Rikagu D/max-1200 diffractometer. Fourier Transform Infrared (FTIR) spectra in the region 4000-200cm^{-1} were acquired by using a P-E 1725 spectrometer and the conventional KBr technique. ^{29}Si and ^{27}Al MAS NMR spectra of the samples were measured with a Bruker MSL-300 NMR spectrometer using one pulse or solid echo sequence at 59.6MHz and 78.2MHz, respctively. Rotors were spun in air at 4-5KHz. The recycle delay of 5s and 0.2s and the pulse width of 2 μ s and 0.6 μ s, and TMS and solution of $AlCl_3$ as external standard references were selected for Si and Al signals separately.

RESULTS

The room temperature XRD patterns of samples show in addition to kaolinite reflections, a peak at d=9.97Å assigned to the 002 reflection of mica in which of highly crystalline attributed muscovite and of poorly crystalline to illite and a peak at d=3.34Å to α-quartz, at d=3.517Å to the 101 reflection of anatase. Between 450 ℃ and 550 ℃, the kaolinites become progressively less crystalline and the intensity of their lines generally decrease. At 500 ℃ for YKB, SK, YK, at 550 ℃ for Y, KK, P1 and at about 600 ℃ for XK, and the lines of the parent kaolinitebassically was eliminated, showing a intense broad background in the 18-28 ° (2 θ). At about 950 ℃ for Y, YKB, P2 and at 1150 ℃ for XK, the cubic γ-alumina gives three broad reflections at d=1.39, 1.98 and 2.37Å, and some faint mullite reflections at d=5.40, 3.42 and 1.52Å. As the firing temperature reaches at 1300 ℃ for sample XK and at 1200 ℃ for others, obviously strong mullite reflections occure and cristobalite in most samples except YK. In the pattern of sample P1 at various temperatures, there are some distinguishing features. The room temperature XRD patterns of samples show a peak at 4.835Å attributed to the 002 reflection of gibbsite. As early as 350 ℃ the gibbsite reflection vanish and three new broad reflections assigned to γ-alumina begin to emerge from the background of the highly crystalline kaolinite reflection. At 500-800 ℃ these diffuse signals and the anatase reflection are more clearly visible. As temperature reaches 1050 ℃, the intensity of the γ-Al_2O_3 signals increase obviously(Fig.1).

FTIR spectra of sample YKB after calcination at different temperatures as an example given in Fig.2. At 450 ℃, the 3600-3800cm^{-1} OH strenching and 1300-1400cm^{-1} lattice vibrations of the sample are significantly decrease. Above this temperature, the former becomes unobservable and the latter becomes some broad bands, exhibiting the characteristic of metakaolinite as being amorphous. In kaolinite, the Al-O bond from AlO_6 octahedral gives a strenching band at ca. 540cm^{-1}[14], the intensity decreases with increasing temperatures so that in the range of 450 ℃ and 850 ℃ this bands is faint. Above these temperatures, this vibration becomes increasingly stronger again. The dehydroxylation of kaolinite results in the growth of a broad band at ca. 810cm^{-1} belong to Si-O strenching and 4-coordinated Al-O strenching which remains up to 1200 ℃. The shoulder band at ca. 670cm^{-1} which was considered as another typical band of metakaolinite[9] disappears above ca. 950 ℃, and then, the weak vibration at 750cm^{-1} emerges in the spectra of samples which was assigned to 4-coordinated Al-O strenching[14]. As temperature increasing to 1200 ℃, the great change of IR spectra of samples appear in the range of 1200cm^{-1} and 400cm^{-1}, developing many new absorption peaks in which the peak at 1160cm^{-1}, 970cm^{-1}, 898cm^{-1}, 741cm^{-1}, 610cm^{-1} and 563cm^{-1} may belong to mullite and the 795cm^{-1} and 610cm^{-1} to cristobalite[15,16].

We give the ^{29}Si and ^{27}Al MAS NMR spectra of three representative sample Y, YKB and KK fired at various temperatures in Fig.3, 4 and 5, respectively. The ^{29}Si spectra of these initial samples consist of a single sharp peak at about -91ppm with FWHH (full-width-at half-height) 2-3ppm except sample YK being 5.6ppm, characteristic of layered silicates and attributed to a Q^3 Si environment. At 500 ℃ for sample Y and at 450 ℃ for YKB and 550 ℃ for KK, the spectra are apparently broadenned towards to high feild besides the sharp peak at -91ppm is retained in Y and YKB. Above these temperature, FWHH of the peak increases to about 20ppm which indicates the metakaolinite being an amorphous and the center of symmetrical signal gradually shifts to -102ppm. At 800 ℃, the peak of sample Y moves to -104ppm. As increasing temperature to 850 ℃, the spectra of these samples begin to become asymmetrical and the center of gravity of the signal moves to about -108ppm assigned to the amorphous silica, showing the main structure of Si-O sheet converts into the Q^4 Si environment, meanwhile, a weak peak at about -87ppm belongging to

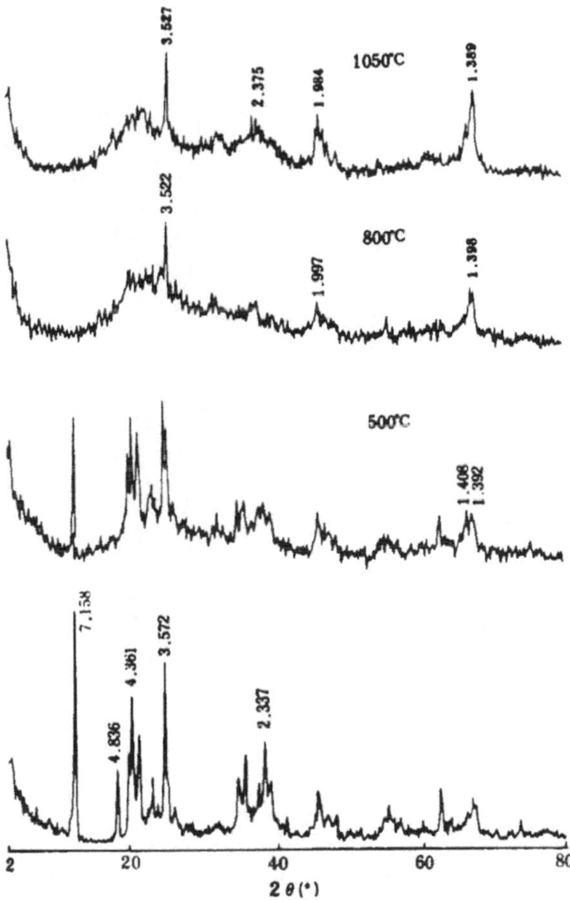

Fig.1 XRD patterns of kaolin P1 heated to indicated temperatures for 1.0h. Apart from the reflection of kaolinite, the reflection at 4.836Å assigned to gibbsite, the one at 3.52Å to anatase and the diffuse reflections at about 1.40, 1.99 and 2.37Å belong to γ-Al_2O_3.

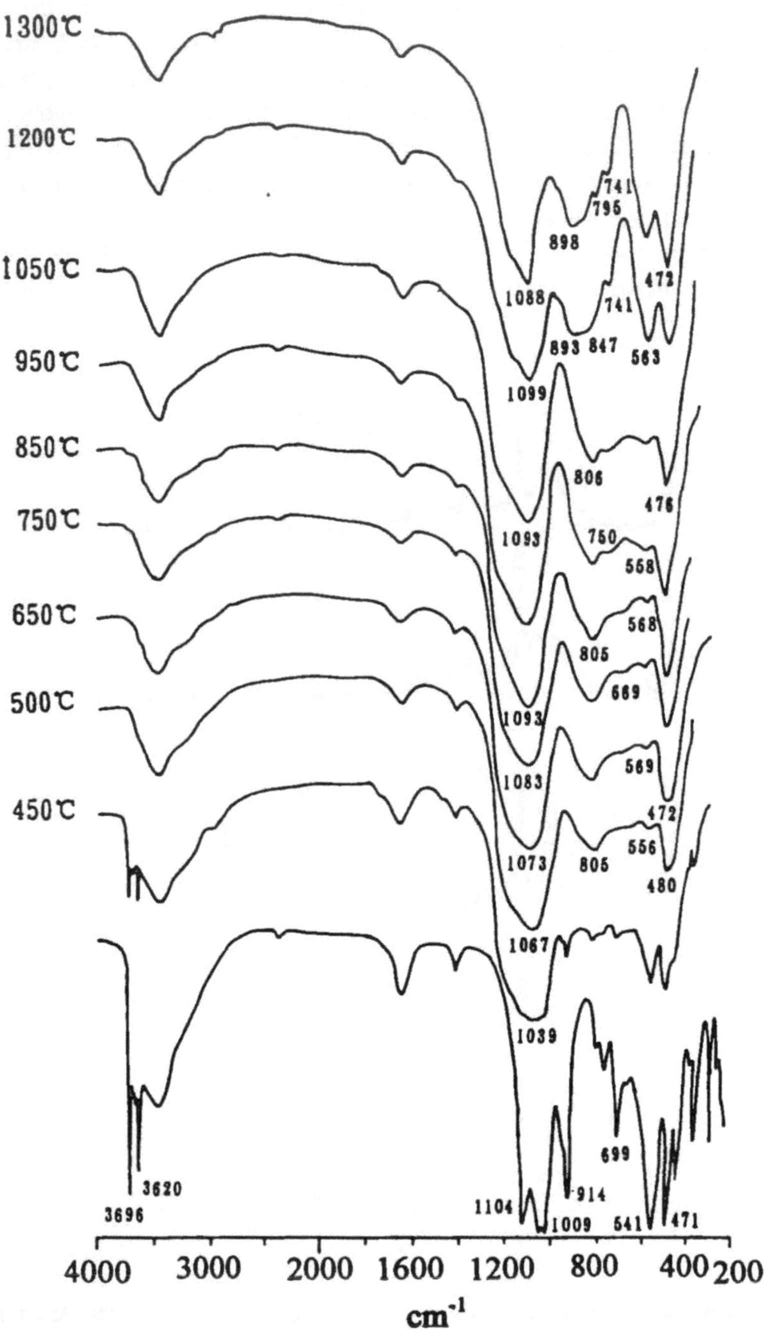

Fig.2 FTIR spectra of kaolin YKB after thermal treatment at different temperatures for 1.0h, showing three time obvious change in the spectra, first at about 450 ℃, secondely at 950 ℃ and thirdly at 1200 ℃. The bottom IR spectrum is of original sample.

mullite (primary mullite) begin to emerge at the low field side of spectra. With furtherly raising temperature, the gravity of spectra shifts to ca. -110ppm and the FWHH decreases which makes the broad lines observed at 850 ℃ being progressively split into the peak of mullite and SiO_2. As heating temperature to 1200 ℃, the obviously various change of the spectra of three samples occure. The spectra of fired sample Y exhibit a very strong ^{29}Si peak at -87ppm which indicate the extensive development of secondary mullite. In the spectra of sample YKB, the intensity of this peak apparently is lower than sample Y. However, the strong signal at -87ppm in the spectra of sample KK does not still emerge, even heated to 1350 ℃.

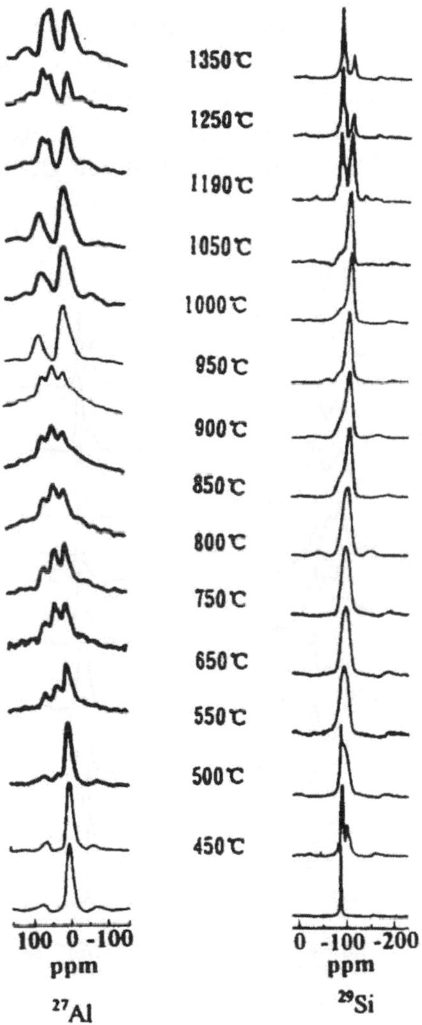

Fig.3 Evolution of the MAS NMR spectra of kaolin Y heated to temperature indicated for 1.0 h using a pulse of 0.6 μ s for ^{27}Al and 2 μ s for ^{29}Si, and spinning at 4-5 KHz on an MSL-300 (ref. aqueous $AlCl_3$ for ^{27}Al and TMS for ^{29}Si). The bottom NMR spectrum is of initial sample.(the same below)

The ^{27}Al spectra of all of the starting samples except YK of which the spectrum contains a weak signal at 70ppm assigned to 4-coordinated Al substituting Si in the Si-O sheet of muscovite, consist of only a single peak at about -3.0ppm assigned to 6-coordinated Al [Al^{VI}]. It is shown from Fig.3, 4 and 5 that the sample prepared from sample YKB, Y and KK at 450 ℃, 500 ℃ and about 550 ℃, respectively, exhibit two other resonances at about 30ppm attributed to 5-coordinated Al [Al^{V}] and at about 57ppm assigned to 4-coordinated Al [Al^{IV}]. Generally, with the dehydroxylation processed, the Al^{IV} and Al^{V} peak are more and more heightenned and the Al^{VI} weakenned. Howevere, the evolution of spectra between different samples has a great difference, for example, the intensity of Al^{V} peak in the spectra of sample Y fired increase much faster than in YKB, whereas, the Al^{IV} peak in the former much slowly than in the later. In the 800-900 ℃ range the signal of Al^{V} of sample Y reaches its maximum. Forming a contrast, the signal of Al^{V} of YKB is still lower than that of Al^{IV}. The another tremendous change has taken place in ^{27}Al spectra of sample YKB and KK at 900 ℃, of Y at 950 ℃, respectively, the Al^{V}

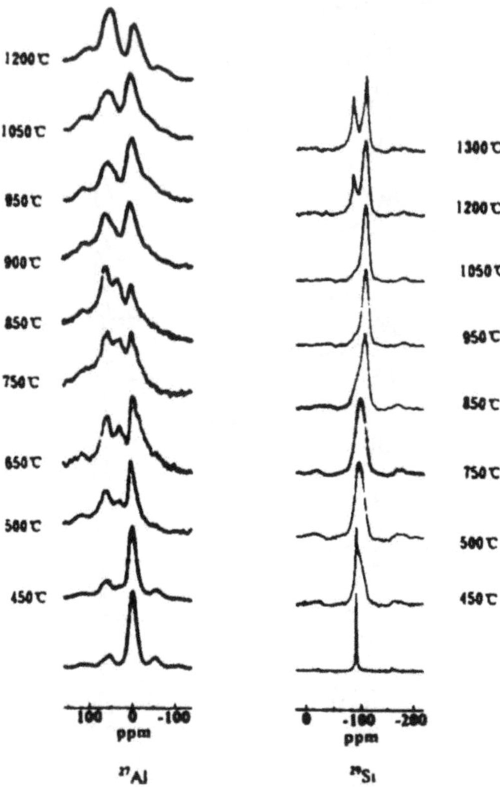

Fig.4 Evolution of the MAS NMR spectra of kaolin YKB heated to indicated temperature for 1.0 h using a pulse of 0.6 μ s for ^{27}Al and 2 μ s for ^{29}Si, and spinning at 4 KHz on an MSL-300 (ref. aqueous $AlCl_3$ for ^{27}Al and TMS for ^{29}Si).

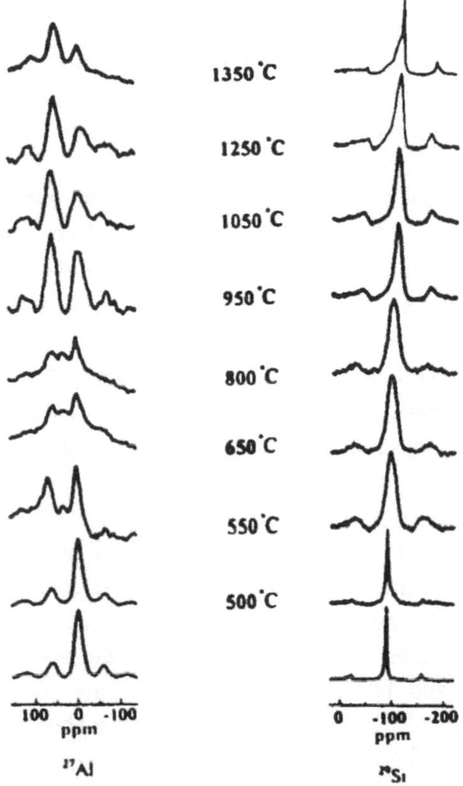

Fig.5 Evolution of the MAS NMR spectra of kaolin KK heated to temperature indicated for 1.0 h using a pulse of 0.6 μ s for ^{27}Al and 2 μ s for ^{29}Si, and spinning at 4 KHz on an MSL-300 (ref. aqueous $AlCl_3$ for ^{27}Al and TMS for ^{29}Si).

signal almost completely disappears and the Al^{VI} signal in sample Y and YKB rise quickly again, showing that the treatment converts all 5- and perhaps some 4-coordinated Al into 6-coordinated. In the spectra of KK, with the further rise of temperature, the Al^{IV} peak at about 56ppm more and more strongthen and the Al^{VI} signal at about 0ppm weaken (Fig.5). As the temperature is increased to about 1200 C for sample Y and YKB, the ^{27}Al spectra display two new Al^{IV} and Al^{VI} signal at ca. 50ppm and -4ppm separately, and the Al^{IV} peak intensity is obviously stronger than Al^{VI}, which is characteristic resonance of mullite.However, the spectra of sample KK did not display the similar change, in spite of that there are similar reflections of mullite with other samples in the XRD pattern.

DISCUSSION

Although the results from XRD and IR show a large variation of the crystalline ordering among these starting samples, the lack of significant difference in the ^{29}Si and ^{27}Al spectra of these samples indicates that short-range silicon and aluminum ordering of these kaolinites is very similar.

The collapse of kaolinite structurw, as detected by our XRD, IR, and ^{29}Si and ^{27}Al MAS NMR, occurs at 450-550 ℃ which is obviously wider than at 480-500 ℃ by Rocha and Klinowski[9]. Recent studies suggested the starting reaction mechanism of dehydroxylation is controlled by diffusion in the kaolinite particles and the diffusion process depend on the defective nature of kaolinite[17] and in some respects on the particle size of sample[13]. To be consistent with this, the collapse temperature of the structure of the sample XK which has the highest crystalline ordering is also the highest at 550 ℃. Howevere, it seems may be suggested that the geological origin of sample may be also a important factor for the callapse temperature of structure due to dehydroxylation. For example, the collapse temperature of sample YKB and SK which are weathered origin at about 450 ℃ is clearly lower than that of not weathered origin, particularly than that of KK at 500-550 ℃, no matter that the crystallinity of KK is much lower than that of YKB and SK.

The structural change of Al atoms in the layered silicates and their treatment products is still a noticeable question. ^{27}Al (I=5/2) is a quadrupolar nucleus and the ^{27}Al NMR spectra of solids can usually be broadened and distorted due to nuclear quadrupole interaction. At present, many people adopt the spectrometer with higher applied magnetic fields and faster sample spinning rate to obtain a real high-resolution solid-state ^{27}Al NMR spectra. However, it is shown from some articles that ^{27}Al nucleus in the different Al sites of a solid have a distinctive quadrupole coupling constant (cq) and there is the differential response of the nucleus in the different sites to the rf pulse. Smith[18] suggested that this effect has been the source of incorrect intensity ratio between the ^{27}Al peaks of various sites obtained in some papers. In order to give a correct intensity ratio of the Al atoms at any kinds of sites for the irradiation pulse, one must use small pulse width or pulse flip angle, for example $\pi/12$[19]. Just owing to the careful choice of experimental parameter, mainly adopting small pulse flip angle, the gained high-resolution ^{27}Al spectra of metakaolinites recorded at 7T by us may compare favorably with the results at 11.7T[8] and at 9.4T[9], clearly resolving the Al^{IV}, Al^{V} and Al^{VI} peak. It can be seen from our results that the Al^{V} is universally present in the metakaolinites even that mixed with a number of impurity mineral bearing-Al. So, we considered that Watanabe et al. did not detected the Al^{V} signal is probably to be due to the unsuitable choice of some experimental parameters, rather than the low applied magnetic field.

Moreover, it was found that the intensity ratio among the three different coordinated Al in spectra may be apparently distinct from sample to sample. Metakaolinite Y and XK of which the original sample belong to sedimentary and alteration origin respectively have similar ^{27}Al spectra, exhibiting the very strong peak of Al^{V}.

Whereas, the spectra of YKB which is weathered residual origin is quite different from that of Y and XK, showing very strong peak of Al^{IV} in the spectra. The results of the simulation of ^{27}Al spectra show the ratio of 4-, 5- and 6-coordinated Al in sample XK at 950 ℃ are 30.5:46.1:23.4, but that in YKB at 850 ℃ 41.3:29.1:29.6 respectively, which is overlaped with the most part of data range given by previous work[8,9,11]. Therefore it should be considered that the different samples, particularly that in geological origin, may have a great influence over the distribution of aluminum species in their metakaolinite.

We measured the acid solubility of Al of the five series of metakaolinite treated at different temperatures (Fig.8). It can be seen from Fig.8 that the acid solubility of Al in the samples increases with increasing calcination temperature, and the acid solubility maximum occurs in the range of 500 ℃ to 900 ℃ except sample XK from 650 ℃ to 950 ℃. Generally, we may parallel the variation of solubility curve with the change of Al coordination state from 6-fold to 5- and 4-fold. Here, it is worth emphasing that the temperature of rapid descent in Al acid solubility of samples at higher temperature is almost consistent with the temperature of Al^{V} vanishing in the samples.

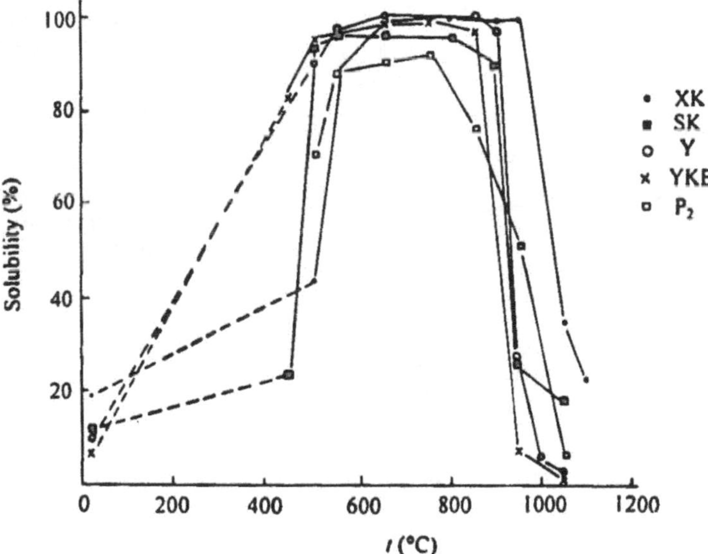

Fig.6 The variation of the acid solubility of Al vs. the calcination temperature of metakaolinites. The sample were treated at indicated temperatures for 1.0 h and the acid soluble experiment of Al of the samples is conducted using the 4mol/L HCl solution at unboiling temperature for 2.0h.

The exothermal peak at about 980 ℃ in DTA curves is a significant characteristic of kaolinite, the nature and significance of which is still debated. Recently, Sanz et al.

deduced that it is associated with the modification of the coordination of Al, which changes from the tetra- or penta- coordination to the more stable octahedral coordination[7]. Fig.7 gives the DTA curves of kaolinite XK heated at various temperatures, showing that the exothermal peak almost disappears at 1000 ℃ at which Al^V in this sample simutaneously disappears. Thinking about the real variation of ^{27}Al spectra of the samples, it can be considered that the exothermal peak at 980 ℃ may be mainly associated with the change from penta-coordinated Al to the stable 6-coordinated Al. Rocha and Klinowski[2] suggested that the penta-coordinated Al may be chosen as the best measure of remarkable chemical reactivity of metakaolnite. According to these results, we can suggest that the Al^V may be used as a sign of the structural change and some properties of the metakaolinite.

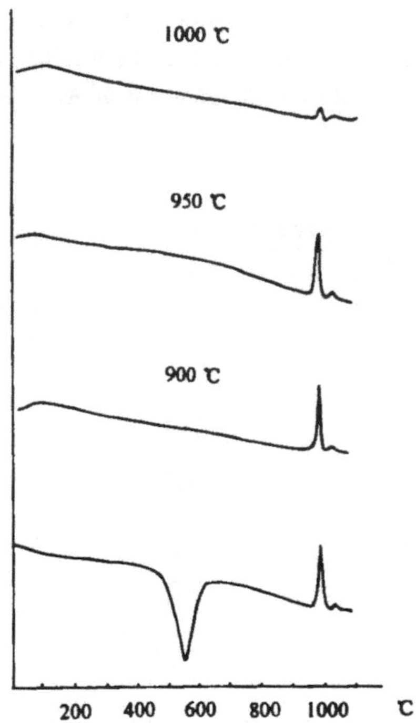

Fig.7 The DTA curves of kaolinite XK after thermal treatment at different temperatures for 1.0 h. The cursve at the bottom is of parent sample and the DTA curves is measured on LCT-thermobalance at 20 ℃/min in air up to 1200 ℃.

The high temperature phase appeared after exotheral peak at about 980 ℃ has attracted much discussion. Earlier ones including the famous one of our country, Zhang Yuanlong (1956) identified the phase as γ-Al_2O_3[20], but in consideration of structure continuity, Brindley and Nakahira (1959) suggested that this phase is a silicon-bearing spinel of composition $Si_8Al_{10.67}\square_{5.33}O_{32}$ rather than γ-Al_2O_3 $(Al_{21.33}\square_{2.67}O_{32})$[21]. According to the formula of the spinel, in the Al-Si spinel the full Si atoms occupy the tetrahedral sites and Al atoms only the octahedral sites, but in γ-Al_2O_3 Al atoms may occupy both tetrahedral and octahedral sites. Based on our spectroscopic data, whether IR or NMR of the samples, there are not only the strong Al^{VI} signal but also a considerable Al^{IV} signal. Moreover, in the all of the ^{29}Si spectra of relatived materials reported yet and our spectra, no any signal at -79.9ppm attributed to Al-Si spinel[5] appear, which is identical with the results from TEM by Sonuparlak et al.[22]. These results may be interpreted as supporting γ-Al_2O_3. The evolution of the X-ray diffraction pattern given in Fig.1 is also considered as a evidence of the high temperature phase being γ-Al_2O_3.

The formation process of mullite is also a problem attracting attention of many people. From our ^{29}Si spectra for most of samples at 850 ℃ and that by Rocha and Klinowski[9] and Massiot et al. [11] as well, it is shown that the metakaolinite begin to transform into the segragated SiO_2 with small amounts of mullite, indicating the formation of the primary mullite is obviously earlier than that of spinel. Whereas in recent studies of calcined kaolinite with XRD, mullite was formed together with pseudoanatase and spinel at 900 ℃ for 8d[23]. The above mentioned difference shows the ^{29}Si and ^{27}Al MAS NMR spectra is more sensitive about the structural variation in the amorphous or microcrystalline systems than XRD.

In previous works, only Brown et al. gave some details of structural change heating one kaolinite sample to 1450 ℃ using ^{29}Si and ^{27}Al NMR spectra[5]. In this work, revealed results show that the extensive development of secondary mullite is at about 1200 ℃ for the most of samples except the XK at 1300 ℃. Although the XRD reflection of these mullites in the various samples have approximately same intensity, the recorded ^{29}Si peak intensity is quite different. According to the difference of that, we may divide the sample into three kinds. First kind include sample Y, XK, P1 and P2 which almost free of any mica minerals, the ^{29}Si spectra of these samples exhibit a very strong peak at -87ppm. The tentative simulation of the spectra of sample Y at 1350 ℃ shows that the intensity ratio between both peaks of mullite and cristobalite is about 4:1. Second kind is of sample SK and YKB containing some mica minerals and the simulation of the ^{29}Si spectra of sample SK at 1300 ℃ shows that the intensity of the mullite ^{29}Si peak only occupy about one third of total signal intensity which is similar with that of Brown et al.. Third kind involves sample KK and YK which contain a large number of mica minerals and the mullite signal in the ^{29}Si spectra of the sample KK at 1350 ℃ represents only 18.1% of observed silicon.

The mineral mullite is a number of the aluminum silicate family with the general composition in $Al_{4+2x}Si_{2-2x}O_{10-x}\square_x$ ($0.2 < x < 0.6$, $\square$ stands for oxygen vacancy) and the structural formula may be represented by $Al_2(Al_{2+2x}Si_{2-2x})O_{10-x}$. The compositional variation is based on the exchange $O^{2-}+2Si^{4+}$ $2Al^{3+}$ which introduces the oxygen vacancies and is accommodated by the development of a superstructure[14,24,25,26]. From the structural formula of mullite, it can be guessed that the amount of Si atoms into mullite structure may be restricted by situation of the aluminum distribution (Al^{IV} and Al^{VI}) in the natural precursor from which the secondary mullite formed. In concert with the change of ^{29}Si spectra of aforementioned samples, the corresponding ratio of Al^{IV} and Al^{VI} in the ^{27}Al spectra increase with the decrease of ^{29}Si signal intensity of mullite, which is associated with the obviously increase of the content of mica minerals as impurity. But, due to the very high spin lattice relaxation time in SiO_2 and adopted recycle time (5s) perhaps more suitable to observe mullite, the aforesaid data about the contribution of Si in mullite may be overestimated and these data can not be used to quantify the stoichiometry of the formed secondary mullite[11]. Nevertheless, we feel that these results have clearly shown the composition or/and structure of mullite formed from the different initial kaolinite sample as the natural precursor to be difference.

CONCLUSION

Based on the abovementioned data, the following conclusions may be made:
(1) The revealed results show once again that ^{29}Si and ^{27}Al MAS NMR is a powerful tool of studing the short-range structure of the thermal transformation products of kaolinites, and may provide more abundant information regarding the structural evolution in kaolinite-mullite reaction series than XRD and IR.

(2) It has been clearly indicated that the evolution of ^{29}Si and ^{27}Al spectra in kaolinite-mullite reaction series of the different samples may greatly differ from each other, and these difference is associated with the geological origin of samples, the crystallinity of kaolinites and other coexisted impurity minerals.

(3) In the metakaolinite, the Al^{V} is universally present and it may be used as a sign of variation of structure and some properties of metakaolinite. So long as careful choice of experimental parameters, ones can distinctly resolve the Al^{IV}, Al^{V} and Al^{VI} signal in ^{27}Al spectra of metakaolinites at 7T applied magnetic field.

(4) The results from various methods show that the high temperature phase occuring after the exothermal peak at 980 ℃ should be γ-Al_2O_3 rather than Al-Si spinel. In the kaolinite-mullite reaction series of the samples studied, the development process of mullite involves a two-step formation. In the first step, the primary mullite in a small amount begin to form directly from the metakaolinite at about 850 ℃ for most samples. In the second step, the secondary mullite is developped in a large amount by the reaction between segragated SiO_2 and γ-Al_2O_3 at about 1200-1300 ℃.

(5) In the cases of calcined kaolinite as natural precursor from which the mullite formed, the composition or/and structure of secondary mullite may be apparently difference depending on the content of mica mineral in these starting sample.

Acknowledgement:

we are grateful to Prof. Yan Yaxiu of China National Building Materials Industry Geological Exploration Center for providing the sample XK and KK.

REFERENCES

1. G. W. Brindley and M. Nakahira. The Kaolinite-Mullite Reaction Series:I. A Survey of Outstanding Problems. *J. Am. Ceram. Soc.* **42:7** 311-314 (1959)
2. J. Rocha and J. Klinwoski. Solid-State NMR Studies of the Structure and Reactivity of Metakaolinite. *Angew. Chem. Int. End. Eng.*, **29:5** 553-554 (1990).
3. R. H. Meinhold, K. L. D. Mackenzie, I. W. M. Brown. Thermal Reactions of Kaolinite Studied by Solid State 27-Al and 29-Si NMR. . *Mater. Sci. Lett.* **4**,163- 166 (1985).
4. K. J. D. Mackenzie, I. W. M. Brown, R. H. Meinhold, M. E. Bowden. Outstanding Problems in the Kaolinite-Mullite Reaction Sequence Investigated by ^{29}Si and ^{27}Al Solid-State Nuclear Magnetic Resonance:I. Metakaolinite. *J. Am. Ceram. Soc.* **68:6** 293-297 (1985).
5. I. W. M. Brown, K. J. D. Mackenzie, M. E. Bowden, R. H. Meinhold. Outstanding Problems in the Kaolinite-Mullite Reaction Sequence Investigated by ^{29}Si and ^{27}Al Solid-State Nuclear Magnetic Resonance:II. High-Temperature Transformations of Metakaolinite. *J. Am. Ceram. soc.* **68:6** 298-301 (1985).
6. T. Watanabe, H. J. Shimizu, K. Nagasawa, A. Masuda, H. Saito. ^{29}Si- and ^{27}Al- MAS/NMR Study of the Thermal Transformations of Kaolinite. *Clay Miner.* **22**,37-48 (1987)
7. J. Sanz, A. Madani and J. M. Serratosa. Aluminum-27 and Silicon-29 Magic-Angle-Spinning Nuclear Magnetic Resonance Study of the Kaolinite-Mullite Transformation. *J. Am. Ceram. Soc.* **71:10** c418-c421 (1988).
8. J. F. Lambert, W. S. Millman and J. J. Fripiat. Revisiting Kaolinite Dehydroxylation: A ^{29}Si and ^{27}Al MAS NMR Study. *J. Am. Chem. Soc.* **111:10** 3517-3522 (1988).
9. J. Rocha and J. Klinowski ^{29}Si and ^{27}Al Magic-Angle-Spinning NMR Studies of the Thermal Transformation of Kaolinite. *Phys. Chem. Minerals.* **17:2** 179-186 (1990).
10. He Hongping, Hu Cheng, Guo Jiugao, Wang Fuya, Zhang Huifen. ^{29}Si and ^{27}Al Magic-Angle-Spinning Nuclear Magnetic Resonance Studies of Kaolinite and Its Thermal Products. *Chinese J. Geochem.* **74:1** 78-82 (1995).
11. D. Massiot, P. Dion, J. F. Alcover and F. Bergay. ^{27}Al and ^{29}Si MAS NMR Study of Kaolinite Thermal Decomposition by Controlled Rate Thermal Analysis. *J. Am. Ceram. Soc* **78:11** 2940-2944 (1995).
12. Guo Jiugao, He Hongping, Wang Fuya, Zhang Huifen and Hu Cheng. The Structural State of Aluminum Atoms and Its Influence on Some Properties of Metakaolinite. *In: Progress in Geochemistry, Contrib. to 30th IGC.* Guangzhou Inst. Geochem. CAS (Ed.) pp. 221-227. Zhongshan Univer. Press, Guangzhou, China. (1996).
13. G. W. Brindley and J. Lemaitre. Thermal Oxidation and Reduction Reaction of Clay Minerals. In. *Chemistry of Clays and Clay Minerals.* A.C.D. Newman (Ed.) pp319-364 Longman Scientific & Technical, London (1987)
14. H. J. Percival, J. F. Duncan, P. K. Foster. Interpretaation of the Kaolinite-Mullite Reaction Sequence from Infrared Absorption Spectra. *J. Am. Ceram. Soc.* **59:2** 57-61 (1974).
15. K. J. D. Mackenzie. Infrared Frequency Calculation for Ideal Mullite ($3Al_2O_3 2SiO_2$) *J. Am. Ceram. Soc.* **55:2** 68-71 (1972)
16. C. H. Ruscher. Phonon Spectra of 2:1 Mullite in Infrared and Raman Experiments. *Phys. Chem. Minerals* **23:1** 50-55 (1996).
17. B. Bellotto, A. Gualtieri, . Artioli, S. M. Clark. Kinetic Study of the Kaolinite-Mullite Reaction Sequence. Part I: Kaolinite Dehydroxylation. *Phys. Chem. Minerals.* **22**, 207-214

(1995).

18. M. E. Smith. Getting More out of MAS NMR Studies of Quadrupole Nuclei. *Bruker Report* **1**,33-35 (1990).
19. D. Massiot, C. Bessada, J. P. Coutures and F. Taulelle. A Quantitative Study of ^{27}Al MAS NMR in Crystalline YAG. *J. Magn. Reson.* **90**, 231-242 (1990).
20. Chang Yuanlung. The Thermal Reactions and the Control of Phase Changes of Kaolinite. In: *Bulletin.* Inst. Geol. Academia Sinica (Ed.) pp4-85, Science Press, Beijing, China. (1956) (in Chinese)
21. G. W. Brindley and M. Nakahira. The Kaolinite-Mullite Reaction eries: III. The High-Temperature Phases. *J. Am. Ceram. Soc.* **42:7** 319-324 (1959).
22. B. Sonuparlak, M. Sarikaya, I. A. Aksay. Spinel Phase Formation During the 980 ℃ Exothmic Reaction in the Kaolinite-to-Mullite Reaction Series. *J. Am. Ceram. Soc.* **70**, 837-842 (1987).
23. J. A. Pask and A. P. Tomsia. Formation of Mullite From Sol-Gel Mixtures and Kaolinite. *J. Am. Ceram. Soc.* **74:10** 2367-2373 (1991)
24. W. E. Cameron. Mullite:A Substituted Alumina. *Am. Mineral.* **62:7-8** 747-752 (1977).
25. J. Angel and C. T. Prewitt. Crystal Structure of Mullite: A reexamination of the Average Structure. *Am. Mineral.* **71:11-12** 1476-1482 (1986).
26. L. H. Merwin, A. Sebald, . Rager and H. Schneider. ^{29}Si and ^{27}Al MAS NMR Spectroscopy of Mullite. *Phys. Chem. Minerals.* **18:1** 47-52 (1991).

Proc. 30th Int'l. Geol. Congr., Vol. 16, pp. 219-232
Huang Yunhui and Cao Yawen (Eds)

A Study of Applied Mineralogy of Ores from Weigang Iron Mine, CHINA

CHEN ZHAOXI
Department of Geology, University of Science and Technology Beijing, China
XU ZHENHONG
China University of Mining and Technology

Abstract

Primary ore and their concentration products from Skarn type Weigang iron mine ore studied by means of Spectral analysis, chemical analysis, opitical microscope and electron probe etc. The primary ores are classified into three categories and eight technological types. Samples of the category one and two are studied for their chemical element association, quantitative mineralogical composition, equilibrium state of iron sulphur element and the state of grain inter-locking, grain size distribution characteristic of exploitable mineral. Twenty-four samples of pyrite have been studied by chemical analysis of single minerals and electron probe, and two of these samples have been studied for their element image of Fe, S and Co. The results suggest that the occurrence form of cobalt in pyrite is regarded as in the form of isomorphism. T Fe value (the total Content of Fe) in the concentrated ore is lower and the T Fe value in the tailings is higher. The lower degree liberation of magnetite is main reason. Additionally the content of the pyrite Can be up to 95% in the concentrated pyrites ores through a simple floatation in laboratory, cobalt content in the concentrates is 0.189%. and It is enough for complex recovery.

Keywords: Iron ore, technological type, occurrence form, cobalt, grain size, liberation, complex recover.

INTRODUCTION

Applied mineralogy is a relatively new subject on which study has just begun. In the past thirty years, as the result of advances in processing method and extensions of raw materials utilization range, the comprehensive utility of mineral resources has been improved to extremely great extent, and Applied mineralogy is increasingly playing an important role in technical improvement of many aspects such as geological exploration, mineral processing and metallurgical processing etc [1,2].

China has held six national symposiums on processing mineralogy since the Association of Process Mineralogy in 1979. Many researchers has been working on it and a number of projects has been finished in the past over 10 years, providing a great deal of useful information and data for the processing renovation of more than 30 metal mineral processing plants. So far a number of studies on many ways of the subject has been reported but the study about the classification of ore process type, especially of Skarn-Iron ore process type has seldom been done[3].

Weigang iron mine has been running for almost ten years, and its final products are mainly iron concentration, sulphur concentration and copper concentration. Its mineral processing plant flow sheet is first flotation-then magnetic separation, separating copper from pyrite by depressing pyrite and floating copper minerals. But because the study of ore process mineralogy has never been done, lack of enough technological basic information and data on several aspects such as characteristics of ore process type, the occurrrence form of cobalt etc, result in many problems during production. For example, indices of concentration and tailing are not meeting the design requirements; copper can not be effectively separated from pyrite, and cobalt element has not been recovered etc[4].

This study will mainly deal with the following: classification of ore processing type; occurrence form of cobalt in pyrite and its complex recover; grain size characteristics of exploitable minerals in ore; liberation characteristics of exploitable minerals in concentrations and tailings; the direction of improving mineral dressing process.

The ore body of Weigang iron mine mainly exists in garnet skarn zone, It can be divided into two minerogenetic zones upper and lower zone according to the structural condition. The lower is the main minerogenitic zone, $1^{\#}$ ore body is the main part of the lower minerogenetic zone and account 90% of the reserves of the whole mining province. Our ores sampling kept consistency with present produce condition of mine. mining of the upper part of -50m level of $1^{\#}$ ore body had been finished. Ore samples are taken at -56m and -100m levels with channel method. The sampling interval is 50 × 10m, and each sample weight is 1 - 2 Kg and total 52 samples are taken.

Basic character of the ores

Composition and texture, structure of the ores

Table1 shows the chemical composition

Table 1 Chemical polyelement analysis result of crude ore

element	TFe	SFe	FeO	S	CaO	Se	Au(g/t)	Ni	Ga	As
content %	38.42	35.47	14.86	3.06	5.08	0.0008	0.021	0.012	0.0017	0.014
element	MgO	Cu	P	Co	Sn	Ti	Ag(g/t)	SiO_2	Al_2O_3	
content %	0.63	0.075	0.146	0.017	0.017	0.0005	0.40	16.73	4.64	

The observation and identification of more than 150 sheets polished sections and thin sections had been finished with microscope and the work of chemical analysis , electron probe analysis . the mineral composition, texture and structure of ores as follows.

Magnetite: it occurs as fine grain massive aggregate or impregnated aggregate, its size is general 0.1~0.3 mm and dissemination grain size is generally than 0.074 mm.

Hematite: it distributes widely in the ores but content is a few and it is the main mineral composition of replaced magnetite.

Pyrite: it is widely found in ores , occur as xenomorphic or hypidiomorphic granular texture, its size is 0.01~2 mm. The pyrites can be divided two stages , the early epoch type occur as fine grain xenomorphic which fill in the gaps of between magnetites and garnets or replace magnetite, the later epoch type mostly are medium xenomorphic -- hypidiomorphic crystal which occur as agglomeration, stockwork aggregate metasomatizing magnetites or gangue minerals.

Chalcopyrite: content in the ores is small, grain size range from 0.02 to 0.08 mm and occur as star-sporadic or impregnate.

Gangue minerals: they are mainly garnet, chlorite, epidote, calsite, quartz and some apatite, feldspar, phlogopite etc., garnet, chorite, epidote, calsite and exploitable metallic minerals are most closely paragenesis, garnet had been chloritization in different degree. The electron probe analysis of chlorite is listed in table2.

Table 2 The result of chlorite chemical composition analysis of EPMA

Composition	Na_2O	CaO	MgO	Al_2O_3	SiO_2	TiO_2	MnO	Cr_2O_3	NiO	FeO	total
Content %	0.42	0.07	14.55	21.07	28.34	0.03	1.58	0.19	0.36	29.61	96.20

Its crystal chemical formula is: $(Fe_{2.4360}, Mg_{2.1323}, Al_{1.1522}, Ni_{0.0303}, Mn_{0.1315}, Cr_{0.0142}$ $Ti_{0.0016})_{5.8981}[(Si_{2.7907}, Al_{1.2093},)_4 O_{10}](OH)_8$.

There are many texture types in the ores which are mainly idiomorphic crystal, hypidiomorphic-xenomorphic crystal texture or metasomatic texture.

The existing state of cobalt in the ores

Cobalt mainly exist in the pyrites and rarely in other minerals such as magnetite, but the occurrence form of cobalt in the pyrite was not clear. In This study twenty four samples of pyrite have been studied by chemical analysis of single minerals and electron probe, the results are listed in table 3. The content of cobalt is not homogenous in the pyrite, and it is different in pyrite of different stages, the content of cobalt is higher in the early pyrite and even the distribution is not homogeneous in the same grain. The content range from trace amount to 0.2%-0.3%, the higest can be up to 7.38%. Two of these samples have been studied for their element image of Fe,

S, Co. Figure 1 is energy spectrum image of cobalt pyrite, photo 1~4 are images of Fe, S, Co element face distribution. These prove the heterogeneity of the distribution of cobalt in the pyrites. The correlation analysis of result of element Fe and Co of pyrite in the table 3 has been studied. We use mathematical model y=a+bx, correlation analysis result: a=45.6616, b=-0.8904, the linear equation y=45.6616 -

Table 3 The result of chemical analysis and electron probe analysis of pyrites

sample	formation of pyrite	analysis result					
		TFe	S	Cu	Co	Ni	As
014		45.66	51.38	0.070	0.222	0.110	0.152
0.20		45.68	51.71	0.046	0.195	0.097	0.154
021		45.58	50.07	0.162	0.265	0.163	0.144
S-2	densely agglomeration and impregnated	45.30	51.44	0.138	0.208	0.132	
S-6	fine grain densely massive and impregnated agglomeration	45.38	50.72	0.177	0.080	0.176	
S-3	fine grain impregnated	45.08	51.35	0.136	0.280	0.203	
S-8	" "	45.63	51.73	0.151	0.337	0.156	
S-10	" "	44.70	51.53	0.029	0.184	0.248	
S-1	fine grain densely impregnated	44.58	50.79	0.113	0.248	0.192	
S-7	medium grain densely impregnated	45.27	51.30	0.110	0.201	0.097	
S-9	impregneted massive	45.42	51.77	0.058	0.268	0.136	
S-5	medium grain pyrites in calsite vein	44.63	50.99	0.026	0.074	0.009	
S-4	" "	44.81	50.99	0.092	0.035	0.006	
S-4-3	" "	44.52	50.54	0.043	0.021	0.003	
	the resulte of elctron probe analysis as follows						
100-5-6-1	medium grain agglomeration pyrites in calsite vein	46.67	53.65	0.04	0.00	0.39	
100-5-6-2	fine grain impregnated	45.94	53.19	0.26	0.33	0.28	
100-5-6-3	hypidiomorphyis crysital in calsite vein	46.63	53.65	0.00	0.00	0.35	
100-5-6-4	impregnated massive	45.71	53.38	0.00	0.02	0.80	
100-5-6-5	" "	45.93	53.43	0.00	0.00	0.91	
56-1-2-1	fine grain impregnated and fine vein in magnetites	47.71	51.40	0.00	0.09	0.04	
56-1-2-2	" " "	46.88	53.60	0.00	0.10	0.04	
56-1-2-3	" " "	40.25	53.66	0.06	5.61	0.09	0.09
56-1-2-4	" " "	40.45	52.95	0.06	5.68	0.04	0.48
56-1-2-5	" " "	39.48	54.11	0.00	7.38	0.04	

0.8904x. The negative correlations relationship between cobalt and Iron in pyrite is obviously. After the crystal chemical formula calculation of the sample 56-1-2-5 in table 3 is $(Fe_{0.8378}, Co_{0.1484}, Ni_{0.0008})_{0.9870}S_2$, which is consistent with chemical formula of standard cobalt pyrite[5]. Hence we say that the occurrence form of cobalt in the pyrites is mainly in isomorphous[6].

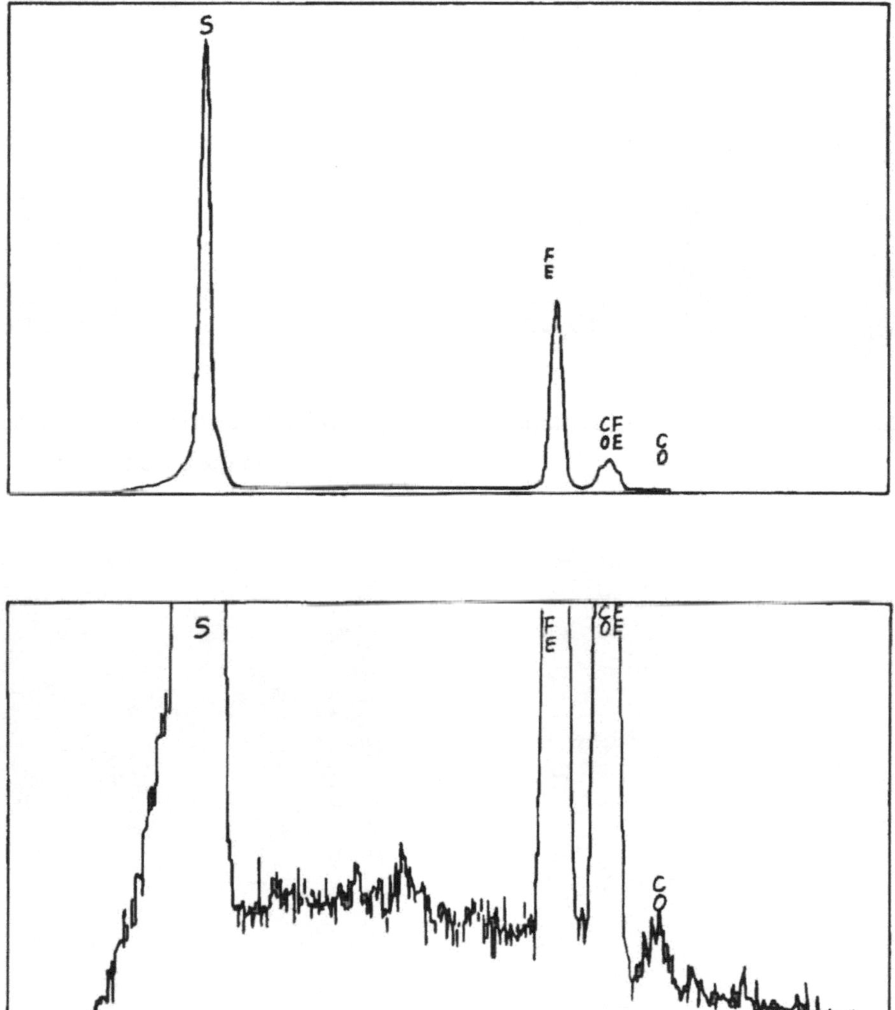

Fig.1. Energy spectrum image of cobalt pyrite, following is enlarged

photo 1 form image of cobalt pyrite(left)

photo 2 face distribution of sulphur in the cobalt pyrite

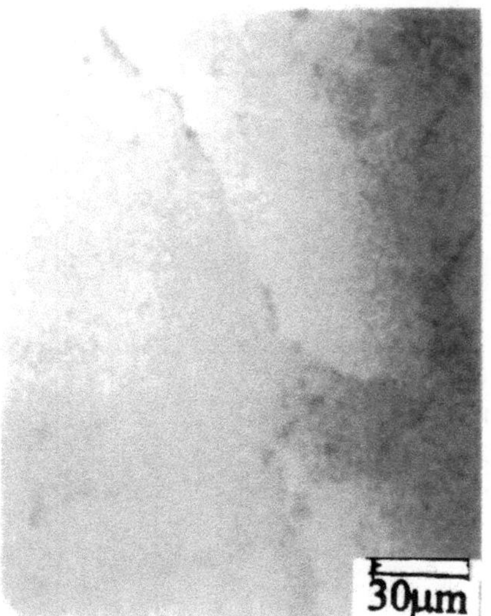

photo 3 face distribution of Iron in the cobalt pyrite

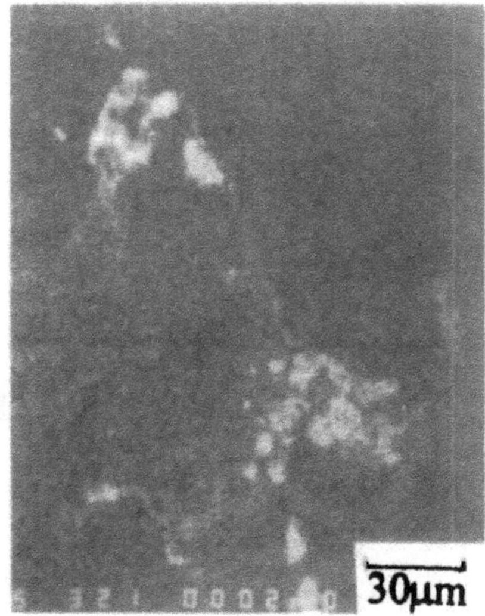

photo 4 face distribution of cobal in the cobalt pyrite

Independent cobalt had not been found in the samples that cobalt content is more than 0.15% by means of many works of SEM and EPMA. According to the research of synthetic pyrite, Co^{2+}, Ni^{2+} occupy the lattice position of Fe^{2+}, the content of cobalt is relatively homogeneous in the pyrites under same physicochemical condition, and is not homogeneoue under different physicochemical condition. So we say that in pyrites of different stages of skarn deposition cobalt and Iron replacement in ismorphously may be very heterogeneous because of polystage mineralization.

Technology type and character of the ores

Table 4 Shows the Technological type of ore

Table 4 Technological types of the ores

type	number	ordinal	name of secondary	associated samples	note
		1	pyrite-magnetite massive ore	-100-7-6, -56-1-2	pyrite ≥5%
		2	hematitized pyrite magnetite - bearing massive ore	-100-3-1, -100-7-2 -100-7-4, -100-7-5 -100-5-5, -100-1-1 -100-4-2, -56-4-4	pyrite <5% Hematite>3%
Mt≥50% magnetite-rich ore	I	3	pyrite-magnetite impregnated masive ore	-100-5-2, -100-5-3 -100-5-6, -100-6-6 -100-5-2, -56-1-3	pyrite ≥5%
		4	pyrite-bearing magnetite impregnated massive ore	-100-6-2, -100-6-4 -100-6-3, -100-6-5	pyrite <5%
		5	Chalcopyrite- bearing pyrite magnetite densely impregnated ore	-100-4-3, -100-4-5 -100-4-4, -100-4-6 -100-5-4, -100-6-7	pyrite ≥5% chalcopyrite≥0 5 -1%
Mt<50% magnetite ore	II	6	pyrite-bearing magnetcte impregnated ore	-100-4-2, -100-5-3 -56-5-4,	pyrite<5%
		7	chalcopyite- bearring hematitized pyrite-magnetite impregnated ore	-100-3-3, -100-7-7 -100-7-8, -56-5-1	pyrite≥5% chalcopyrite ≥1% hematite>5%
sulphide ore	III	8	magnetrte- bearing pyrite disseminated ore	-100-4-7, -100-7-1 -100-4-6, -100-5-1	magnetite<5% pyrite5-10%

chemical anlysis, mineral content of first and second kind in the type I and Type II

Chemical composition and mineral content shows table 5,6

Table 5 Chemical analysis of associated samples of the ores

ore type	TFe	SFe	FeO	SiO_2	P_2O_5	Cu	Co	S	CaO	MgO	Al_2O_3
1	51.09	44.78	20.63	5.45	0.69	0.25	0.047	8.86	3.94	2.18	1.88
2	52.35	50.95	20.90	8.61	0.62	0.08	0.017	2.44	8.22	2.12	3.36
II	44.81	41.90	18.80	9.78	0.13	0.12	0.029	4.26	11.18	1.81	3.42

Note, FeO amount doesn't include Iron from sulphide.

Table 6 Mineral composition and mineral content of the ores

ore type	magnetite	pyrite	hematite	chalcopyrite	sphalerite	chlorite	calsite	apatite	quartz	garnet	epidote
1	49.85	16.17	3.12	0.55	0.45	14.38	7.25	1.72	1.55	-	4.94
2	54.20	4.37	8.41	0.23	-	13.59	3.607	1.57	0.65	11.13	2.17
II	48.27	8.196	2.912	0.219	-	8.37	12.65	0.131	0.55	13.34	5.36

Note: Hematite includes pseudomorphic hematite, specularite and goethite, limonite etc. chalcopyrite includes bornite etc

Table 7 Balonce calculation of Iron

name of mineral		magnetite	pyrite	hematite	chalcopyrite	chlorite	epidote	total
Iron Content		72.4	46.69*	69.94	30.52	21.33*	13.73	
Content of mineral	1	49.85	16.19	3.12	0.55	14.38	4.94	
	2	54.20	4.37	8.41	0.23	13.59	2.17	
	II	48.27	8.196	2.912	0.219	8.37	5.36	
distribution of Iron	1	36.09	7.56	2.18	0.17	3.07	0.68	49.75
	2	39.096	2.04	5.88	0.07	2.90	0.30	50.286
	II	34.95	3.83	2.04	0.07	1.78	0.74	43.41
distribution ofIron	1	72.54	15.20	4.38	0.34	6.17	1.37	100.00
	2	77.75	4.06	11.69	0.14	5.77	0.59	100.00
	II	80.51	8.82	4.70	0.16	4.10	1.71	100.00
distrilution of	1	72.54	87.74	92.12	92.46	98.63	100.00	
accumulator	2	77.75	81.81	93.50	93.64	99.41	100.00	
	II	80.51	89.33	94.03	94.19	98.29	100.00	
balamce coefficient	1	chemical analysix TFe=51.09% K=49.75/51.09 =0.974						
	2	chemical analysix TFe=52.35% K=50.286/52.35 =0.960						
	II	chemical analysix TFe=44.81% K=43.416/44.81 =0.969						

Note: the Iron content in pyrite and sphalerite the value of eletron probe analysis, the others are theoretical value

The table 5,6 indecate that there are notable difference between different types of ores for their element contente of Fe, S, Co and Cu, and the benefical recovery composition enrichment have been done through the division of tochnological types.

The balance calculation of element Fe.S shows table 7,8,9.

Table 8 Balance culculation of sulphur

Ore type	name of mineral	Sulphur content	content of mineral	distribution of sulphur	distribution ratio	accumulator	balance coefficient
I	pyrite	53.45	16.19	8.65	96.22	96.22	chemical analysis value 8.86%
	chalcopyrite	34.92	0.55	0.19	2.11	98.33	K=8.99/8.86 =1.015
	sphalorite	32.60	0.45	0.15	1.67	100.00	
	total			8.99	100.00		
2	pyrite	53.45	4.37	2.34	96.69	96.69	
	chalcopyrite	34.92	0.23	0.08	3.31	100.00	K=2.42/2.44 =0.992
	total			2.42	100.00		
II	pyrite	53.45	8.196	4 38	98.43	98.43	K=4.45/4.26 =1.044
	chalcopyrite	34.92	0.219	0 07	1.57	100.00	
	total			4 45	100.00		

Table 9 Theoretical index of Iron in the ore

ore type	name of manerals	out put	grade	recovery	loss
I	magnetite	49.85	72.4	72.54	27.46
	hematite	3.12	69.94	76.92	23.08
	pyrite	16.19	46.69	92.12	7.88
	chalcopyrite	0.55	30.52	92.46	7.54
2	magnetite	54.20	72.4	77.75	22.25
	hematite	8.41	69.94	89.44	10.56
	pyrite	4.37	46.69	93.50	6.50
	chalcopyrite	0.23	30.50	93.64	6.36
II	magnetite	48.27	72.4	80.51	19.49
	hematite	2.912	69.94	85.21	14.79
	pyrite	8.196	46.69	94.03	5.97
	chalcopyrite	0.219	30.52	94.19	5.81

The occupancy of Iron are 76.92%, 89.44% and 85.21% separately in the oxides of magnetite etc of the first and second kind of type I and type II ; in sulphides the occupancy of Iron are 15.54% 4.2% and 8.9% separately. All recovery of magnetite is the theoretical recovery of Iron which are 72.54% 77.75% and 80.51% respectively. Hematite etc. can not be recovered. The occupancy of Iron in the sillicate minerals are 7.54% , 6.36% 5.81% respectively, and tyey are theoretical grade of Iron in the tailings .

grain size character of exploitable minerales

The statistical measurement of dissemination grain size of exploitable minerals in two types of ore with linear method has been done (table 10). In ores of type I magnetites are mainly medium grain[7]of which grain size 0.074~1.168mm is 88.24%and 0.589~0.147mm is up to 61.61%. In ores of type 2 magnetite finer than type 1, of which 0.074~1.168mm is up to 82.91%, 0.037~1.168mm is up to 94.84%. The grain size of pyrite distribuite between 0.074~1.168mm is up to 86.64%and 91.05%. Grain size of chalcopyrite which +100 mesh in type 1 and type 2 are 45.36%and 22.02% seperately, +200 mesh are 58.76% and 42.71%. Under present level of ground product,-200 mesh amount up to 60%, Grinding level of magnetite is lower, grinding level of pyrite is reasonable.

Table 10 Dissemination grainness character of the exploitable minerales

mesh	size grade (mm)	type	magnetite		pyrite		hematite		chalcopyrite	
			distribution ratio%	accumu -lator%	distribution ratio%	accumu -lator%	distribution ratio%	accumu -lator%	distribution ratio%	accumu -lator%
+14	>1.168	I	5.00	5.00						
		II	3.41							
-14~	1 168~	I	10.72	15.72	17.90	17.90			24.74	24.74
+28	0.589	II	12.41	15.82	11.83	11.83				
-28~	0.589~	I	29.83	45.55	25.22	43.12			8.25	32 99
+48	0.295	II	20.92	36.74	32.87	44.7	13.11	13.11		
-48~	0.295~	I	31.79	77.34	21.15	64.27	13.54	13.54	12.37	45.36
+100	0.147	II	32.66	69.4	30.90	75.6	32.79	45.90	22.02	22 02
-100~	0.147~	I	15.90	93.24	22.37	86.64	29.80	43.34	13.40	58.76
+200	0.074	II	16.92	86.32	15.45	91.05	29.51	75.41	25 69	47.71
-200~	0.074~	I	6.25	99.49	11.59	98.22	36.34	79.68	36.08	94.84
+400	0.037	II	12.00	98.32	8.38	99.43	21.31	96.72	42.20	89 91
-400	<0.037	I	0.51	100.00	1.78	100.00	20.32	100.00	5.15	100.00
		II	1.69	100.00	0.57	100.00	3.28	100.00	10.09	100.00
+200		I	93.24		86.64		43.34		58.76	
mesh content	>0.074	II	86.32		91.05		75.41		47.71	

technological mineralogy analysis of final concentrate and tailing products.

technological mineralogy analysis of iron concentrate

Table 11 Indices of Iron in the final Iron concentrates and tailings

final products	TFe_2O_3	SFe_2O_3	FeO	Fe_2O_3
Iron concentrate	88.90%	88.27%	25.47%	60.60%
tailing	23.27%	21.53%	4.95%	17.78%

Table 12 mineral constitue and balance calculation of iron in the iron concentrate and tailing

name of mineral	content of mineral %		Iron % content	distribution of Iron %		distribution ratio %		distribution of accumulator %	
	C	T		C	T	C	T	C	T
magnetite	83.22	5.42	72.4	60.25	3.92	94.75	26.87	94.75	26.87
hematite	2.39	4.21	69.94	1.672	2.947	2.63	20.20	97.38	47.07
pyrite	0.52	5.22	46.69	0.238	2.42	0.37	16.59	97.75	63.66
chalcopyrite		0.41	30.52		0.12		0.82		64.48
chlorite	5.18	9.47	21.33	1.105	4.13	1.74	28.31	99.49	92.79
epidote	2.37	7.50	13.73	0.325	1.05	0.51	7.21	100.00	100.00
garnet	1.35	40.92							
calsite	2.85	12.46							
quartz	2.12	4.40							
total	100.00	100.00		63 59	14 587	100 00			
balance coefficient	Kc=63.59/62.17=1.023 Kt=14.587/16 27=0 8970								

The total content of iron in the concentrated ore is 62.17% which is much lower than the normal reguired minimum of 65%, and the TFe value in the tailings is 16.27% higher than the normally required one which is 10-11%. Table 12 is about the mineral composition and mineral content and balance calculation of iron in the concentrates and tailings, magnetites is 83.22% of Iron concentrate, occupancy of iron is 94.75%. Table 13 is the degree of liberation and character of grain size distribution of exploitable minerals in the concentrated ore and tailings, the degree of liberation of magnetite in the iron concentrates is 70.81%. The magnetite is 5.42% in the tailings, the independent magnetite crystal in the tailings is 86.68%, distributes between 0.295~`0.037mm up to 92.48%, +200 mesh up to 62.64% of all independent magnetites. The iron grade of iron concentrates is lower, because: 1. the degree of liberation of magnetite is lower (the normally value is more than 80%); 2. about 13.67% gangue minerals is left in the concentrated ore, they are higher; 3, there are 5.42% magnetite crystals in the tailings.

Table 13 Liberation character of the magnetite in iron concentrate

size grads mesh (mm)	liberation distrbution			Locked 3/4			1/2			1/4		
	content %	ratio %	accumulator %	content %	ratio %	accumauator %	content %	ratio %	accumauator %	content %	ratio %	accumulator %
+4.8 >0.295	0.38	0.53	0.53									
-48~+65 0.295~0.208	1.07	1.51	2.04	0.60	2.69	0.53	9.86	9.06	9.06			
-65~+100 0.208~0.147	2.64	3.72	5.76	3.11	13.92	16.61	1.04	17.61	26.67	0.52	49.37	49.37
-100~+150 0.147~0.104	8.66	12.23	17.99	5.59	25.07	41.68	1.20	20.38	47.05	0.16	15.87	65.24
-150~+200 0.104~0.074	16.67	23.54	41.53	5.30	23.74	65.42	1.60	27.21	74.26	0.16	15.71	80.95
-200~+400 0.074~0.037	25.84	36.50	78.03	7.14	32.01	97.43	1.33	22.64	96.90	0.20	19.05	100.00
-400 <0.037	15.55	21.97	100.00	0.57	2.57	100.00	0.18	3.10	100.00			
total %	70.81	100.00		22.31	100.00		5.88	100.0		1.04	100.0	

technological mineralogy analysis of sulphide concentrate

Table 14 Mineral composition and amount of sulphide concentrate

mineral composition	pyrite	magnetite	hematite	chalcopyrite	garnet	chlorite	epidote	calsite	quartz	total
content %	78.52	12.34	1.53	0.54	0.47	3.00	1.23	1.53	0.93	100.00

The degree of liberation of pyrite in the sulphide concentration is 92.2% , they are good , the content of magnetite in sulphide concentrate is higher which is up to 12.34%, they are mainly single magnetite crystal, so the seperated process may need to be improved. But chalcopyrites content is low , so the process of depressing sulphur and floating copper should be thinked with ecconomy efficiency.

complex utilization of cobalt

The content of cobalt is 0.154% in sulphide concentraete, but the pyrite crystals can be up to 95% in the concentrated pyrites ores after simple floating seperation the content of cobalt is up to 0.189%, these cobalt have not been recovered yet. In fact the process of recovering cobalt by roasting sulphated cobaltiferous pyrite is mature and the changing rate of cobalt is 70-80%(fig2. the flow sheet of roasting sulphuted of cobalt - bearing pyrite). So the complx utilization of cobalt in sulphide concentrates of Weigang mine is practical. According to the present production the 27ton cobalt metal can be extracted per year from 25 thousand ton pyrite concentrates of which content of cobalt is 0.154%, the changing rate of cobalt is 70%. That is to say that $ 665, 400 is got per year. It is suggested that using the existing facilities of the sulphuric acid plant of Weigang mine should get certain economic benefit.

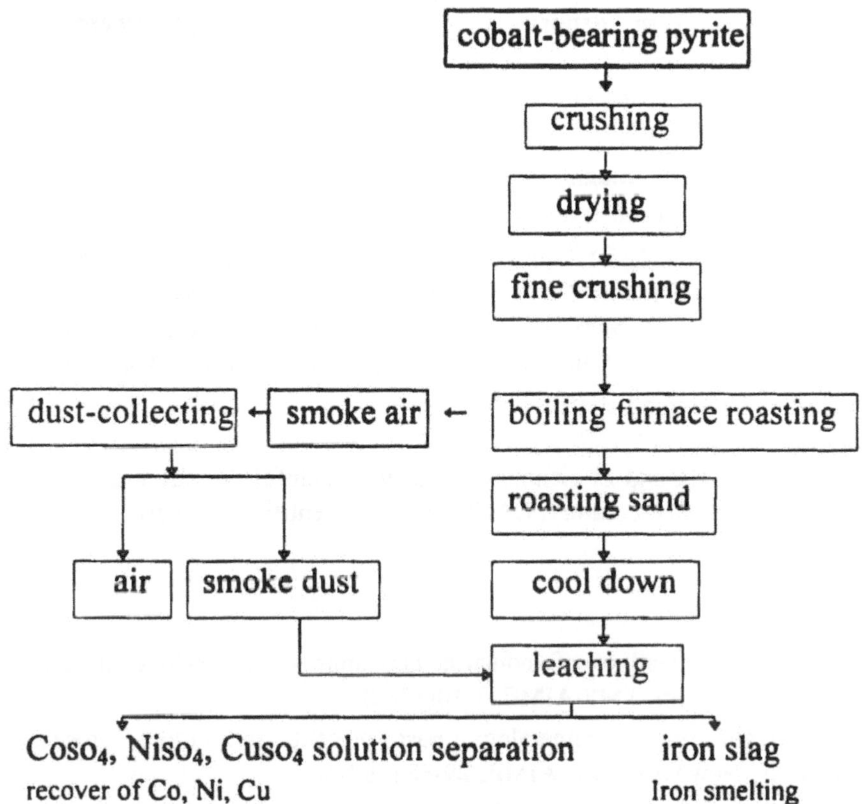

Fig.2. The flow sheet of roasting sulphuted of cobalt pyrite[8]

CONCLUSIONS

The primary ores of Weigang iron mine are classified into three categories and eight technology types in the light of content of major exploitable minerals (magnetite, pyrite, chalcopyrite and hematite etc). There are notable difference between different type of ores for their elemental content of Fe, S, Co,Cu and also a notable difference of their state of grain inter-locking and grain-size distritubion characteristice. S, CO, Cu is up to 8.86%, 0.25% and 0.047%.

Magnetite ore of the category one is medium coarse grain, the category two is mainly medium grain. The results provided a fundamental information for the techinical improvement of ore mining and ore dressing.

The existing state of cobalt in the pyrite was not completely clear. This study suggests that the occurrence form of cobalt in the pyrites is mainly in isomorphous. The content of cobalt is higher in the early pyrite and even the distribution is not homogeneous in the same grain, We say that in the pyrites of different stages of skarn deposition cobalt and iron replacement in ismorphously may be very heterogeneous during the polystage mineralization of the skarn-type iron deposit . In order to recover cobalt, cobalt-bearing pyrite is our main target.

Degree of liberation of the magnetite in the iron concentrate is only 70.81%, but the independent magnetite crystals in the tailings is higher, so we suggest that -200 mesh should be up to 70% to inprove magnetite separation process, and reduce the lost of independent magnetite crystals and rich locked particle.

The content of cobalt is 0.154% in sulphide concentrate it is suggested that using the existing faclities of the sulphuric acid plant of WEIGANG mine should get certain economic benefit.

Acknowledgemeute

We thank prof. Ren runfu and Li qianmaɔ etc for their helpful. We are grateful to yu jing etc of electron prob laboratory, CAGS, for their identification work.

References

1. G. C. Amstutz. Recollections of industrial appliation of mineralogy, process mineralogy II New york TMS-AIME, 3-10(1982).
2. Richard & Hagni. process mineralogy: past, present, and Future. process mineralogy II New york TMS-AIME, 29-38 (1982).
3. Zou Jian, review and outlook on the advancement of mineral proceesing technology in China's metallurgic mines, China mining magajine, vol 1.No.2 74 ~76 (1992) .
4. The exploration prospecting report of Weigong Iron mine, the third team of bureau of geology and mineral exploration, MMI (1974).
5. Wang pu etc, systematic mineralogy, Geological publishing house, 342-343 (1985).
6. G. A. Klotove. Cobalt deposite, China Industrial press, 103,109 (1965).
7. Xu guofeng. ore microscopy, geological publishing house, 166-168 (1986)
8. Quan hongdong, chemical treatment of mineral, metallurgic industrial press, 104(1984).

Proc. 30th Int'l. Geol. Congr., Vol. 16, pp. 233-242
Huang Yunhui and Cao Yawen (Eds)

CISMMI: a Computer Information System for Microscopic Mineral Identification

M.M. BOLDYREVA,[1] V.K. PETROV,[2] and B.N. POPOV[2]

[1] Dept. of Geology, St.Petersburg State University, St.Petersburg, 199034, Russia
[2] State Optical Institute, St.Petersburg, Russia

Abstract

CISMMI is a computer system for identification of ore minerals in a polished section using spectral reflectance curves in visible spectrum from 400–700 nm with some additional qualitative properties of minerals which can be observed with polarized microscope. The system comprises a series of programs for IBM compatible personal computer. The measured and calculated values are compared with those in a data file that contains data on over 640 minerals. The determination parameters are as follows: calculated reflectance value and degree of anisotropic rotation at 550 nm and coefficient of similarity of the curves. Every next parameter reduces the number of minerals to a limited set of most similar ones. In the output final list, the minerals are ordered by descending of their similarity to the tested mineral. For every mineral from the list one can call addition information on chemical formula and admixtures, symmetry, ranges of Vickers hardness number, internal reflexes, the colour values, degree of anisotropic rotation and mineral association. More then 300 samples of 58 minerals were tested and approximately in 90% cases the right name of the mineral was among the first three names of the result list, and in 50% cases the right name was in the top.

Keywords: computer, identification, ore minerals, visible spectrum, spectral reflectance curves

PREFACE

CISMMI (a Computer Information System for Microscopic Mineral Identification) is a computer supported system for ore mineral identification in polished sections which uses reflectance spectra in visible light. These spectra are distinctive and very informative features of minerals, and they provide a useful and objective criterion for mineral identification. In the process of computer identification, a measured spectral curve of an unknown mineral is compared with the set of pattern curves collected in the data file.

Some computer supported systems for microscopic identification of ore minerals have been offered: NISOMI [1], PYRITE [5], POLYFIT and POLYORT [4], DIAG [3], MOMI [2]. Quantitative optical characteristics are the definitive criteria in all of these systems but the similarity of spectral reflectance curve is a main criteria in the three last systems. CISMMI is the most recent and elaborate implementation of this concept. It was developed in the State Optical Institute (GOI, St.Petersburg) as modification of DIAG system. Since 1992, CISMMI has been used in the Ore Microscopy Course at the St.Petersburg State University.

DATABASE SYSTEM

CISMMI program package contains several blocks: database system, diagnostic programs, block of colour characteristics analysis and utilities (menu-driven procedures to change database, to store new reflectance data, to print and plot data, etc.).

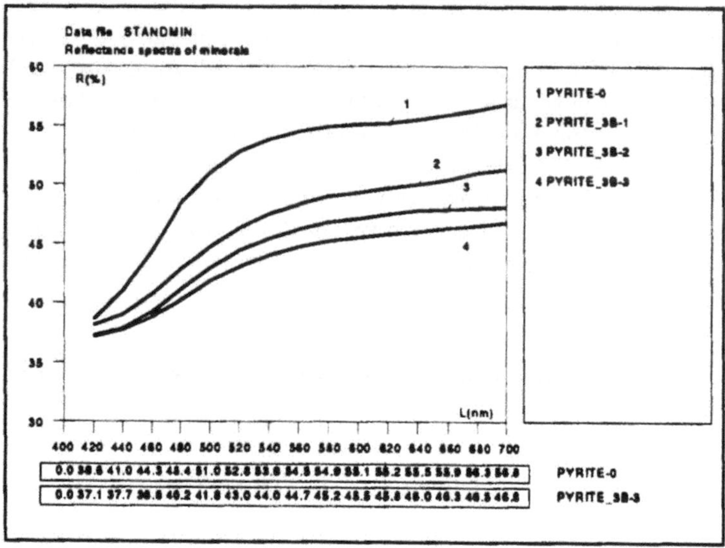

Figure 1. Reflectance values and spectral reflectance curves of Pyrite and its Sb-varietes. (**1** – 0%, **2** – 2.0%, **3** – 11.9%, **4** – 17.0%.)

Database system contains characteristics of more than 640 isotropic and anisotropic minerals. First of all, there is the information on reflectance values measured at interval of 20 nm in the visible spectrum from 400 to 700 nm for each mineral and its varieties and the spectral reflectance curves based on them (Fig. 1). Simultaneously the reflectance spectra no more than 10 samples can be shown at the monitor. Such optical properties as reflectance and degree of anisotropic rotation at 550 nm are calculated. Block of colour characteristics analysis determines the chromaticity diagram for each mineral using reflectance spectra and calculates colour coordinates (x, y), luminance (Y,%), a spectral purity (P, %), the dominant wavelength (L,nm) for A, B, C-sources of light (Fig. 2). The data file with additional information (**Fig.** 3) contains some general data: symmetry, internal reflexes, ranges of Vickers hardness number (VHN), chemical formula, chemical admixtures and mineral association.

	A	B	C
X	60.14	53.16	51.90
Y	54.57	54.22	54.06
Z	16.43	38.73	53.33
m	131.1	146.1	159.3
x	0.459	0.364	0.326
y	0.416	0.371	0.339
z	0.125	0.265	0.335
L,nm	582	575	572
P,%	14.89	16.21	16.48

Figure 2. Calculated colour values of Pyrite-0 (data file STANDMIN) for **A**, **B**, **C**-source of light. Here Y,% — luminance, P,% — spectral purity, L,nm — dominant wavelength, x, y — colour coordinates.

Additional information

Name	Pyrite-Sb-3	Pyrite-0
Symmetry	Cubic	Pyrite-Sb-1
Internal reflexes	--	Pyrite-Sb-2
VHN	0415-0925	Pyrite-Sb-3 <--
Chemical formula	$Fe(S,Sb)_2$	
Chemical admixtures	Sb(17.3%), As(0.16%), Cu(0.90%)	
Associate minerals	antimonite, berthierite, chalcopyrite,ullmannite, antimony	

Figure 3. Additional information on Sb-pyrite, data file STANDMIN.

The database is stored in two files — STANDMIN ("Standard" mineral file), which contains above mentioned data on 472 mostspread minerals [3], and SHUMIN with reflectance spectra and colour data on 170 rare minerals [5]. Large set of reflectance spectra for each mineral from the database were analyzed, and individual variation limits were determined. The pattern reflectance spectra and the limits are stored in the database. Data can be selected and transferred to any new user created files. System is opened for updating reflectance data on new minerals. It is also possible to create new reflectance data files using samples of interest (Fig. 4) with inserting new data of additional information for each mineral.

Data file : BISMUTH	(continued)	Data file : FAHLORE
Aikinite-167	Galenobismutite-146	Fahlore-1
Berryite-159	Galenobismutite-348	Fahlore-10
Berryite-302	Galenobismutite-358	Fahlore-11
Bi-Fahlore	Galenobismutite-361	Fahlore-12
Bismuth-229	Gladite-236	Fahlore-13
Bismuth-231	Goongarite-148	Fahlore-14
Bismuthinite-143	Ikunolite-172	Fahlore-15
Bismuthinite-238	Joseite-A-309	Fahlore-2
Bismuthinite-239	Joseite-B-331	Fahlore-3
Bismuthinite-243	Krupkaite-284	Fahlore-4
Bismuthinite-355	Lindstromite-234	Fahlore-5
Cosalite-145	Matildite-155	Fahlore-6
Cosalite-350	Matildite-MSF-10	Fahlore-7
Cuprobismutite-270	Pavonite-282	Fahlore-8
Dognachkaite-323	Shirmerite-285	Fahlore-9
Emplectite-241	Shirmerite-286	
Emplectite-244	Wittichenite-234	
Emplectite-275		

Figure 4. Newcreated files of spectral data for minerals of interest.

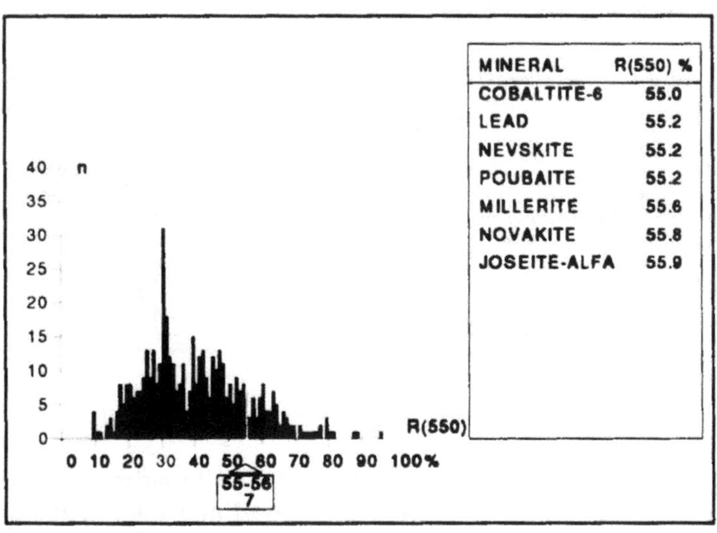

Figure 5. Histo R-mineral distribution for reflectance value at wavelength of 550 nm for data file STANDMIN. The marker beneath the scale shows the inverval of reflectance value and the number of minerals.

For every data file, the histogram of reflectance value of minerals at wavelength of 550 nm is presented, and the user can see the number of minerals in the required range of R and the list of these minerals (Fig. 5).

In design of the database, we followed the existing world practice and experience and collected an extensive information on properties of ore minerals. Currently, the database is in correction, and new data files and additional information on rare minerals (file SHUMIN) are developed at our University.

DIAGNOSTIC PROGRAMS

Diagnostic programs contain procedures for mineral identification by comparison of reflectance spectra of tested and reference minerals at 400–700 nm. Measured reflectance values with intervals of 20 nm for the tested mineral are transferred to computer from keyboard (Fig. 6). During computer identification, the similarities between measured and stored reflectance curves can be determined. The determinative parameters for comparison are: calculated reflectance at 550 nm, degree of anizotropic rotation ($A\% = (R1 - R2)/R1$), and coefficient of similarity of the curves defined as $\delta = \sum_1^{m-1}(R_i'1 - R_i'2)/(m-1)$, where $R_i'1$ and $R_i'2$ are first derivatives of compared curves, and m is the number of measured points in the spectra.

NAME OF MINERAL :
M-1

R1(550) = 23.95

Remark:
for isotropic minerals
assignment is used
R2 = 0 .

L(nm)	R1(%)	R2(%)
400	0.0	0.0
420	0.0	0.0
440	19.1	0.0
460	20.1	0.0
480	20.9	0.0
500	21.9	0.0
520	22.7	0.0
540	23.6	0.0
560	24.3	0.0
580	25.1	0.0
600	25.8	0.0
620	26.3	0.0
640	26.9	0.0
660	27.4	0.0
680	27.8	0.0
700	28.3	0.0

Figure 6. Updating measured reflectance value of unknown mineral M-1 from the keyboard. The shown at left reflectance value R1 was calculated by the program for the wavelength 550 nm.

Every next parameter reduces the number of minerals to a limited set of most similar ones. As a result, reflectance curves of the examined mineral and the closest ones and their names are displayed on the screen (Fig. 7). To facilitate visual comparison of spectral curves in cases when many minerals are proposed for identification and plot is not legible the user can reversibly remove some curves from the plot. Graphic

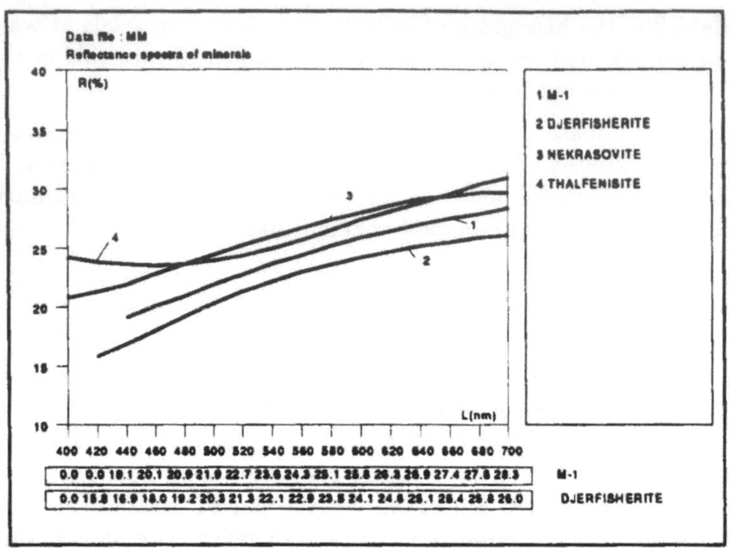

Figure 7. Graphic presentation of results of identification of unknown mineral M-1.

Protocol of identification

Sample M-1
R1(550) = 28.9 %

List of minerals

N	Data file	Mineral	Class	VHN
1	STANDMIN	DJERFISHERITE	1	0155-0193
2	STANDMIN	NEKRASOVITE	1	0260-0285
3	STANDMIN	THALFENISITE	4	0146-0149

Figure 8. The log of identification: the minerals are ordered by descending of their similarity to the tested mineral M-1. Taking into account its microhardness, it was identified to be Djerfisherite.

presentation is followed by a log of identification (Fig. 8). In the output final list, the minerals are ordered by descending their similarity to the tested mineral. For each mineral, it contains its ordinal number, name of the file, name of the mineral, class of similarity (0 means the highest similarity), and microhardness. For anisotropic minerals R1 spectra are taken into account. If in such cases R2 spectra correspond to the same class or better asterisk is used.

EFFICIENCY

Efficiency of CISMMI was determined by the use of data for minerals that were treated as unknowns. More than 300 samples of 58 minerals were tested, and approximately in 90% cases the right name of mineral was among the first three names

Table 1. Test of bismuth minerals from collection

N	Name of mineral	N species	N samples	Position in the list 1	2	3	4
1	Bismuthinite*	2	8	3	2		
2	Galenobismutite	2	17	14	3		
3	Cosalite	1	7	6	1		
4	Goongarite	1	4	4			
5	Matildite	1	4	2	1	1	
6	Berryite	2	5		1	3	1
7	Aikinite	1	6	4	2		
8	Ikunolite	1	6	5		1	
9	Bismuth native	1	2	2			
10	Wittichenite	1	1	1			
11	Lindstromite	1	2	2			
12	Gladite	1	2	2			
13	Emplectite	1	4	1	3		
14	Dognachkaite	1	4	4			
15	Cuprobismutite	1	3	1	2		
16	Shirmerite	1	1	1			
17	Joseite-A	1	5	5			
18	Joseite-B*	1	7	4			

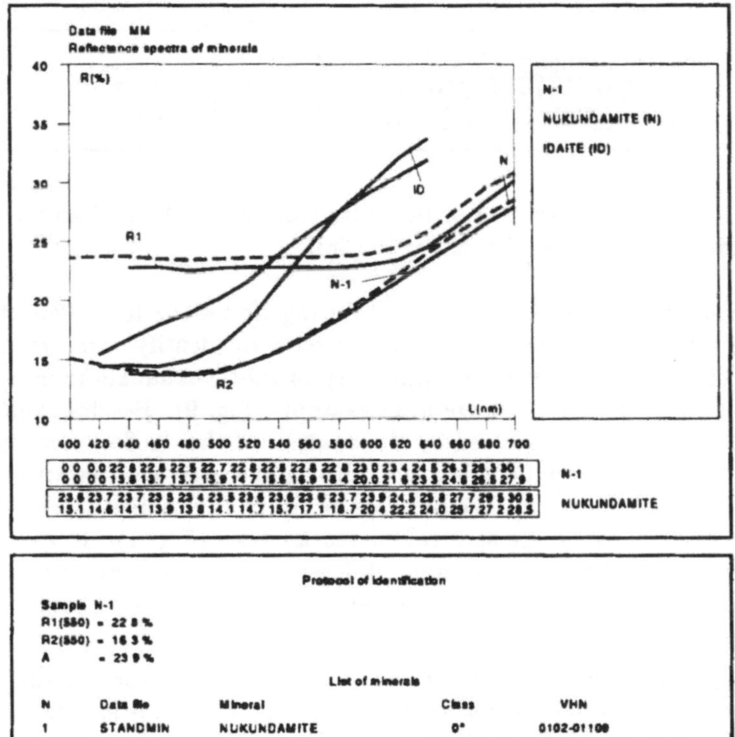

Figure 9. Identification of a rare mineral Nucundamite in sulphide deposits Mir, Mid-Atlantic Ridge, 26° N. The spectra of Idaite are shown for comparison.

of the result list and in 50% cases the right name was in the top.

Test for 88 samples of 18 bismuth minerals from the St.Petersburg State University collection are shown in Table 1. 60 samples (68%) were attributed to the top in the list, and 81 samples (90%) were attributed to the first three positions. 6 samples were not attributed to any positions, due to minimal anizotropic rotation of mineral grains.

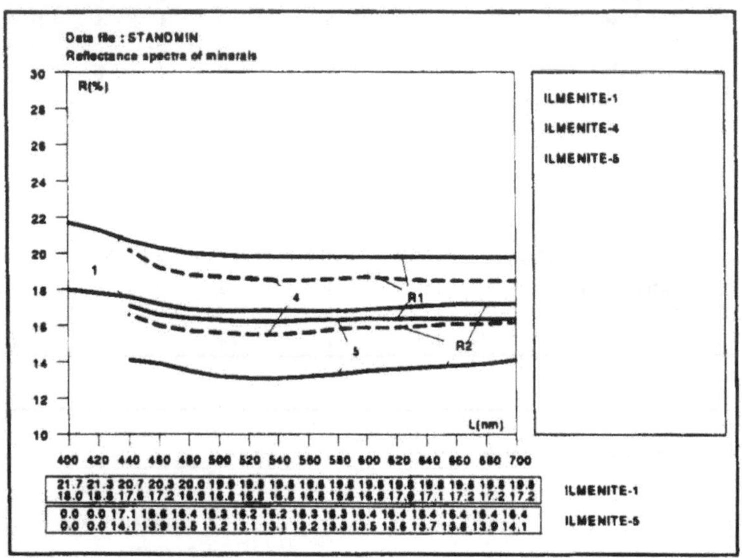

Figure 10. Reflectance spectra of Ilmenites from kimberlites (4 and 5) — dependance of MgO, (**1** – 0%, **4** – 5.6%, **5** – 16.6%).

New applications of CISMMI were found during its testing in St.Petersburg State University. First of all, the system allows ones to identify rare ore minerals at the stage of their optical research. Discovery of rare Nukundamite in the sulphide deposits of the Mid-Atlantic Ridge is an example (Fig. 9). Besides, composition of some minerals can be determined by the analysis of their reflectance spectra. For instance, MgO-rich Ilmenites (Fig. 10) from kimberlites are characteristic of the diamond-bearing kimberlite. The presence of Sb in Pyrite also can be determined with high probability by the spectral curves analysis (Fig. 1). It was found that CISMMI allows ones to detect a minor difference in the spectral curves of Fahlore (Fig. 11). Relations between spectral reflectance and Fahlore composition were determined (Fig. 12).

Considering our experience with CISMMI application, we suggest that success of the mineral identification depends on the reliability of diagnostic parameters, size of database, and precision of stored information.

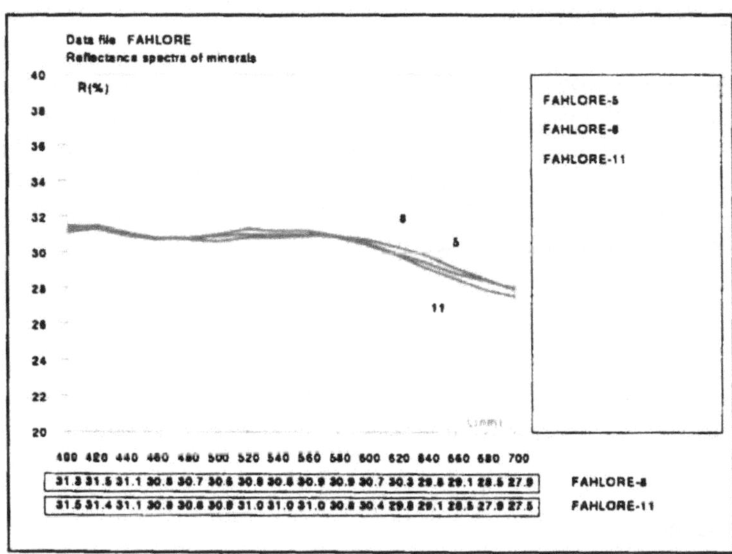

Figure 11. The most similar reflectance spectra of Fahlore achieved by the CISMMI system.

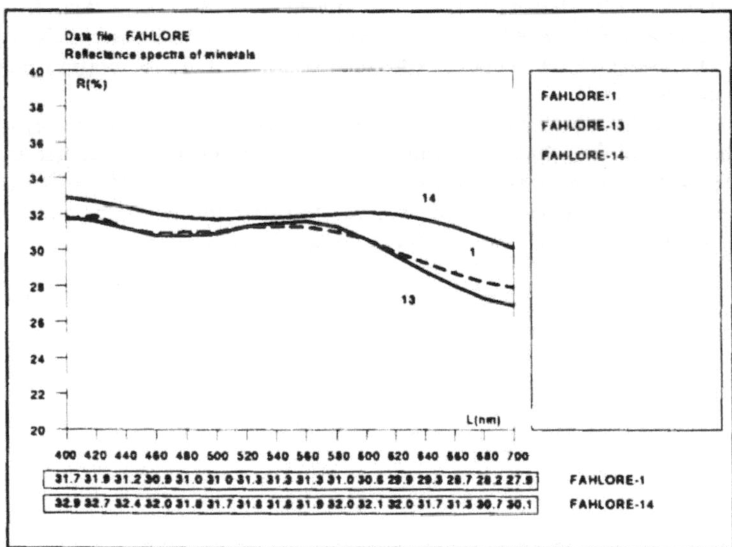

Figure 12. Influence of Bi on reflectance spectra of Fahlore, (**1** – 1.3%, **13** – 0.0%, **14** – 16.9%).

COMPUTER SYSTEM

The CISMMI system consist of a number of subroutine structured computer programs and associated data files for IBM compatible personal computer with EGA, VGA type of monitor and operation system MS DOS 3.30 or higher. The System is programmed in TURBO PASCAL. It requires 60–100 Kbytes per a data file with 400–500 reflectance curves. CISMMI is easy to operate. It is called by its name and starts by file `bankdir.exe` or `cismmi.bat`.

Acknowlegements

The authors express their gratitude to Joseph Romanovsky and Natalia Ovchinnikova for their help in preparation of the final version of the paper.

REFERENCES

1. B.P. Atkin and P.K. Harvey. Nottingam interactive system for opaque mineral identification: NISOMI. Transactions of Instn. of Mining and Metallurgy. **88**. 1325–1327 (1979).
2. H.-J. Bernhardt. MOMI, Computer-supported System for Microscopic Identification of ore Minerals by means of qualitative and quantitative properties. The 15th General Meeting of the IMA, 28 June-3 July, 1990, China. Abstracts. **1**, 240–242 (1990).
3. T.N. Chvileva, M.S. Bessmertnaya, E.M. Spiridonov, A.S. Agroskin, G.V. Papaian, P.A. Vinogradova, S.I. Lebedeva, E.N. Zavialova, A.A. Filimonova, V.K. Petrov, L.R. Rautian, O.L. Sveshnikova. Handbook for identification of ore minerals in reflected light. Nedra, Moskow (1988).
4. V. Kovachev, S. Strashimirov and D. Damianov. Testing of Polyfit and Polyort programs for development of data bank of standard reflection spectra of ore minerals. Annual of the Higher Inst. of Mining and Geology (1986–1987). Sofia. **33** (II), 41–48 (1987).
5. N.I. Shumskaya, L.I. Iliana, I.T. Milovsorova. A system for identification of ore minerals by means of crystal optical data. Nedra, Leningrad (1979).

Proc. 30th Int'l. Geol. Congr., Vol. 16, pp. 243-254
Huang Yunhui and Cao Yawen (Eds)

Computer Graphics of Crystal Shape

FENNA ZHOU
Editorial Department of Journal , Wuhan University of Hydraulic and Electric Engineering , Wuhan 430072 , P . R . China
WENGUI WANG
Faculty of Earth Sciences , China University of Geosciences , Wuhan 430074 ,P. R .China
LIPING CHEN
CAD Center, Huazhong University of Science and Technology , Wuhan 430074. P.R.China

Abstract

According to the principle of V . Goldschmidt's gnomonic projection and computer graphics , a new system of computer graphics of crystal shape is presented in the paper . By getting the data from the database of directly inputting information of geometry and topology of crystal shape , the figure can be displayed on the monitor or draw by plotter . The system can draw not only the ideal shape of crystal shape , but also the practical shape of crystal , quasi-crystal , and operate the figure with the geometrical and projective transformation .

Keywords: crystal shape , gnomonic projection , computer graphics

INTRODUCTION

There are many manual methods to draw the crystal shape which are crystal axis framework , gnomonic projection and so on . Although the gnomonic projection is more accurate than the others and not complex , it is sorry for the manual manner to spend two much time and energy of the operator . Actually , it is very difficult to get an exact figure for complex twin crystal with the manual methods . By the development of computer technologies , many drawing tools based on computer have been proposed . Dowth E . in U .S .A . has developed a system called SHAPE; Nakamuta Y . has presented his CRYSTAL UTILITY in Japan ; Prof . Shen J . C . has developed "3D color display system of crystal shape" in China . Pointing to ideal crystal , all these above is related to the solution of the crystal plane . Unfortunately , in some situation it is difficult to get the important parameter——the distance between center and plane . Utilizing the polar coordinates and the coordinates of the vertices on the crystal top view from the crystal measuring , the method in this paper can draw crystal shape with gnomonic projection on computer . Because of its simplicity , this method is suitable to get the crystal figure of the practical crystal even before the kind of the crystal is recognized . Avoiding the great computation of the crystal plane equation , the process only needs computing the 2D coordinates of the points .

PRINCIPLE OF CRYSTAL SHAPE GRAPHICS BASED ON GNOMONIC PROJECTION

The top view of crystal shape is an important element in crystal shape graphics .It is the vertical view of crystal. Here is the Dicar coordinate system shown in figure 1. There is the relation that the original point of crystal axis coordinates coincides with the one of the Dicar coordinate system and the directions of the crystal axises should be close respectively to the ones of the *X* axis , *Y* axis and *Z* axis of the Dicar coordinate system as possible. According to demands ,crystal can be projected in the direction of *a* axis or *b* axis , *c* axis . The *a* axis(or *b* , *c*) of crystal can be keep with the *Z* axis when placing the crystal with the relations above .

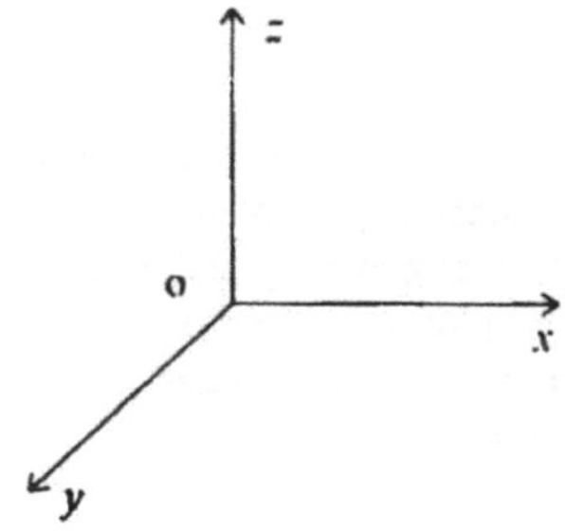

Figure 1 . The Dicar coordination system

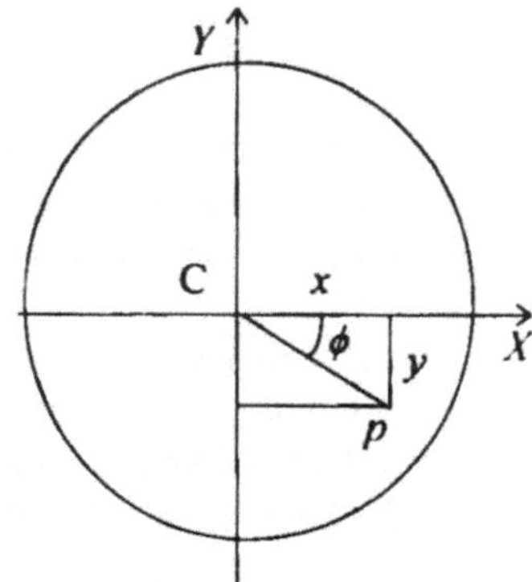

Figure 2 . The relation of (ρ , ϕ) and (x , y) . (ρ , ϕ) is the spherical coordination ; (x , y) is Dicar coordination ;p is the gnomonic projection of the crystal plane.

The relation between spherical coordination (ρ, ϕ) and Dicar coordination (x, y)is shown in figure 2 . The configuration of the crystal plane can be represented as:

$$x=R \cdot \tan\rho \cdot \cos\phi \qquad (1)$$

$$y=R \cdot \tan\rho \cdot \sin\phi \qquad (2)$$

here R is the radius of the base circle .

With the top view , the 2D coordinates of crystal corner in a certain projective direction can be known while the 3D ones unknown . In order to find the 3D coordinates to draw the crystal , it should be combined the crystal plane gnomonic projection and top view . Only with normal axis projection , it is impossible to reach it .

In figure 3 , *M* is the center of the projection sphere and *F* , *G* are the projection of the sphere points *F'* , *G'* on the gnomonic projection plane π . The crystal zone line *Z* is the intersection of the crystal zone plane *Z'* and π . The projection of the drawing plane *L'* on the π become the conduct line *LL*. *MS'* is the intersection of *L'*and *Z'*. The projection of *MS'* on the plane π is *S*. *MP'* is the crystal zone axis of *Z'* representing the direction of the practical crystal edge. *MQ'*is the vertical projection of *MP'* on *L'* . There is $MQ' \perp MS$. Let $DM \perp LL$ of which the point *D* is the intersection . Rotating *L'* around the *LL* while *S* is fixed until coinciding with π, *M* is to reach the *W* . The *W* is called the corner point where *SW=SM* , *DM=DW*. When it reaches π , *MQ'* become *QW* . Because of the rotation of the drawing plane about *LL* , there are $QW \perp SW$, $DW \perp LL$, where *QW* represents the projective direction of the practical crystal edge on the drawing plane .

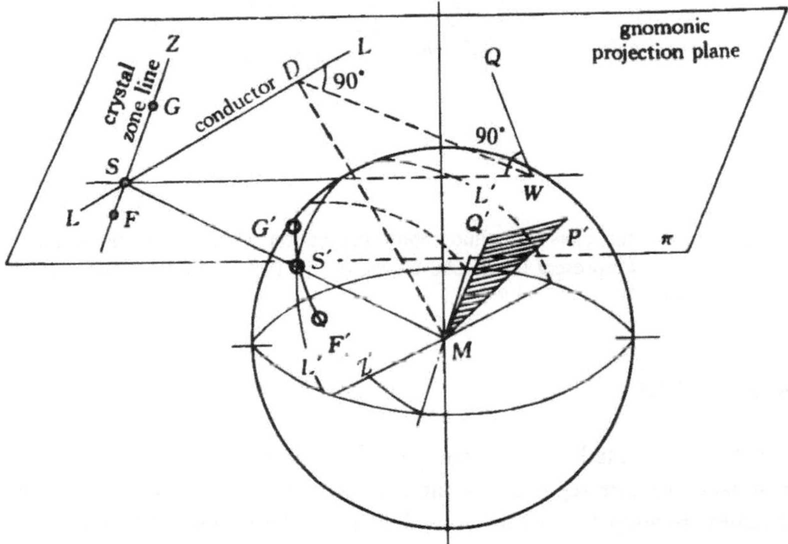

Figure 3 . The diagram of principle of crystal drawing transferred from gnomonic projection . *M* indicates the center of the projection sphere ; *F* , *G* are the projection of the sphere points *F'* , *G'* on the gnomonic projection plane π; *S* is the projection of *MS'* on π; *MP'*represents the direction of the practical crystal edge; *MQ'*is the vertical projection; *W* is the corner point; *QW* represents the projective direction of the practical crystal edge .

What decided above is just the direction of crystal edge in crystal figure . The length of crystal edge is controlled by the top view . In gnomonic projection , when the direction of crystal edge is known , the length of it can be computed with the lines which is perpendicular line of the conduct line through the corner of the relative edge . Shown as in figure 4 , the crystal edges *ad* , *be*, *cf* of the cylinder planes *abed*, *acfd*, *bcfe* is perpendicular to the plane *P* (just like a gnomonic plane) . For these cylinder planes , to drawing the vertical projection about the *P* and *Q*, it will get vertical view *a'b'c'* on *P* .

On Q , the normal projection can be gotten which are $a''b''c''d''e''f''$. The $a''d''$, being on the plane $a''aa'$, is perpendicular to AB(conductor). The projection of the other edges is similar to that above . By these measures , every point on top view can get its relative point of crystal figure on the perpendicular line of its conductor .

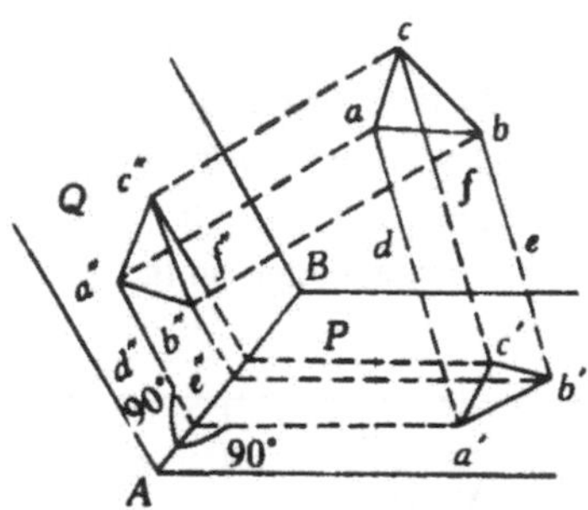

Figure 4 . Relation of top view and clinographic projection. P is the gnomonic plane; Q is the drawing plane; *ad*, *be*, *cf* represent the crystal edges; $a'b'c'$ is the vertical view; $a''b''c''d''e''f''$ is clinographic projection .

DATA STRUCTURE

In order to store the crystal shape on computer for all kinds of operations, it is necessary to design a data structure representing the crystal shape , In this me3thod , single chain and three tables are adopted . Organized with a chain , face table , ine table and point table, recording the crystal shape, represent the topological and geometric relations of the face , line and point .

In the crystal shape drawing , the whole graph is the combination of the outline of the top and bottom of crystal which are separated . When designing the data structure , it firstly abstracts the top view as a 3D drawing (takes the up top view as the up of the crystal and takes the bottom top view as the bottom of it) . For the up and bottom top view , it orders face , line and point and designs the variable arrays respectively .

According to the watch direction , point , line , face are numbered from No 1 . If there is inner loop , its number will be arranged behind the number of the point , line , face of general loops . By a virtual line , the inner loop is connected with the boundary outline of crystal plane which the inner loop exists . Here point , line and face represent corner , edge and plane of a crystal respectively . Figure 5 explains the ordering rule .

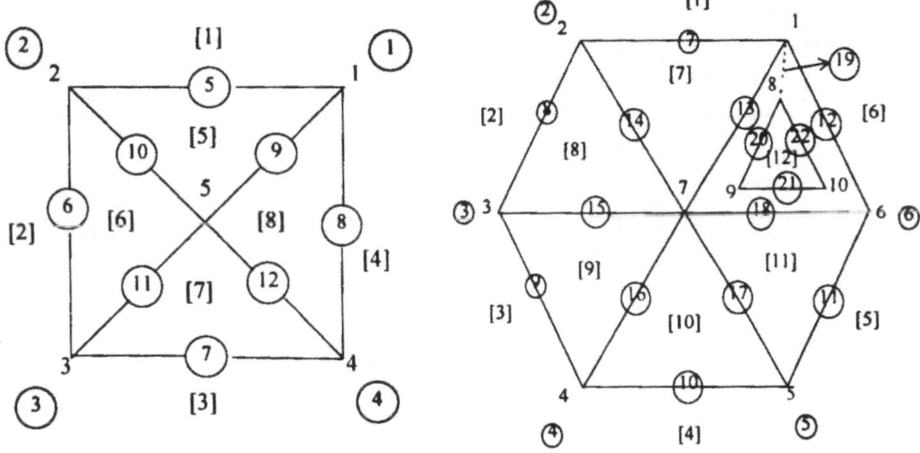

Figure 5. The numerical order of top view . 1 , 2 , ··· number of the point; ① , ② ··· number of the line; [1] , [2]··· number of the face .

The employed Variables and Arrays:

(a) $a\% = \begin{cases} 1 & \text{drawing the up of the crystal ,} \\ 2 & \text{drawing the bottom of the crystal;} \end{cases}$

(b) $as\% = \begin{cases} 0 & \text{the bottom - top view is not symmetric ,} \\ 1 & \text{the up - top view is the same as the bottom - top view ,} \\ 2 & \text{the bottom - top view is the reflection of the up about center point;} \end{cases}$

(c) $d\% = \begin{cases} 1 & \text{convex crystal ,} \\ 2 & \text{concave crystal;} \end{cases}$

(d) *dp* is the length of the last vertical edge of the outer loop;

(e) *ioi*(*I* , *J*) is a array which depicts the characters of the inner loop, *ioi*(*I* , 1)is the number of the inner loops;

(f) *hs*(*I*) is the number of crystal plane;

(g) *nl*(*I*) is the number of crystal edge;

(h) *np*(*I*) is the number of crystal corner;

(i) *nsv*(*I* , *J* ,*HK*) is the number of the crystal edge on crystal plane and relative order number of the crystal plane , *K*=0 , it is the number of the crystal edge of the J-th plane;

(j) *lp*(*I* , *J* , *K*) is the ordering number of the two polar points of the J-th edge .

(k) v(*I* , *J* , *K*) is the values *x*(*y*) of the J-th points;

(l) *rp*(*I* , *J* , *K*) is the spherical coordinates of the J th plane;

In the arrays above , *I* can be 1 or 2 which means the up top view or the bottom . *nsv* , *lp* , *v* constructed practically three related tables (face table , line table and point table) . Figure 6 shows the single chain and three tables data structure in which it can be found the topological relations among plane , edge , and points .

	sum of edge		sum of inner loop	
1	3		0	
2	3		0	
3	3		0	
4	3		0	
5	3		0	
6	3		0	
7	3		0	
8	3		0	

	loop table		
1	1	5	2
2	2	6	3
3	3	7	4
4	4	8	1
5	5	10	9
6	10	6	11
7	11	7	12
8	9	12	8

	plane (ρ,ϕ)	
1	90	270
2	90	180
3	90	90
4	90	0
5	45	270
6	45	180
7	45	90
8	45	0

	edge table	
1	1	1
2	2	2
3	3	3
4	4	4
5	1	2
6	2	3
7	3	4
8	4	1
9	5	1
10	5	2
11	5	3
12	5	4

	vertex table	
1	5	5
2	1	5
3	1	1
4	5	1
5	3	3

Figure 6 . Single chain and 3 table structure . The face table consists of the loop table and the crystal plane (ρ,ϕ) table ; the line table is the edge table : the point table is the vertex table

HIDDEN LINE ELIMINATION

The shading of the convex crystal is simple . But for the concave crystal such as some twin crystal , the crystal with growth hillock , it is complex . For the shading of the convex crystal , the gnomonic projection of the crystal practically are the one of the norm of the plane . By this property , the visibility of the plane can be decided on the gnomonic plane . As shown in figure 7 , *R* is the gnomonic point of the drawing plane on the gnomonic plane , meanwhile *R* keeps on the extend line of the *DW* and *a* , *b* , *c* are respectively the gnomonic points of the plane . Giving the projection of crystal plane on the conductor , the projection of the plane which intersect with the drawing plane must drop on the two sides of the conductor . For these planes located in the up of the crystal , the gnomonic point and the *R* are just at the two sides of the conductor when the angle of the outer norm line of the plane to the viewing line is bigger than 90 . For example , the *a* in the figure is invisible; The gnomonic point and the *R* are just at the same side of the conductor when the angle of the outer norm line of the plane to the viewing line is smaller than 90 . For example , the *c* in the figure is visible . When the angle is equal to 90 , gnomonic point is just on the conductor while the crystal plane degenerates as a line , for example the *b* in the figure . For those planes located in the bottom of the crystal , the projection of the plane is of the inner norm line and the visibility is opposite to these above .

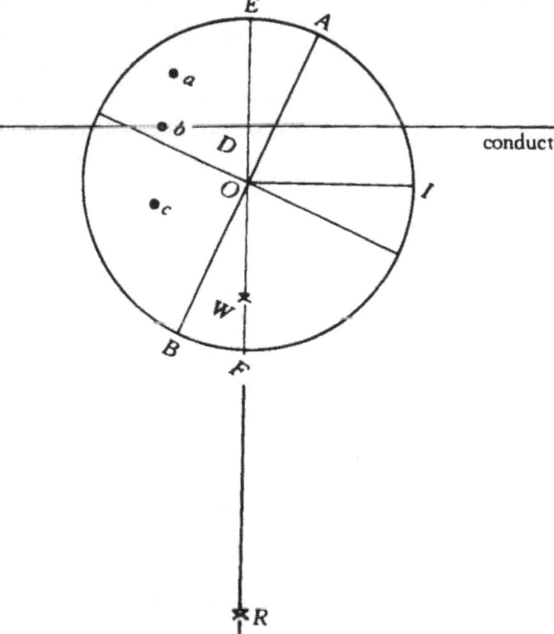

Figure7 . The visibility estimation of crystal faces . *R* is the gnomonic point of the drawing plane; *a* , *b* , *c* are the gnomonic points of the crystal plane .

For the shading of the concave crystal , it is necessary to use many technologies such as the max min test , the depth identification , the certification of the hidden line and the deletion of the hidden line .

(1) max min text

It is used to identify whether the line segment on the projective plane overlap part of the polygon or not . $T_{x\,\max}$, $T_{x\,\min}$, $T_{y\,\max}$, $T_{y\,\min}$ represent the location and volume of the minimal projective rectangle of the polygon on the drawing plane . $P_{x\,\max}$, $P_{x\,\min}$, $P_{y\,\max}$, $P_{y\,\min}$ represent the ones of the line segment . The line segment on the projective plane mustn't overlap the minimal projective rectangle of the polygon when it doesn't overlap the polygon . In this situation , it is satisfied with one of the four inequalities:

$$P_{y\,\max} < T_{y\,\min} , \quad P_{y\,\min} > T_{y\,\max} ,$$
$$P_{x\,\max} < T_{x\,\min} , \quad P_{x\,\min} > T_{x\,\max} .$$

When the line segments overlap the minimal projective rectangle , not any inequalities above will be satisfied when it is possible for the line segment to overlap the polygon .

(2) the depth identification

The values about the depth are not given directly because the top view and the crystal figure are taken in 2D plane . But , indirectly , it can identify the depth by the relation between every crystal plane , edge on the top view and the drawing plane: move the top view to hold the top view and the corner(view point) occurs in the same side of the conductor , then find out the crystal edge and crystal plane which overlap each other in the crystal figure and compare the relation of the edges , plane and the conductor . For example , if the edge just locate between crystal plane and conductor , the crystal plane must overlap the edge so that it is necessary to take intersective operation about the edge and the outline of crystal plane . If not , the edge must be visible so that it is not necessary to call the intersective operation .

(3) the certification and deletion of the hidden line
It firstly compute intersection of line segment to the outline of the polygon to label the start point and the end point that divide the segment into invisible segment and visible segment .
For line segment L_1 , L_2 , its parametric equation is as:

$$\begin{cases} x = x_1 + (x_2 - x_1) \cdot s \\ y = y_1 + (y_2 - y_1) \cdot s \end{cases} \quad (0.0 < s < 1.0)$$

$$\begin{cases} x = x_3 + (x_4 - x_3) \cdot u \\ y = y_3 + (y_4 - y_3) \cdot u \end{cases} \quad (0.0 < u < 1.0)$$

here , (x_1, y_1), (x_2, y_2) are the coordinates of the points of L_1 , (x_3, y_3), (x_4, y_4) are that of L_2 . Parameters s , u can be gotten by solving the equations:

$$s = \frac{\begin{vmatrix} x_3 - x_1 & x_4 - x_3 \\ y_3 - y_1 & y_4 - y_3 \end{vmatrix}}{\begin{vmatrix} x_3 - x_4 & x_2 - x_1 \\ y_3 - y_4 & y_2 - y_1 \end{vmatrix}}, \quad u = \frac{\begin{vmatrix} x_3 - x_1 & x_2 - x_1 \\ y_3 - y_1 & y_2 - y_1 \end{vmatrix}}{\begin{vmatrix} x_3 - x_4 & x_2 - x_1 \\ y_3 - y_4 & y_2 - y_1 \end{vmatrix}}$$

if

$$\begin{vmatrix} x_3 - x_4 & x_2 - x_1 \\ y_3 - y_4 & y_2 - y_1 \end{vmatrix} \neq 0,$$

there is only one solution for s , u . But for the intersection which is in the two line segment currently , there is another conditions that are $0.0 < s < 1.0$ and $0.0 < u < 1.0$.

After computing the intersections , the sum of intersection can used to decide the visibility of the line segments:taking a vertical line through the center point of the line segment , if the sum of intersection is singular , it means the segment is invisible , if it even , it means the segment is visible . The union of invisible segment can be given when all face comparison are taken which may hide the segments meanwhile the visible segments are identified . Consisting of the visible segment , the shading crystal figure will be given finally .

THE DRAWING MODULE

The modular programming structural design about the crystal shape adopted the top-bottom strategy . The main module is used to control systematic running . Some sub module will be introduced respectively in the following .

Gnomonic Projection

It identifies the gnomonic plane firstly and then takes gnomonic projection about crystal plane . The three parameters R, ω, OD are necessary for the gnomonic plane where R, taking values as 5 generally , is the radius of the base circle , ω is the angle of the projection of viewing line on gnomonic plane to the X axis; OD , $OD=R \cdot \tan 9°28'$ generally , is the distance from the center of the circle to the conductor . R, ω, OD determine practically the projective direction of crystal . the coordinates of the vertical point from the center of the circle to the conductor:

$$x_D=OD \cdot \cos\omega, \quad y_D=OD \cdot \sin\omega$$

the slope of the conductor: $d_k = \tan\omega$

the equation of the conductor: $y= d_k \cdot (x - x_D) + y_D$

the coordinates of corner point $W(x_W, y_W)$

$$x_W = \left(\sqrt{OD^2 + R^2} - OD\right) \cdot \cos(270° + \omega)$$

$$y_W = \left(\sqrt{OD^2 + R^2} - OD\right) \cdot \sin(270° + \omega)$$

With eq. (1) , (2) , the spherical coordinates of crystal plane is projected into gnomonic plane . It transforms the spherical coordinates into Dicar coordinates .

The Top View

Firstly it holds the up top view and the bottom top view in the same direction that make the line between the centers of up top view and bottom top view perpendicular to the conductor . It must be satisfied with:

$$(y_{c2} - y_{c1}) / (x_{c2} - x_{c1}) = -1 / d_k .$$

here , (x_{c1}, y_{c1}), (x_{c2}, y_{c2}) are the coordinates of the center of up-top view and bottom-top view respectively . Secondly , it verifies top view because the configuration of the top view is not consistent with that of the gnomonic projection of the spherical coordination of crystal plane . In another words , the edge direction in top view is not consistent with the direction of the line . This line is perpendicular to the line that connects the two gnomonic points which is transferred from the gnomonic projection of the two plane sharing the edge . The processing is as the following:

l_1 is a general edge . (x_{v_1}, y_{v_1}), (x_{v_2}, y_{v_2}) are coordinates of two points of l_1 ; (x_{f_1}, y_{f_1}), (x_{f_2}, y_{f_2}) are coordinates of two gnomonic points of two crystal plane sharing this l_1 .

the slope of l_1 : $k_{l_1}=(y_{v_2} - y_{v_1}) / (x_{v_2} - x_{v_1})$

the slope angle of l_1 : $\beta_1 = \arctan k_{l_1}$

the slope of f_1f_2 : $k_{f_1f_2} = (y_{f_2} - y_{f_1}) / (x_{f_2} - x_{f_1})$

the slope angle of f_1f_2 : $\beta_2 = \arctan k_{f_1f_2}$

the angle of f_1f_2 to l_1 : $\alpha = \beta_2 - \beta_1$
Rotating top view with α by the rotating matrix , the new coordinates of top view V_r can be gotten :

$$V_r = V \cdot M_r$$

$$M_r = \begin{bmatrix} \cos\alpha & \sin\alpha & 0 \\ -\sin\alpha & \cos\alpha & 0 \\ x_c \cdot (1-\cos\alpha) + y_c \cdot \sin\alpha & y_c \cdot (1-\cos\alpha) - x_c \cdot \sin\alpha & 1 \end{bmatrix}$$

here , (x_c , y_c) is the coordinates of the center of top view .

Drawing Crystal Figure

(1) the determination of the direction of crystal edge

(x_f, y_f), (x_g , y_g) are coordinates of two gnomonic points of two crystal plane sharing line l.
the equation of f_g : $y - y_f = [(y_g - y_f) / (x_g - x_f)] \cdot (x - x_f)$
the equation of the conductor : $y - y_D = d_k \cdot (x - x_D)$
Solving the equations above , it gets the coordinates of intersection S

$$x_S = \frac{\begin{vmatrix} y_f - x_f \cdot (y_g - y_f)/(x_g - x_f) & 1 \\ y_D - d_k \cdot x_D & 1 \end{vmatrix}}{\begin{vmatrix} -(y_g - y_f)/(x_g - x_f) & 1 \\ d_k & 1 \end{vmatrix}}$$

$$y_S = \frac{\begin{vmatrix} -(y_g - y_f)/(x_g - x_f) & y_f - x_f \cdot (y_g - y_f)/(x_g - x_f) \\ -d_k & y_D - d_k \cdot x_D \end{vmatrix}}{\begin{vmatrix} -(y_g - y_f)/(x_g - x_f) & 1 \\ d_k & 1 \end{vmatrix}}$$

the slope of the line , perpendicular to the connection of point S and corner point W , is k_f,

$$k_f = -(x_W - x_S) \quad (y_W - y_S)$$

(2) the algorithms about the coordinates of the corners of crystal figure

After obtaining the direction of the edge by the gnomonic projection , the length of the edge is limited by the perpendicular line of the conductor . The conductor is through two points of the related edge in the crystal figure . The processing is as following:
Through a end point of the first non pillar edge J , it draws a line perpendicular to the conductor . Taking any point (x_1 , y_1) on this new line as the coordinates of corner of 1st point of the edge J, it can determine the location of the starting point of the crystal figure . the equation of the edge J:

$$y = k_f \cdot (x - x_1) + y_1 \qquad (3)$$

Through 2nd point (x_v , y_v) of the edge J, drawing a line perpendicular to the conductor , the equation of this line is:

$$y = -1 / d_k \cdot (x - x_v) + y_v \qquad (4)$$

By solving eq . (3) , (4)the coordinates of 2nd point of the edge J can be obtained . Taking the coordinates as the 1st point of next edge , similar to above , 2nd point of new edge can be determined. Repeating the steps above, the point of all edges can be computed . Exactly , the computation about points of the edge in crystal figure is based on the condition that the first point is known so that it is necessary to check if any point of the edge has been resolved before the computation . If there is , the unknown point can be referred with the known point . If the two point are all known , not any computation about intersection is demanded .

(3) hidden line elimination and drawing

After the computation of points of all edges , it still identifies the visibility about these edges and output the shaden crystal figure at last .

(4) transformation about the draft

According to demands , it can take the top view and crystal figure moving , scaling , rotating and projecting with which it can give out many different configurations about the crystal top view and figure .

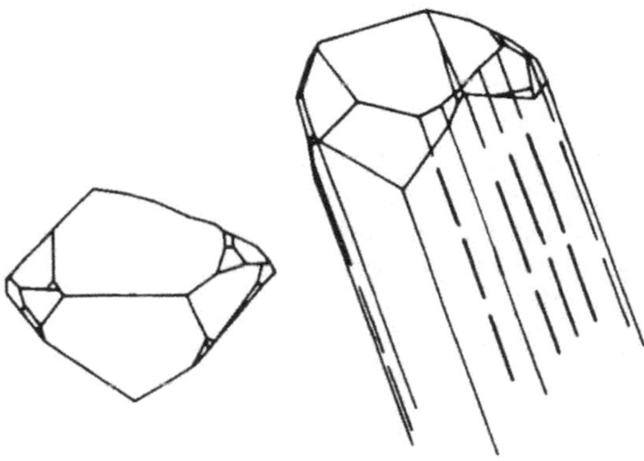

Figure 8 . Practical drawing of orpiment . The right of figure is top view and the left is gnomonic projection .

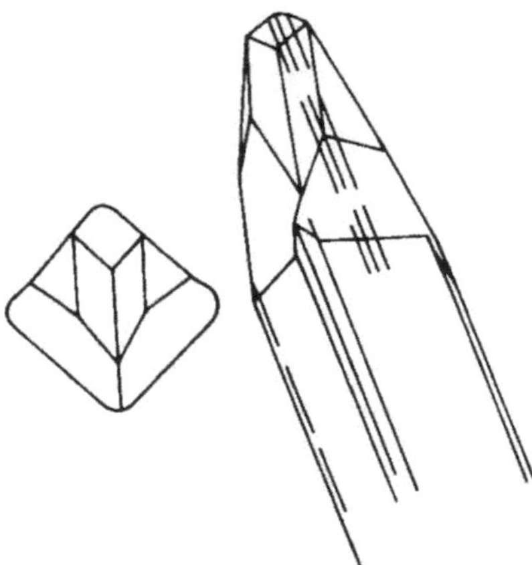

Figure 9 . Practical drawing of aegirine . The right of figure is top view and the left is gnomonic projection .

Based on the INTEL series , the CSGS (Crystal Shape Graphics System) has been developed according to the principles above . Figure 8 , 9 are respective the practical shape of orpiment found in Shimen and aegirine discovered in Kuaian . This system modernizes the crystal measuring graphics , avoids the manual complex operation , saves time greatly and promotes the precision . It is an advanced tool for crystal shape research .

REFERENCES

1 . E . Dowty . Computing and drawing crystals shapes . Am . Mineral . 65 , 465 ~ 471(1980)

2 . Y . Kanazawa and Y . Endo . Drawing of crystal and twin figures . Mineral Journal . 10 , 279 ~ 295(1981)

3 . Shoupeng Zhou . Handling hidden line of concave polyhedron with edge displaying routine . Journal of zhejing Univ . 4 , 35 ~ 48(1982)

4 . Jinchan Chen . 3D color displaying system for crystal shape . wuhan(1985)

5 . Jiagang Sun . Computer Graphics . Qinghua Univ . Publishing House , Beijing(1987)

6 . Fachen Jiao . Computer Drawing. Tianjing Univ . Publishing House , Tianjing(1987)

7 . Y , Nakamuta . Utility programs for the description of crystals using a persional computer . Mineral Journal . 20 , 71 ~ 74(1991)

8 . Zhizhong Peng and Wengui Wang . General Crystallometry . Geological Publishing House , Beijing(1992)

Proc. 30th Int'l. Geol. Congr., Vol. 16, pp. 255-263
Huang Yunhui and Cao Yawen (Eds)

Melt Inclusions in Eclogites from High-Pressure and Ultra-High Pressure Metamorphic Belt in the Dabie Mountains, China

HAN YUJING , ZHANG ZEMING and LIU RONG
Faculty of Earth Sciences ,China University of Geosciences, Wuhan 430074, China

Abstract

The eclogites mostly occur as lenses or layers intercalated in the metamorphic supracrustal rocks in the Dabie Mountains, China. Melt inclusions have been discovered for the first time in the eclogites. Petrographical, microthermometric and Raman spectroscopic studies have been made of the inclusions.

The melt inclusions are isolated within omphacite, kyanite and coesite. The homogenization temperatures for the melt inclusions generally vary between 820 and 950 °C. Some inclusions contain two phases, a silicate glass phase and a gas phase in which the dominant gas is CO_2, and some three phases with an aqueous solution (15.9 ~ 22.0 Wt%NaCleq). The composition of glass phase from a kyanite eclogite(Sample 1) is close to its host mineral kyanite. H_2S and OH^- detected by Ramma analyses are respectively molecules and ions in monomer [SiO_4] networks. The glass phase from coesite-bearing eclogite (Sample 4) corresponds to a type of frame [SiO_2] network structure, with a composition close to the host mineral coesite. Thus the host minerals possibly crystallized from the melts of corresponding compositions, and the melts are likely of anatexis origin.

Keywords: Melt inclusion, eclogite, high-pressure and ultra-high pressure metamorphism, Dabie Mountains, China

INTRODUCTION

The Dabie high pressure (HP) and ultra-high pressure (UHP) metamorphic belt lies in the eastern part of the Qinling-Dabie orogenic belt which is sandwiched between the North China and Yangtze continental blocks. It can be divided into three belts from south to north: an epidote blueschist facies belt, a high pressure eclogite facies belt and an ultra-high pressure eclogite facies belt(Fig.1). The eclogites occur as lenses and layers intercalated in the metamorphic supracrustal rocks of the Dabie Mountains. The protoliths of the eclogites are tholeiitic basalts in the majority of cases and calc-alkaline basalts in the minority, representing an island arc environment [15]. They have undergone polyphase metamorphism and deformation. Peak metamorphic P-T conditions of HP and UHP eclogites were 450~650°C at 1.5GPa and 650~900°C at 3.0GPa respectively[16,15,17]. In order to obtain information on the fluids present during the metamorphism and deformation, a study of fluid inclusions has been made in HP and UHP metamorphic rocks from representative areas. Melt inclusions have been discovered for the first time in HP and UHP eclogites. Here we report on petrographic, microthermometric and Raman spectroscopic studies of the melt inclusions.

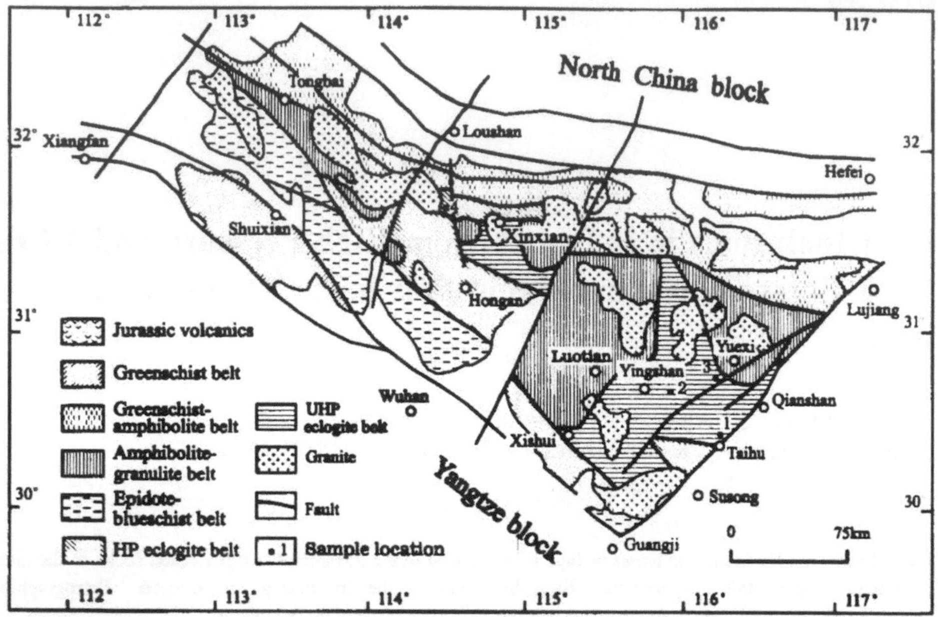

Figure 1. Simplified geological map of high-pressure and ultra-high pressure metamorphic belt in the Dabie Mountains. Sample location: 1, Shima; 2, Mifengjian; 3, Bixiling; 4, Xiongdian.

PETROGRAPHY OF MELT INCLUSIONS

Samples were selected from representative areas (Fig.1) of the HP and UHP eclogites: kyanite eclogite at Xiongdian(sample 1), coesite-bearing eclogite at Mifengjian(sample 2), eclogite (sample 3) and coesite-bearing eclogite(sample 4) at Shima, and kyanite eclogite at Bixiling (sample 5).

Sample 1, with the assemblage garnet + omphacite + kyamite + phengite + quartz, occurs as lenses within garnet-amphibole schists. Banding on a mm to cm scale is quite common and is defined by garnet-rich layers alternating with omphacite-rich layers. The melt inclusions have a liquid phase existing between the glass and gas phases, thus they are fluid-melt immiscible inclusions[5,3, 11,12] (Fig. 2). They are isolated in kyanite crystals and their morphologies are negative crystal shapes or irregular, ranging from 15 to 30μm in size.

Sample 2, containing garnet + omphacite with minor kyanite, phengite,coesite (quartz) and rutile, occurs as lenses within biotite gneisses. Symplectic textures are developed in this rock. Fluid-melt immiscible inclusions have been found in omphacite. They have negative crystal shape and range from 8 to 20μm in size.

Sample 3, consisting of garnet + omphacite + muscovite + epidote + rutile and having well developed symplectic texture, occurs as layers or lenses within biotite gneisses. The melt inclusions are negative crystal shaped and isolated in omphacite, ranging from 8 to 13μm in size. These inclusions are also fluid-melt immiscible inclusions.

Sample 4, having garnet + omphacite + coesite (quartz) + rutile, occurs as lenses within biotite gneisses. One melt inclusion is olive-shaped, 25μm long and isolated in coesite within garnet that shows well developed radial fractures (Fig. 2).

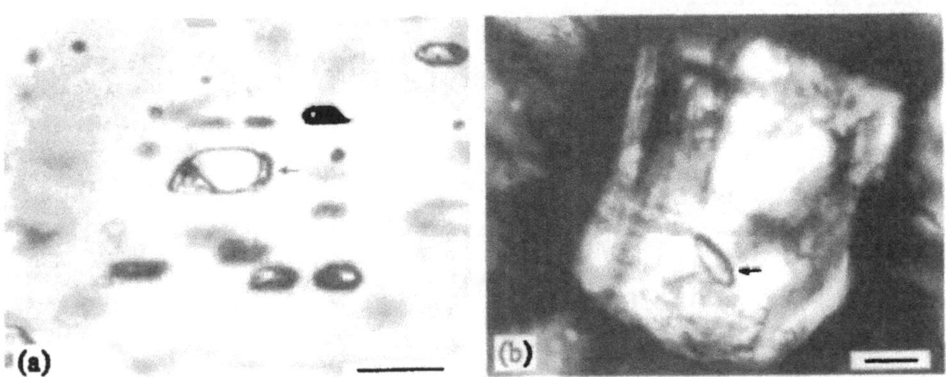

Figure 2. Photomicrographs of melt inclusions hosted in kyanite (a) and coesite within garnet (b). Scare bar is 15μm in (a) and 25μm in (b).

Sample 5, composed of garnet + omphacite + kyanite + phengite +quartz + rutile, occurs as lenses within biotite gneiss intercalated with bands of garnet-peridotite. The melt inclusions consisting of glass and gas phases are isolated in omphacite or garnet. They range from 10 to 25μm in size and have negative crystal shapes or irregular morphologies.

MICROTHERMOMETRY OF MELT INCLUSIONS

Microthermometric analyses of the melt inclusions were performed using a Leitz 1350 heating stage. The homogenization temperatures for the melt inclusions vary between 820 and 950°C (Table 1).

Table 1 Homogenization temperature of melt inclusions

Sample No.	Rock	Host mineral	T_i(°C)	T_h(°C)
1	kyanite eclogite	kyanite	790	850
2	coesite-bearing eclogite	omphacite	830~870	890~950
3	eclogite	omphacite	790~830	820~890
4	coesite-bearing eclogite	coesite	*	*
5	kyanite eclogite	omphacite	790~830	850

T_i, Initial melting temperature; T_h, Homogenization temperature.
* Heating up to 600°C led to bursting of the melt inclusion.

MICROANALYSES OF MELT INCLUSIONS

Microanalyses for individual inclusions were made to investigate the composition of melt inclusions, using a Ramanor U1000 Spectrometer, fitted with a spectra physics Ar ion Laser (514.5nm line) at Xi'an Insititute of Geology and Mineral Resources. The results are shown in Figures 3 to 7 and Table 2.

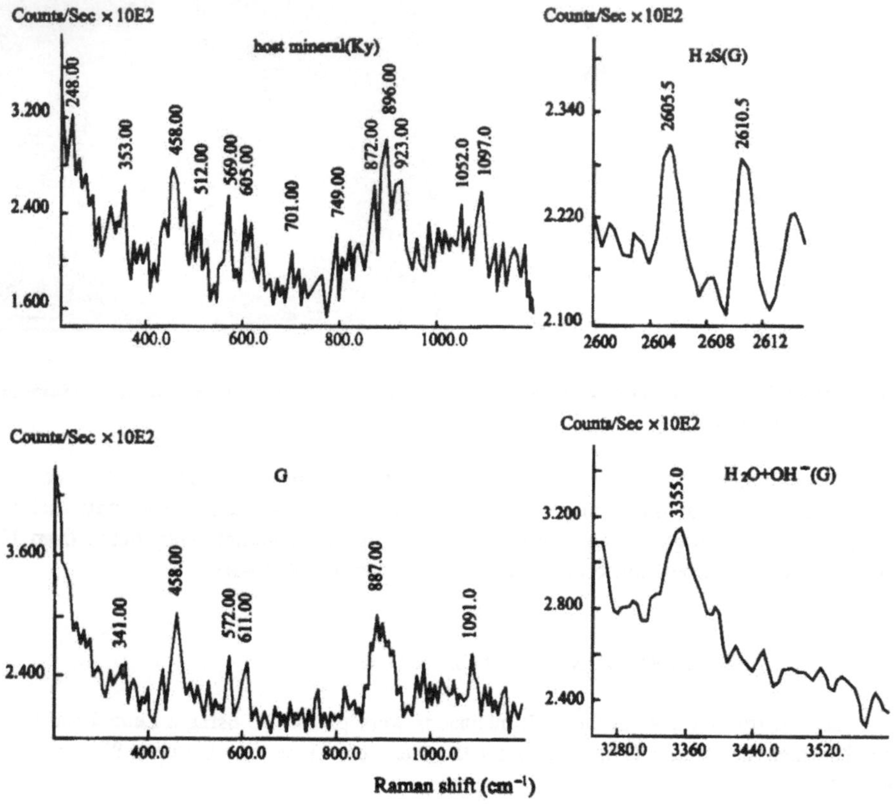

Figure 3. Raman spectra of glass phase (G) in melt inclusions and host mineral kyanite(Ky) from sample 1.

There are different types of polymeric molecular network structure in the glass phase [8,13,14,7] of the melt inclusions: 1) A devitrified glass phase with monomer [SiO_4] network(Fig.3); 2) A predominantly monomer [SiO_4] network, with subordinate dimer [Si_2O_7] networks (Figs.4, 5); 3) A devitrified glass phase with a pure frame [SiO_2] network and minor dimer [Si_2O_7] networks (Fig.6) and 4) A chain Si-O network (Fig.7). It shows that the glass phase are silicate glass phase in the samples. Raman spectra of some host minerals were measured for comparison with those of glass phases. In HP eclogite from sample 1, the Raman spectrum of glass phase is similar to that of the host mineral kyanite[6] (Fig. 3). It shows that the composition of glass phase is close to kyanite, i.e., it is a transitional phase from glass phase to kyanite, H_2S and OH^- [9] detected by Raman analyses are respectively molecules and ions in monomer [SiO_4] network. In UHP eclogite sample 4, the composition of glass phase is also close to its host mineral coesite(incipiently-transformed coesite)[2,4,14] (Fig. 6).

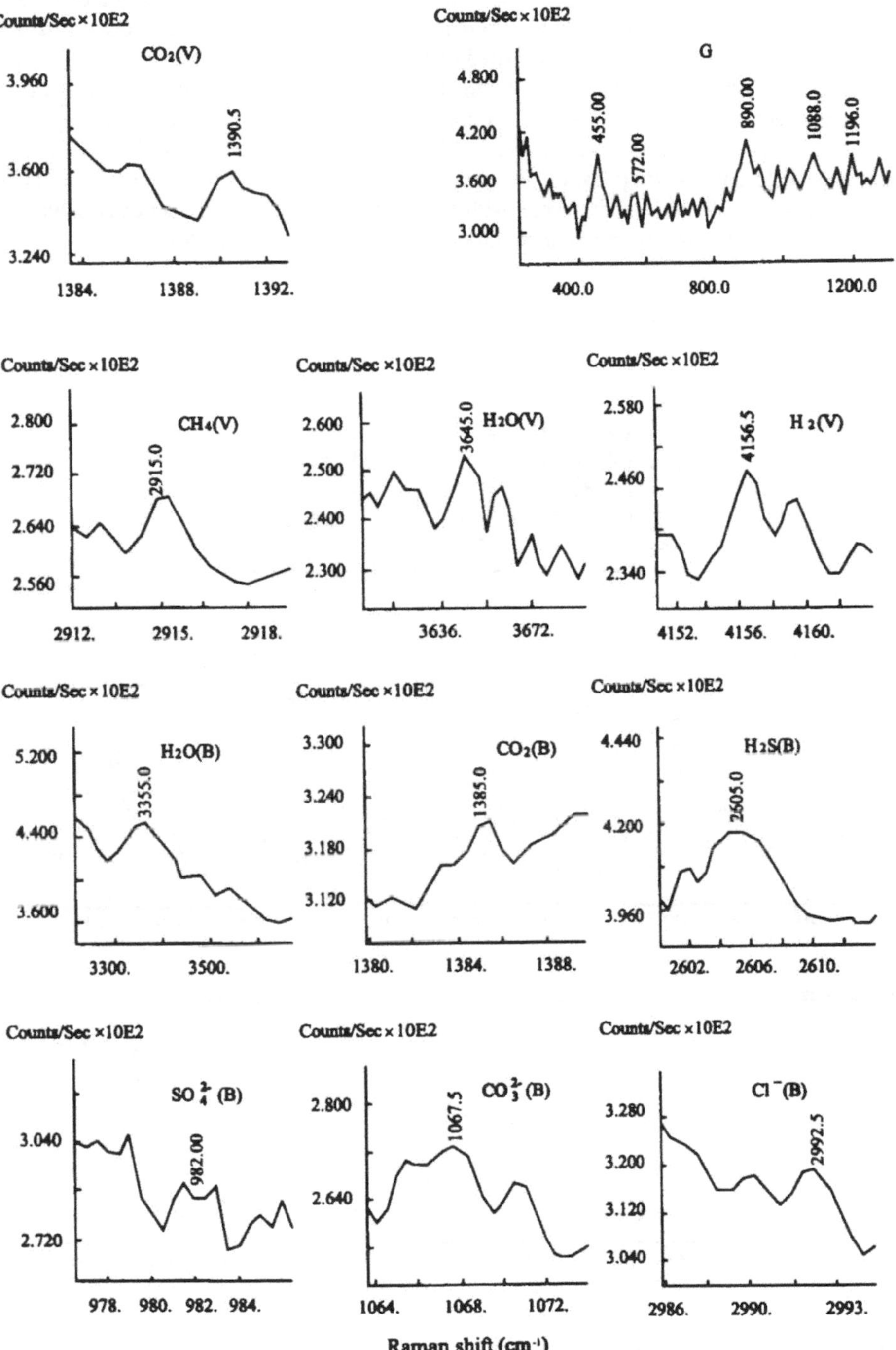

Figure 4. Raman spectra of melt inclusion from sample 2. G, glass phase; V, gas phase; B, Liquid phase.

Table 2 shows that CO_2 is the dominant species in the gas phase of the melt inclusions, ranging from 57.0 to 85.1mol%. A majority of samples contains H_2O(13.4~28.5mol%) and CH_4(5.2~14.9mol%), mixed with H_2, H_2S or SO_2 in very few samples. The liquid phase existing in fluid-melt immiscible inclusions is composed of H_2O and CO_2. The content of H_2O ranges from 65.3 to 80.0mol% and CO_2 from 12.7 to 20mol%. H_2S was only detected in sample 2. In the aqueous solutions the main species of cations are Na^+, K^+, Mg^{2+} and Ca^{2+}, among them Na^+ being the most dominant and ranging from 1.31 to 1.96mol/l. The ratios of Na^+/K^+ fall in the range of 4.90~6.55 and Ca^{2+}/Mg^{2+} ratios between 0.50~0.71. The dominant anion is Cl^-, ranging from 2.21 to 3.0 mol/l, sometimes mixed with CO_3^{2-}, HS^- or SO_4^{2-}. The aqueous solutions belong to the H_2O-NaCl system and their salinities are 15.9 ~ 22.0 wt% NaCleq.

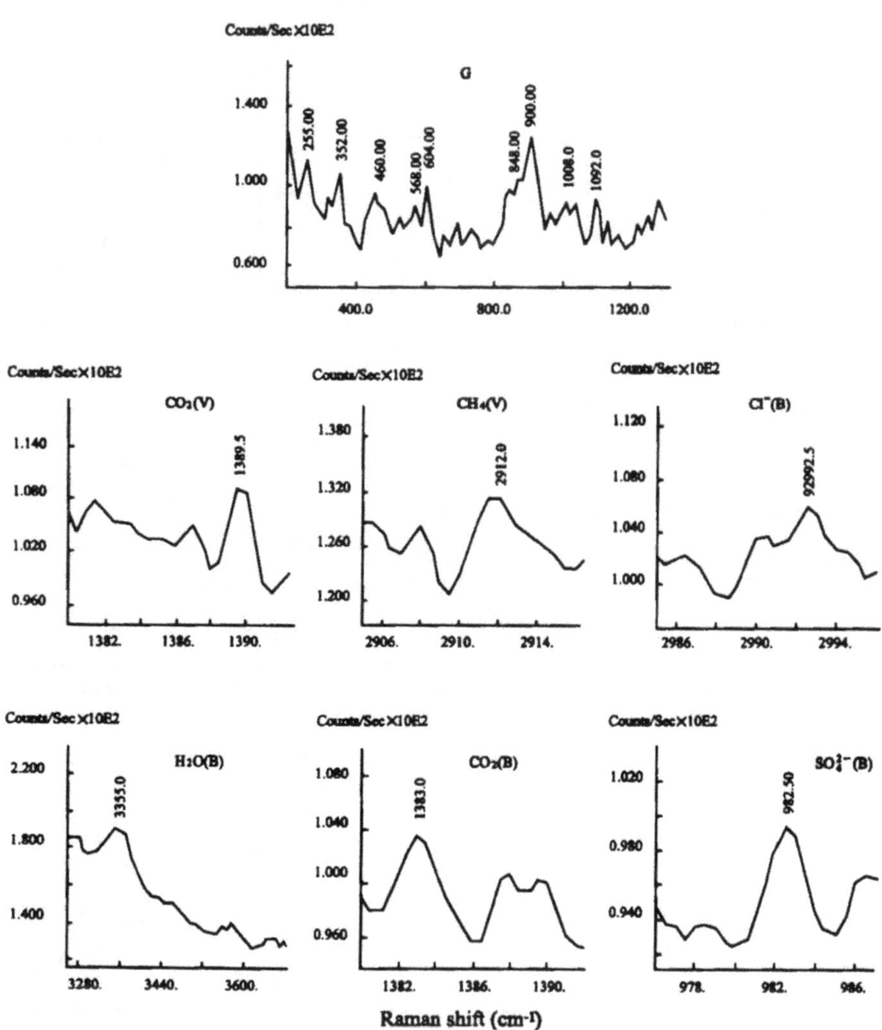

Figure 5. Raman spectra of melt inclusion in omphacite from sample 3. The meanings of G, V, B are the same as those in Figure 4.

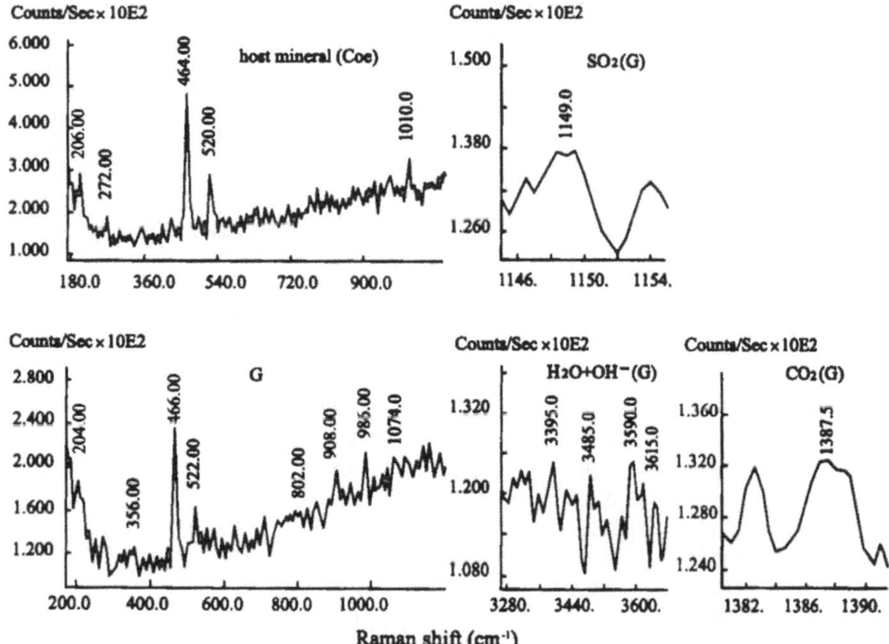

Figure 6. Raman spectra of melt inclusion and host mineral coesite(Coe) from sample 4.

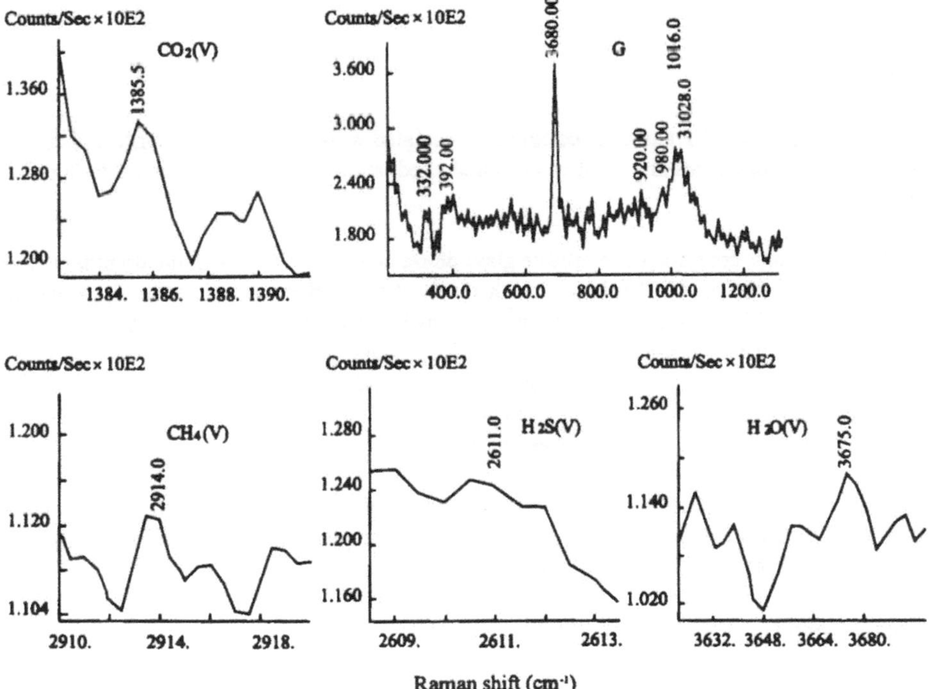

Figure 7. Raman spectra of melt inclusion from sample 5.

Table 2. Raman data on gas phase and liquid phase in individual melt inclusions

Sample No.		1	2	3	4	5
Gas phase	CO_2	72.2	57.0	85.1	66.3	68.9
(mol %)	CH_4	11.4	5.2	14.9		8.9
	H_2O	13.4	28.5		16.7	17.6
	H_2S					4.6
	H_2		9.3			
	SO_2	3.0			17.0	
Liquid phase	H_2O	65.3	73.1	80.0		
(mol%)	CO_2	34.7	12.7	20.0		
	H_2S		14.2			
Liquid phase	K^+	0.30	0.40	0.20		
(mol/l)	Na^+	1.50	1.96	1.31		
	Ca^{2+}	0.20	0.25	0.25		
	Mg^{2+}	0.40	0.45	0.35		
	Cl^-	2.51	3.08	2.21		
	CO_3^{2-}	0.17	0.13			
	HS^-	0.15				
	SO_4^{2-}			0.25		
Salinity(wt%NaCleq)		17.6	22.0	15.9		

Analysed by Xu Peicang and Wang Zhihai

CONCLUSIONS

1. Melt inclusions from HP and UHP eclogites are isolated within omphacite, kyanite and coesite crystals. their morphologies are negative crystal shaped or irregular, ranging from 8 to 30μm in size.

2. Melt inclusions, consisting of a silicate glass phase and a gas phase with the dominant CO_2, have been discovered in UHP eclogites with rarely developed symplectic texture; while the fluid-melt immiscible inclusions have been found in HP eclogite and UHP eclogite with well developed symplectite. They are composed of silicate glass, gas and liquid phases. The major component of gas phase is mainly CO_2, and the liquid phase belongs to the H_2O-NaCl system with a fairly high salinity of 15.9~22.0 Wt% NaCleq . The formation of sympectite in eclogites may be related to the presence of the aqueous solution.

3. The fact that the composition of glass phase is close to that of its host mineral shows that the host mineral possibly crystallized from melts of similar composition and the melts are likely of anatectic origin related to subduction[1,8].

Acknowledgement

The authors are grateful to Xu Peicang and Wang Zhihai for help with analyses by Laser Raman

Microscope. Chen Ziying is thanked for the careful reviews of the manuscript and for useful suggestions. Thanks are also to Roger Mason for improving the English. This work was supported by Doctorate Special Foundation from the Nationa Education Committee of China (Project No. 9349101).

REFERENCES

1. T. Anderson, H. Austrheim and E. A. J. Burke. Mineral-fluid-melt interactions in high-prssure shear zones in the Bergen Arcs nappe complex, Caledonides of W. Norway: Implications for the fluid regime in Caledonian eclogite-facies metamorphism. *Lithos* **27**, 187-204 (1991).
2. H. Boyer, D. C. Smith, C. Chopin and B . Lasnier. Raman microscope (RMP) determinations of natural and synthetic coesite, *Phy. Chem. Minerals* **12**, 45-48 (1985).
3. L. S. Hollister and M. L. Crawford (eds). *Short course in fluid inclusions: applications to petrology. Mineral. Assoc.* Canada (1981).
4. K. J. Kingma and R. J. Hemley. Raman spectroscopic study of microcrystalline silica, *Am. Mineral.* **79**, 269-273 (1994)
5. H. Lu, B. Li, K. Shen, X. Zhao, T. Yu and J. Wei. *Fluid inclusion geochemistry*, Geological Publishing House, Beijing (1990) (in Chinese with English abstract).
6. P. McMillan and B. Piriou. The structures and vibrational spectra of crystals and glasses in the silica-alumina system, *J. Non-Crystalline Solids* **53**, 279-298 (1982).
7. P. McMillan and B. Piriou. Raman spectroscopec studies of silicate and related glass structure--a review, *Bull. Mineral.* **106**, 57-75 (1983).
8. B. O. Mysen, D. Virgo and C. M. Scarfe. Relations between the anionic structure and viscosity of silicate melts--a Raman spectroscopic study, *Am. Mineral.* **65**, 690-710 (1980).
9. B. O. Mysen, D. Virgo, W. J. Harrison and C. M. Scarfe. Solubility mechanisms of H_2O in silicate melt of high pressures and temperatures: a Raman spectroscopic study, *Am. Mineral.* **65**, 900-914 (1980).
10. P. Philippot. Fluid-melt-rock interaction in mafic eclogites and coesite-bearing metasediments: Constraints on volatile recycling during subduction, *Chem. Geol.* **108**, 93-112(1993).
11. E. Roddar. *Fluid inclusions. Reviews in Mineralogy 12.* Mineral. Soc. America (1984).
12. T. J. Shepherd, A. H. Rankin and D. H. M. Alderton. *A practical guide to fluid inclusion stydies.* Blackie, Glasgow(1985).
13. P. Xu, W. Wei, R. Li, Y. Wang and X. Zhang. A study of molecular network structure in magmatic glass phases and industry silicates glass phases and its application, *Northwest Geosci.* **14**, 63-120 (1993) (in Chinese with English abstract).
14. P. Xu, R. Li, Y. Wang, Z. Wang and Y. Li. *Raman spectoscopy in geosciences.* Shanxi Press of Science and technology, Xi'an (1996)(in Chinese with English abstract).
15. Z. You, Y, Han, W. Yang, Z. Zhang, B. Wei and R. Liu. *The high-pressure and ultra-high pressure metamorphic belt in the East Qinling and Dabie Mountains, China.* China University of Geosciences Press, Wuhan (1996).
16. Z. Zhang, Z. You, Y. Han, W. Yang and B. Wei. High-pressure and ultra-high pressure metamorphic belt in Tongbai-Dabie Mountains, central China, *J. China Univ. Geosci.* **6**, 139-145 (1995).
17. Z. Zhang, Z. You, Y. Han and L. Sang. Petrology, metamorphic process and genesis of the Dabie-Sulu eclogite belt, Eastern Central China. *Acta Geologica Sinica* **9**, 134-156 (1996).

Proc. 30th Int'l. Geol. Congr., Vol. 16, pp. 265-276
Huang Yunhui and Cao Yawen (Eds)

Study On The Silicate Melt Inclusions In Accessory Minerals Of Various Igneous Rocks In China

Li Zhaolin
(Dept. of Geology. Zhongshan University, Guangzhou, 510275, China)

It is well known that the forming temperature of magmatite can be determined by means of homogenizing melt inclusions in its rock forming minerals. The homogenization temperature are mainly acquired from high temperature heating stage homogenization and quenching methods. But the homogenization temperatures thus determined, due to the wide range of temperature variation during the rock forming process, usually do not perfectly reflect the whole spectrum of the temperature range, especially in the case here there were no melt inclusions in rock forming minerals (e. g. quartz in some granites and diorites) and only gas liquid inclusions that could be found used. Actually the homogenization temperatures of gas liquid inclusions can only represent the low limits of rock forming temperature. In order to get initial crystalling temperatures of magmatic rocks, one should determine the homogenization temperatures of melt inclusions in accessory minerals that crystallized from magma in the early period of intrusion. From 1979 to 1981, we succeeded in developing the quenching oil immersion method, and homogenization method by means of which homogenization temperature of melt inclusions in the accessory minerals in the magmatic rocks can be obtained and phase change of melt inclusions under thermal states may be studied and the effection is good. It provided new technology and had great significance in advancing for the study of rock- forming temperature of magmatic rocks.

In the passing decade, we have done our best to study on the determination of rock-forming temperature and magma composition by means of melt inclusions in accessory minerals (Zircon, apatite) of igneous rocks, and probed into the forming physicochemical conditions of some granites, diorites, porphyry, volcanic rocks, kimberlites and sediments of modern marine volcanic eruption in southern, central, southwestern and northwestern China and good results have been described for this method .

I. Determinational methods of rock — forming temperature by melt inclusions in accessory minerals

1. Homogenization methods : the two — side polished section or accessory minerals with smooth crystal face (zircon, aptite etc) contained melt inclusions were first selected and then put it on the high — temperature heating stage (Leitz — 1350, from west German; LGHS — 1, from China). Since the melt inclusions principally belong to dry

system and gas spreading velocity in the high viscous magma melt is quite slowly , so the experiment of melt inclusions homogenization required temperature risen slowly and equal temperature was kept for stage. The temperature was allowed to rise little faster , such as 10— 20℃/min. , when it had not been raised to 700℃. In this range of temperature, some appearances of heat expanding and gas expanding along solid fissure will take place, but there is no solid dissolved. The temperature was allowed to rise slowly (2—5℃/min.) , when it had already been raised to higher than 700℃. And after every increment of 100℃, it was kept constant for 20 minutes to several hours. So it normally took n—24h for a complete run up, usually when the temperature is higher than 700℃ ,the solid will start to melt. This could be recognized with the help of the contraction of gas, edges and corners of solid phase become blunt, and the temperature at that time is called initial melt temperature, that is glass or crystalline start to be soften or melten. With the rising of temperature or prolong of heating time, the volume of melt expand and gas contract until the whole inclusions were full of the melt, and reached homogenization condition. The temperature at that time is just homogenization temperature. This method is objective, reliable, accurate, and can be used to understand the relation between the phase equilibrium of melt and temperature during the experiment.

2. The quenching—oil immersion method

This method is used to determine the rock—forming temperatures of melt inclusions in accessory minerals and it is a new method to determin the crystalline temperatures of early period magamtic rocks . Because accessory minerals are usually small and impossible to preserve in preparing the two —side polished thin sections, special method should be developed if the inclusion in these minerals are to be studied, a few of grains of the separated transparent and semitransparent accessory minerals, such as zircon, apatite and titanite etc, were put on a slice of glass and covered with a small thin glass cover ;then some drops of immersion oil, the refractive index of which was less than that of mounted minerals, were dropped on the sample, and then the slice with the minerals were to be examined with petrographic microscope. The purpose of dropping in the immersion oil was to reduce the reflective light on the coarse surfaces of the minerals, and get a clear picture of the melt inclusion (Photo 1). After detailed being observed, the samples were cleaned with alcohol and put into a silicon tube for heating up with in auto —furnace (LGHQ—1 type) rise temperature slowly and keep equal temperature stage for quenching. At the desired temperature, the silicon tube quenched in water and accessory mineral samples were take out for the oil immersion examination again to see whether the phase change in them had take place. If the phase in the inclusions changed, the temperature at the phase changing is the initial melting temperature .Continue to the process of heating and quenching until the inclusion reached homogenization condition , thus we get the homogenization temperature. The advantages of this methods are: simple facilities, and easy operational process. It can be regarded as a valuable new method in studying rock—forming temperatures at present.

II. The features of melt inclusions in accessory minerals

based on the phases and rations of these phase presented in the various kinds of accessory minerals in igneous rocks, such as zircon, apatite, diamond, carborundum, hypersthene. these inclusions can be classified into four groups:

(1) Single phase solid inclusion (C_{gl}, C_{Fe}, A_{gl}, C_{Si}, C_P, C_C)

(2) Two phase melt inclusion(A_{gl} + V , A_{gl}+ A_{Fe}, A_{gl} +nV, A_{gl} + C_{Si}, C_{Si} + V , C_{Si} + C_{Fe}, C_{Fe}+nV, A_{gl}+ nC_{Si},)

(3) Multiphase melt inclusion (C_{Si} +A_{gl} + V, A_{gl} +C_{Si} +C_{Fe} +nV, A_{gl} +C_{Si} +V, A_{gl} +C_{Si} +nV, C_{Si}+V , A_{gl} + C_{Fe} +V)

(4) Unmixing phase melt inclusion (A_{gl} +A_{Fe} , A_{gl} +nC_{Si} + nC_{Fe} +V, A_{Si+Fe}+A_{gl} +V)

The traits of four groups are as follows:

1. Single phase solid inclusions contain only one phase , which may be amorphous or crystalline silicate, or magnetite and graphite. The apatite solid inclusions usually existed in zircon of granite, suggests that phosphate separate out earlier than that of zircon in the magma .Carbonaceous (perhaps graphite) and silicate melts usually appeared in the diamonds of kimberlite also. Solid inclusions are usually n —10 μ m in size, which had been mechanically captured by the accessory minerals, and indicated that the include material had formed earlier than host minerals. In the heating— up process, these inclusions usually did not change their phases contained, and can't be taken as the sample for determination temperature.

2. Two phases melt inclusions include the following types:

(a) Amorphous two phases melt inclusions (photo 2): they are composed of glass and gas, the range of size is n—60 μ m .the gaseous phase ratio is in the range of 1—25% and can indicate the viscosity of magma .these kind of inclusions usually distributed in the accessory minerals of volcanic and subvolcanic rocks and sediments of submarine volcanic eruption. There exist relations between the phase features of inclusion and the dramatic dropping of temperature and pressure in magma. In addition, the amorphous two phases melt inclusions also distributed in the accessory minerals (zircon, pyrope)of some granites and kimberlites. The shape of this typical inclusions in zircon is ovoid, tubular or irregular, with its long axis paralleled to the C axis of zircon crystal. The inclusions of pyrope are composed of amorphous silicate and material riching in iron (A_{gl} + C_{Fe}). The silicate of inclusion is about the range of 80% , but the gaseous phase is very little, and the higher magma density was dedicated by the sieve—like gaseous phase distributed in the silicate. There are some relations between this kind of inclusion existed in plutonite and immiscible magma or melt resulted from the dropping temperature.

(b) Crystalline two phases melt inclusion (photo 1): this kind of inclusion consists of crystalline and gaseous phase , and mainly distributed in zircon of granites, granodiorites and quartz monzonites. It is usually big, ranging from 3 × 6 to 20 × 60 μ m, with one individual as large as 142 × 5 μ m, from this kind of inclusion, the gaseous phase is in the range of 10—40%. The shape of this kind of inclusion, which usually situated in the center of the host minerals, is tubular, and its long axes are paralleled to the C axis of zircon crystal, and also about 1/4 to 4/5 of the length of zircon crystal housing them (photo 1). It is obvious that their shapes are controlled by the way the host minerals grow. small inclusions (long axes is less than 10 μ m), with one of two phases (C_{Si}, C_{Si}+V), are usually located in the cone area of zircon crystals or near their surface. The inclusions located in the center of zircon are mainly three phases (G_{gl} +C_{Si}+V)with only a small portion of two phases inclusions. This indicates that the sensitivity of granitic magma, in which the zircon crystallized, had decreased with time. in some zircon crystals, the phase appearances of melt inclusions differed greatly from each other locally, even they are in the same growth zone of single zircon crystal, and the crystalline and amorphous melt inclusions are usually coexisting. This indicates that the magma was heterogeneous and immiscible during some periods of time. In the pyrope, the inclusions are consist of C_{Si} +C_{Fe} ,gaseous phases are very little and difficult to observe. It suggests that the viscosity of magma is quite high. In addition, a few inclusions of this kind have been discovered in the hypersthene of sediments of submarine volcanic eruption.

3. Multiphase melt inclusions are mainly found in zircons of granites, quartz monzonites, porphyry granites and granodiorites, most of them are three phase (Photo. 3).The gas percentage of these inclusions varies from 3% to 15%, with a mean of 5%. This shows that the magma included in multiphase inclusions is denser than that in two phase one, as the denser the magma inclusion, the lower the gas ratio in it. The gaseous phase is distributed in the same way as the mentioned types. The shape of this kind of inclusions mostly assumes the form of tube and negative crystal, occasionally oval, with their long axis paralleled to the C axis of zircon, and those with long axes perpendicular to the C axis of zircon are in the minority. Their sizes range from 6 × 8 to 20 × 60 μ m , usually 15 × 20 —24 × 32 μ m, with axis 3/5 —1/2 the length of the host mineral . Large individuals are usually distributed in the center of the crystal ; and bigger the crystals, the larger the inclusion it contains. According to statistics, the amount of large

inclusion (larger than 20 μm) is only 1/5 — 1/10 (or less than) of small ones. The crystalline material is mainly colorless silicate, or iron-riched material. The inclusions are composed of C_{Si} + C_{Fe} + A_{gl} in pyropes of kimberlite, and the gaseous phase are scattered in the silicate. and scarcely appeared as single gaseous phase. This suggests that features of magma are lower volatile component, higher viscosity and immiscible.

4. Immiscible phase melt inclusions are composed of gaseous phase and powdery iron — riched material which well — distributed in the glass (photo 4) , and distributed in some mineral. such as: hypersthene in the sediments of submarine volcanic eruption , pyrope of kimberlite and zircon of some granites. Immiscible amorphous melt inclusions are brown and black , silicates ranging of 60— 70% in inclusions and gaseous phase distributed as sieve — like in glass sometimes . This kind of inclusions entrapped iron — riched immiscible silicate magma at the speedy or temperature dropping . Silicate melt in inclusions boiled and gaseous phase concentrated relatively in the heating —up process .

III The rock — forming temperature determination of melt inclusion in accessory minerals of various igneous rocks (table. 1 , Fig. 1)

1 Granite : The melt inclusions in zircons of granites and quartz porphyry , which related to the deposits of rare earth elements (REE) , tungsten , tin and gold , began to fuse at 700 — 900 ℃ and homogenized at 780 — 1180 ℃ . The homogenization temperature of REE — bearing granite (Raoping) is at 1020 — 1180 ℃ , tin — bearing (Gejiu) at 800 —1200 ℃ , mostly 925 — 1056 ℃ ; tungsten — bearing granite and quartz porphyry at 950 —1000 ℃ ; auriferous granites (Shandong) at 780 — 980 ℃ , among them , the temperature of auriferous granite is lower than that of former . The temperature of early stage granite is higher than that of later in the same district , such as Gejiu , early stage of Youchahuand Masong porphyry biotite at 800 — 1200 ℃ , 840 — 1056 ℃ on the average , second stage of Senxianshui middle grain biotite granite at 800 — 1200 ℃ , 966 ℃ on the average , late stage of Laoka fine — middle grain biotite granite at 800 — 1000℃ , 925℃ on the average . there are some difference of melt inclusions homogenization temperature in zircon among various period granites . The melt inclusions of Shanxi granites began to fuse in the temperature range of 800 —950 ℃ , and homogenized at 950 —1100 ℃ . In the same district , the homogenization temperature in the range of early Proterozoic era granite (Erxianzi magmatic complex) are 950 — 1000 ℃ , late proterozoic era (Hannan magmatic complex) and late Yenshan period (Dongjiankou magmatic complex) is 950 — 1100 ℃ ,with a few samples higher than 1100 ℃ . This means that crystalline temperature of younger granites is slightly higher than that of older granite .

2 Intermediate — acid rocks: The intermediate —acid intrusive rocks related to iron —copper deposits are widely distributed in the middle —lower reached of Yangzi river. Melt inclusions of intermediate—acid intrusive rocks in Ningzhen district began to fuse in the temperature ranges of 640—1020℃, and homogenized at 740—1080℃.The various faces belts exist various homogenization temperature of melt inclusion in zircon,

Table. 1 Phase characteristics and homogenization temperature of melt inclusions in accessory minerals

Sample NO.	host mineral	Phase ratio in melt inclusion (%)	Initial melting temp. (℃)	Homogeniz—tion temp. (℃)	Type of melt inclusion
G1—25—12/1	zircon	C_{Si}=82; C_{Fe}=8; V=10 C_{Si}=87; C_{Fe}=2; V=8 C_{Si}=88; C_{Fe}=2; V=10 C_{Si}=92; V=8	850 850 850 850	950 950 950 950—1000	Crystalline Crystalline Crystalline Crystalline
G2—30/2—2	zircon	G=75; V=25 C_{Si}=90; V=10 C_{Si}=60; V=40	800 850 850	850 950—1000 950	Amorphous Crystalline Crystalline
G2—140/1—2	zircon	G=60; C_{Fe}=10; V=30 C_{Si}=75; V=25 C_{Si}=67; G=8; V=25	950 850 950	1050 1000 1000 5%Cryst	Amorphous Crystalline Crystalline
G14—173/2—2	zircon	C_{Si}=85; G=15; V=10 C_{Si}=89; G=3; V=8 G=60; V=40	950 850 800	1100 1050 950	Crystalline Crystalline Amorphous
L—1	zircon	G=85; V=15	900	1050	Amorphous
L—3	zircon	G=75; C=15; V=10	950	1100	Crystalline
L—4	zircon	G=80; V=20	800	1000	Amorphous
GE 1	zircon	C_{Si}=70; C_{Fe} =5; V=10	700	940	Crystalline
83—T11	pyrope	C_{Si+Fe} =90; C_{Si}=10	600	950	Crystalline
W02	zircon	A_{Si}=50; V=35; C=15	700	980	Amorphous
W04	zircon	A_{Si}=30; V=25; C=45	750	820	Crystalline
H81—1	zircon	A_{Si}=20; V=40; C_{Si}=40	800	840	Crystalline
H81—2	apatite	A_{Si}=90; V=10	700	740	Amorphous
H12	zircon	C_{Si}=50; C_{Fe} =30; V=20	1020	1080	Crystalline
83—86	zircon	C_{Si}=40; A_{Si}=10; V=40	700	1000	Crystalline
83—10—1	zircon	C_{Si}=65; A_{Si}=10; V=35	700	950	Amorphous
7205/93—101	hy	A_{Si}=600	600	850	Amorphous
7205/G—17—101	hy	A_{Si+Fe} =85; V=15	700	1100	Amorphous
HN—01	Sapphire	A_{Si}=85; V=15	785	1125	Amorphous

G1—25—12/1—granite of Erxianzi, G2—30/2—2—middle grain hornblend granodiorite of Hannan, G2—140/1—2—quartz diorite of Hannan, G14—173/2—2—biotite diorite of Dongjiangkou, L—1—coarse grain granite of Lianhuashan, L—3—quartz porphyry of Lianhuashan, L—4—volcanic breccia of lianhuashan, GE—1—porphyry biotite grantie of Gejiu Masong, 83—T11—kimberlite of Mengyin, W02—granite of Linglong, Shandong, W04—granodiorite of guojialing, Shandong, H81—porphyry granodiorite of Suzhou, Jiangsu, H12—quartz Monzonite porphyry of Suzhou, Jiangsu, 83—86 and 83—10—1—alkaline syenite of Saima, Liaoning, 7205/93—101 and 7205/G—17—1—sediments of submarine volcanic eruption from Okinawa, HN—01—basalt of Penglai, Hainan

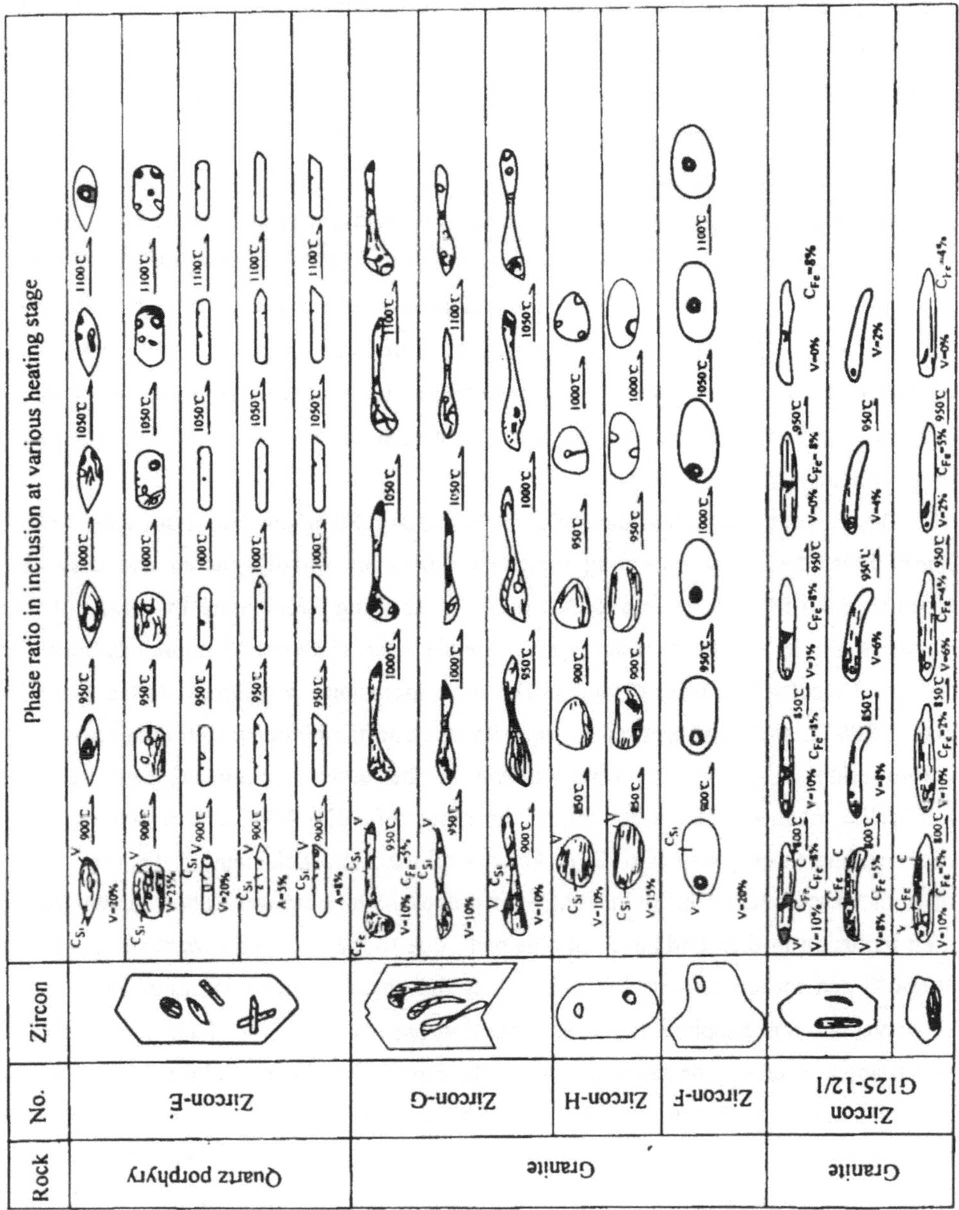

Fig. 1 Phase-assemblage variation of heated melt inclusion in accessory minerals

such as Gaozi — Xiaoshu complex, marginal faces quartz syenite diorite porphyry at 900 — 1080 ℃, transitional faces granodiorite porphyry at 880 — 900 ℃, central faces porphyry granodiorite at 740—840℃. As results mentioned above, the forming temperature dropped progressively form the margin to the center. In addition, late stage quartz diorite porphyry homogenized at 720℃, which is lower than that of early stage mentioned above. The forming temperatures of zircon and apatite in the granodiorite porphyry, both of them are at 900℃, but in porphyry granodiorite and quartz monzonite porphyry, zircon homogenized at 840—1080℃, which is higher than that of apatite at 900—740℃. this means that phosphate is actively than that of zirconium silicate in intermediate—acid magma.

3. Alkaline rocks: Crystalline melt inclusions were extensively distributed in zircon of syenite from Saima,Liaoning the inclusions consist of C_{Si}+V, C_{Si}+A_{gl}+V(C_{Si}=45—80%, A_{Si}=10—15%. V=20—40%). They are rectangle, ovoid triangle, and their sizes range form 20 × 20—20 × 40 μ m. The melt inclusions start to fuse at 740—840℃ and homogenized at 950—1000℃ in the heating—up process. The refractive index of crystalline minerals in inclusion is lower than that of host minerals. The complex is magmatic genesis.

4. Basic rocks: Abundant diamonds in Shandong Mengying are associated with the kimberlites, their rock—forming mineral olivines are usually altered and easy to be weathered, so it is impossible to determine the forming temperature by means of those minerals. But by using of the accessory minerals, such as pyrope, diamond, carborundum, the good results had been obtained. The melt inclusions in pyropes started to fuse at 500 —700℃, and homogenized at 800—950℃, and the homogenization temperature of carborundum is at 600℃. Carborundum is the accessory mineral separated from kimberlite magma when it cooling to the range of 500—600℃.There are two phase (Csi+Ac)and three phase (Csi+CFe+Ac) crystalline melt inclusions in diamond and no obvious phase change in the heating—up process. The homogenization temperature of melt inclusion in diamond have not obtained, due to the diamond crystal starting to fuse at 600℃ and carbonization at 800—900℃, so that there is much work to be done.

Hyacinth and sapphire are usually found in basalts form East Chain (Hainan Penglai and Jilangsu Liuhe). The sapphires in Penglai basalt contain crystsalling and amorphous melt inclusions which consist of A_{gl}+V, A_{gl}+V+C, C_{Si}+A+V. The size of inclusions range between o.n—3.6 μ m and the ratios of gas range between 5—10%. In addition, the fluid melt inclusion also found, which are composed of A_{gl}+L_{CO_2}+V (V=5-10%,L_{CO_2}=20-30%, A_{gl}=60—70%). The most inclusions began to fuse at 760—825℃and homogenized at 1125—1265℃, which is higher than that in rocks(860—1230℃)

5. Sediments of modern submarine volcanic eruptions:

In order to probe into the temperatures of modern submarine volcanic eruption , the phase features and forming temperature melt inclusion in hypersthene and quartz contained in the sediments of volcanic eruption from Okinawa trough , were studied by us for the first time in 1986. The types of melt inclusion are complicated and composed of two phases . three phases and immiscible phases (C_{Si} , A_{gl} +C_{Fe} +V . A_{gl1}+ A_{gl2} + V . A_{gl1} +A_{gl2} +L_{CO_2}+V . A_{Si+Fe} +A_{gl} +V) (photo 5) , and there exists a fluid inclusion in the heating —up process , melt inclusion began to fuse at 500 — 800℃ , and homogenized at 700 —1100℃. which is lower than that of terrestrial basalts (890 — 1450 C) . The samples from the northeastern part of the geosyncline gave out relatively low forming temperature of 750 — 950 ℃, mainly 800 — 950 ℃, samples from the middle east area gave out higher forming temperature , i.e. 900 — 1060 ℃, while samples from middle south area gave out forming temperature between 800-1000℃. Although there are some temperature changes from the northeast to middle south area , the changes are very small , and only 50 ℃ , which displays the features of fissure eruption and is in accordance with the geology of geosyncline . in addition , the content of material riching in iron of the melt inclusions in hypersthene usually reaches 10 — 30 % , and also indicates the magma is a basic one .

IV. The chemical composition of melt inclusion in accessory minerals

In order to inquire into the chemical characters of early stage magma , we analysised the composition of melt inclusion in zircon of Gejiu granite . The inclusion is crystalline melt type (Csi +V) . in which the gaseous phase is 20% . The inclusions were heating — up , they homogenized at 1050 ℃ . The samples were treated specially and then took spectrum analysis by 650 —H type electronic microscope . According to the experiment results (table . 2) .the content of P2O5 ,ZrO2 . TiO2 , and Cr2O3 , in melt inclusion is

Table. 2 Results of electronic microprobe analysis of melt inclusions in zircon

No. Of sample	Position	composition (%)										
		SiO_2	Al_2O_3	MgO	CaO	Na_2O	K_2O	TiO_2	Cr_2O_3	ZrO_2	P_2O_5	FeO
1	M	52.89	35.9	0.68	1.44	0.76	3.03	1.25	0.45	1.27	0.36	1.93
2	E	65.66	18.85	0.31	1.16	1.18	3.63	0.58	0.21	5.89	1.78	0.74
7	Zc	34.55	2.02	0.80	0.15	0.86	0.22	0.22	0.29	43.96	16.6	0.30
ACMI		66.83	15.93	0.53	1.13	1.48	2.83	0.55	0.45	6.79	2.43	0.97
ACZc		33.97	1.42	0.72	0.05	0.79	0.07	0.08	0.097	46.4	16.29	0.01
ACGZc		67.32	15.31	0.62	1.96	3.05	5.57	0.46	0.00	0.00	0.20	3.68

M —middle part of melt inclusion . Zc —Host mineral zircon. E —Edge part of mel inclusion . ACMI —Average content in melt inclusion . ACZC —Average content in host mineral Zircon. ACGZc—Average content in granite containing zircon.

higher than that of the host rocks , but the content of Na2O, K2O .CaO , FeO. SiO2 , is lower . Near the margin of the inclusion , the SiO2 content is higher than that in the center , and the feature of differentiation is obvious , K2O/Na2O ratio is slightly higher than that of host rocks . P2O5 .ZrO2, Cr2O3 is obviously higher . This gave some proves of apatite ,zircon and chromite etc. separated from the early period of magma , based on the statistics of 10 sample compositions , the early period compositions of magma are heterogeneous .

The aforementioned result indicates that using the silicate melt inclusions in accessory minerals to study rock -forming physical and chemical conditions of igneous rocks , we can get not only the initial rock-forming crystallization temperature , but also a lot of significant information about phase equilibrium conditions and material components of the initial magma system during initial crystallization of magma . It is very useful to research the present diagenetic theories of magma and go far into the relations between magmatism and movement of mantle material . It is a great breakthrough that taking advantage of the silicate melt inclusions of accessory minerals to research the temperature and magmatic components on the present research technology of inclusions in minerals . The way needs only simple equipment and is easy to operate and fine on result . It may be widely spread and used in production , science research and education .

It is important to study the silicate melt inclusions in accessory minerals. However , it is also practical on the research of astronomy science, meteorites, lunar rocks , sedimentary metamorphic rocks and alluvial sandy minerals or so .

I am grateful to prof. Yang Rongyong ,Yang Zhongfang , Niu Hechai , Dr. Zhou Fengying and my postgraduate Li Wen for their help to this paper .

Reference

1. Ermakov , N. P. , 1972 Geochemical system of inclusions in minerals . Nedra press , Moscow (in Russian)

2. Sobolev V S and Kostyuk V . P etal. , 1975 Magmatic crystallization based on a study of melt inclusions " Nuka " Press , Novosibirsk (in Russian)

3.Li Zhaolin etal. , 1978 Preliminary study on inclusions in mineral of the Jilin meteorite shower Proceeding of Jilin meteorite shower , Science Press, Beijing: 139—144(in Chinese,English abstract).

4. Ermakov, N.P.and dolgov. Y.A.,1979, Thermobaro geochemistry. Nedra Press, Moscow, (in Russian)

5. Li Zhaolin, wu Qizhi etal., 1982 The determination of formation temperature of the igneous rocks from some areas. Scientia Geologica sinca, 1:80—86.

6. Rodder. E., 1984 Fluid inclusions. Mineralogical Society of America Reviews in Mineralogy, 12:644.

7. Li Zhaolin, Chen Jianlin etal., 1986 The physiochemical conditions for the sedimentation of Okinawa submarine geosyncline volcanic products . Proceedings of Symposium on Geology of Continental Margins

(abstracts) Sponored by Nanjing University . October 20th November 2th 1986 .Nanjing. People's republic of China 166 169

8 Liu yinjun . Li Zhaolin etal . 1986 Study of melt inclusion in some basalt from eastern China Geochemistry . 5 (2) 108 121

9 Li Zhaolin . 1987 A new method of rock forming temperature determination of melt inclusions in accessory minerals of igneous rocks Scientis Sinica (Series B) XXX (9) 986 993

10 Li Zhaolin Pang Xuebin etal 1992 A study of geochemical characteristics of volcanic rocks in some seamounts of the South China Sea Process to 29th IGC Geological Publishing House . Beijing 121 123

11 Li Zhaolin 1992 Rock forming temperature determination applying melt inclusions in accessory minerals in granite Petrogenesis and Mineralization of Granitoids Proceedings of 1987 Guangzhou International Symposium Science Press .Beijing 209 220

12 Touret J 1993 CO transfer between the upper mantle and the atmosphere temporary storage in the lower continental crust Terra Research Terra Nova . 4 87 98

13 Li Zhaolin Yang Zhongfang 1996 Rock forming ore —forming temperatures of Lianhuashan tungsten deposit Guangdong province China Journal of Geochemistry . Vol 15 No 3 PP 239 248

Note: The project supported by National Science Foundation of China, Special Foundation of Scientific Research for Doctorial Subjects of Institution of Higher Learning and The Frontal Projects in Basic Research of Zhongshan University.

Plate

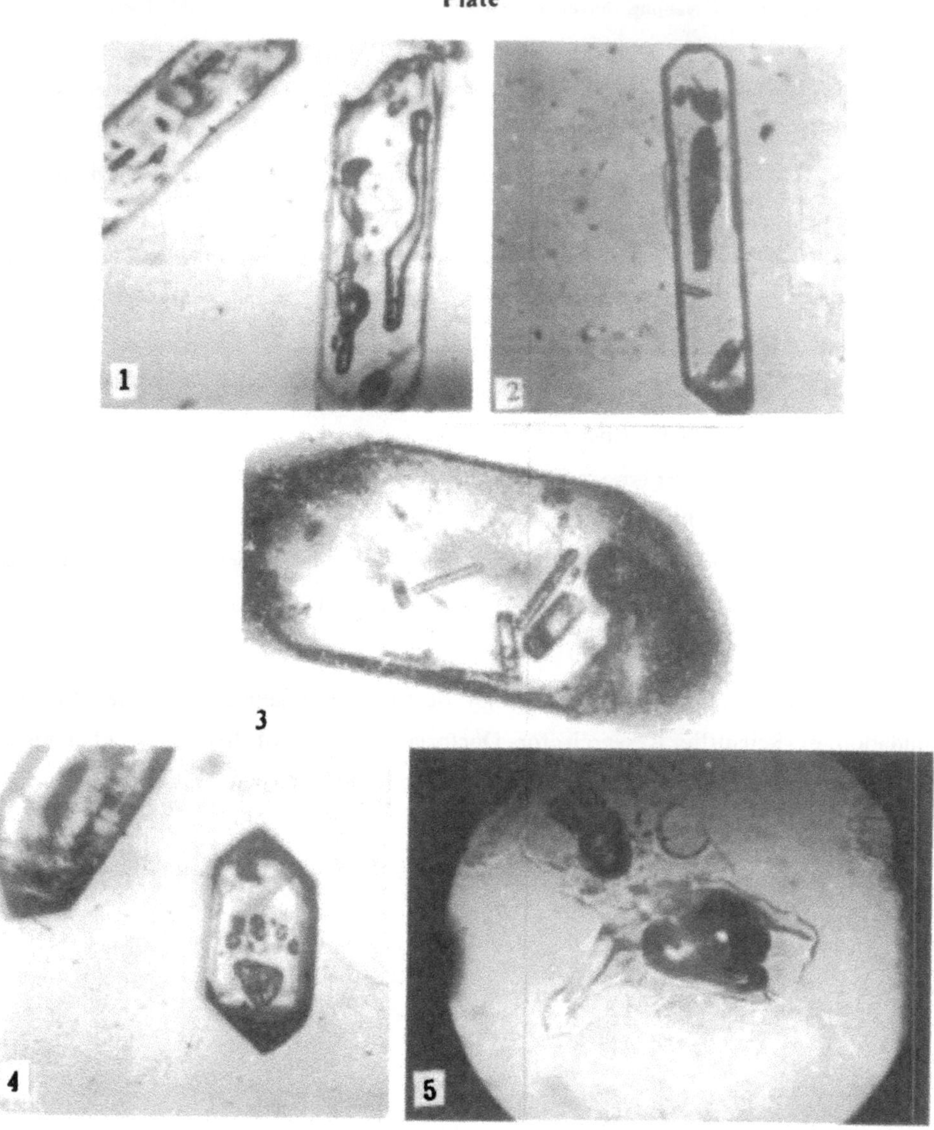

1.Two phase melt inclusion(C_{Si}+V) in zircon in Gejiu granite, × 250.

2.Multiphase amorphous melt inclusions(A_{Si}+V, A_{Si}+C_{Si}+V) in zircon in Dongjiangkou biotite diorite, × 250.

3.Multiphase melt inclusions(C_{Si}+C_{Fe}+V, C_{Si}+V) in zircon in Shandong granite, × 250.

4.Immmiscible phase melt inclusions(A_{Si}+nC_{Si}+nC_{Fe}+V) in zircon in Hannan granite, × 250.

5.Melt inclusion(A_{Si}+C_{Si}+V) in volcanic glass in Okinawa trough, × 200.

Proc. 30th Int'l. Geol. Congr., Vol. 16, pp. 277-297
Huang Yunhui and Cao Yawen (Eds)

Geology and fluid inclusions studies on Keketuohai No. 3 rare-element pegmatite, Xinjiang, Northwest China

Huan-Zhang Lu [1&2] Zhonggang Wang [2]
1. Science de la terre, Université du Québec à Chicoutimi, Quebec, G7H 2B1, Canada
2. Institute of Geochemistry, Academia Sinica, Guiyang, Guizhou, 550002, P. R. China

ABSTRACT

The Keketuohai No. 3 pegmatite, the largest in China, located in Fuyun county, Xinjiang province, being mined for Li, Be, Nb and Ta contained spodumene, beryl, columbite, lepidolite and pollucite. The pegmatite occurs in a gabbro which lies above a biotite granite. It consists of 10 zones (in the probable sequences of formation): I) graphic zone with an assemblage of quartz, feldspar and beryl; II) coarse grained albite zone, the Be ore body; III) massive microcline zone, with beryl; IV) muscovite and quartz zone, with Be, Nb, and Ta mineralization; V) cleavelandite and spodumene zone with Li, Nb and Ta mineralization; VI) quartz-spodumene zone rich in Li, Nb and Ta; VII) muscovite and albite zone with Ta and Cs mineralization; VIII) albite-lepidolite zone Li, Ta, Rb, Cs ore body; IX) quartz-pollucite zone with Cs mineralization; and X) massive microcline and quartz core, rich in Rb and Cs.

Fluid, fluid-melt and melt inclusions studies in spodumene, beryl, tourmaline and coexisting quartz from different zones show five types of inclusion: 1) melt inclusion, 2) fluid-melt inclusion, 3) H_2O-CO_2, 4) Halite bearing and 5) aqueous inclusions. Melt inclusions are found in zones I and III, and melt-fluid inclusions (consist of melt and fluids, the fluids are mostly H_2O and CO_2, in an inclusion) occur in zones I, III and VI. Aqueous inclusions occur in all zones. Most H_2O-CO_2 inclusions are found from Zone VI to X. Halite bearing inclusions coexisting with aqueous inclusions occur mostly in zones IV to VII. Microthermometric data show that homogenization temperatures (Th) of melt inclusions ranges from 700 to 850°C. Th of melt-fluid inclusions ranges from 299 to 465°C, and that for aqueous, halite bearing and H_2O-CO_2 inclusions ranges from 220 to 404°C. The salinity of aqueous inclusions range from 5 to 9 wt.% NaCl. equiv., which is relatively low. The salinity of halite bearing and some fluid melt inclusions range from 28.50 to 32.00 wt.% NaCl equivalent. The Th H_2O-CO_2 inclusions range from 220 to 311°C and the pressure calculated from H_2O-CO_2 and CO_2 inclusions is from 1800 to 2500 bars.

The results indicate that the No.3 pegmatite was formed during the transition from magmatic to subsolidus hydrothermal fluid conditions, occurring probably in three stages. Stage 1 was magmatic; stage 2 required a processes of immisciblity to produce melts and fluids; the final stage was predominantly hydrothermal fluids. Li, Be, Nb, Ta and Cs mineralization occurred throughout this transition.

keywords: pegmatite, Keketuohai, fluid, fluid-melt and melt inclusions, immisciblity, magma, hydrothermal fluids

INTRODUCTION

Most geologists recognize that an important genetic link exists between granite and the formation of pegmatites. How pegmatites are related to larger masses of granitic magma is as enigmatic today as it was almost half a century ago [7]. Hypotheses for the generation of pegmatites include subsolidus vein fillings related to dilatant strain rate [4], in situ anatexis in response to regional geothermal gradients [12] derivation from largely consolidated granite bodies[2,3] and formed during magmatic-hydrothermal transition [6, 7, 11]. This paper is a study of the Keketuohai No. 3 pegmatite to test the genesis and how to form the Be, Li Nb and Ta mineralization.

The Keketuohai No. 3 pegmatite, the largest in China, is located in Altay district, Fuyun county, Xinjiang province, NW China (Fig.1). The No. 3 pegmatite was discovered in the 1940s and was mined by the former Soviet Union. In 1951, the Soviet-Union -China Nonferrous and rare elements Co-op was established and put in charge of development and systematic exploration in Altay area. In 1955, all capital in this co-op was transferred it to China and thus the company's name Changed to Xinjiang nonferrous metal Co-op. Production, during this period, was on a small scale and all products were transported to the Soviet Union. On 1960, large scale of geological survey had been undertaken and Keketuohai mine company formed to operate the No. 3 pegmatite. The authors have studied this pegmatite since 1979, with research projects on the mineralogy and geochemistry, and revisited the mine in 1992. There are numerous geologists including ones from the former Soviet Union worked in this mine and they call it "Mongolia No. 3 pegmatite". It is a wrong name because this pegmatite is located in the Xinjiang province of China, not in the Mongolia. Zhou et al. (1986) describe the mineralogical and structural zones of this pegmatite and Wang (1993) did comparative studies between the No.3 pegmatite and others located in the Keketuohai area. However very little attention was paid on the magmatic and fluid inclusions aspect and specially on the genesis of the pegmatite.

GEOLOGY OF NO.3 PEGMATITE OF KEKETUOHAI

Regional geology of No. 3 pegmatite

Tectonically, there are two subtectonic plates: the Siberia and Kazakhs plates in the Altay area (Fig. 1). The No. 3 pegmatite located in the south margin of Siberia plate and in the central part of NW-SE Altay Caledonian- Hercynian fold belt [15,]. For detail about the tectonic, magmatism and metamorphism of the Altay area, which it can be found in the paper of Zhang et al., 1990 and Fig.1.

The Regional geology of Keketuohai: The lithology of this region including Carboniferous sandstone, siltstone, and limestone, Devonian sandstone, limestone and volcanic clastic rocks, and Ordovician sandstone, slate and shale. The Caledonian intrusions intruding into the Devonian and Ordovician strata, are norite-amphibolite, tonalite- granodiorite, biotite-plagioclase granite, biotite granite and tourmaline muscovite granite, most are

I type granite with minor S type ones. The Hercynian intrusions comparing to the Caledonian ones are small in size and intruding into the Caledonian plutons, consist of gabbro, diorite, norite, biotite granite, amphibolite-biotite two feldspar granite, biotite granite, two mica granite and muscovite granite, most are S type granite with minor amounts of I type granite. The pegmatites in this region are related to the Hercynian porphyritic biotite granites, and two mica granites. In the Altay area, there are 7 pegmatite districts (Fig.2, including the Keketuohai district) containing about over 1,000 pegmatites but the most important one is the Keketuohai district containing not only the No.3 pegmatite, the biggest, mineralized pegmatite, but also another 24 pegmatites. The Keketuohai district is about 7 km^2 and consists of Ordovician sandstone, slate and shale and intrude by Hercynian gabbro, porphyritic biotite granite, biotite granite and muscovite and two mica granite.

In the Keketuohai district, the pegmatites, among of them 11 of the 25 pegmatites containing Li, Be, Nb and Ta mineralization. The No.3 pegmatite intruding into a gabbro, strikes 310°, dipping SW at an angle of 80-90° in the upper part of the ore body and 10-25° in the lower part. It forms a cupola shape body in the upper part and low angle body in the lower part, and is seated in a gabbro in the close vicinity (deep part) of a granite body. Both the gabbro and granite are Hercynian in age. An granite dyke cut the gabbro and the pegmatite cuts both. The granite is characterized by higher contents of Be, Li, Nb, Ta, Rb, Cs and F than that of Clark value, but no economic significance. Up to date, diamond drilling and open pit exploration has intersected the pegmatite approximately 2000 meters along strike and 1500 meters down dip; the thickness ranges from 20 to 60 meters. The pegmatite occurs as a elliptic cylinder with a long diameter of 250 m, and short diameter of 150 m at surface (Fig. 3a). The contact of the pegmatite with the gabbro is knife sharp, and with several inclusions of gabbro in the pegmatite. The alteration zone of pegmatite is restricted to a narrow zone not exceeding 3 meters inside the gabbro. The alteration mineral assemblages include Li-bearing amphibole, Cs-bearing biotite, muscovite, tourmaline and fluorite. The Li, Rb, Cs and F elements are not only concentrated in the pegmatite but also in the alteration zone. Since the mining have been mining over a half century, the internal structural zones of No.VIII and IX described in the following section, have been mined out. In this study, all the samples were from our field work in 1992 except for zones No. VIII and IX are which from our previous work.and Since these two zones only contribute about 0.1 % of volume, it should not be as important as other zones.

Internal structure zones of No. 3 pegmatite

The No.3 pegmatite is mined for Li, Be, Nb and Ta contained in spodumene, beryl, and pollucite. As shown by earlier investigators [15, 17], the No. 3 pegmatite has a complex symmetry concentric zoned structure. On the basis of mineral composition, texture, and location within the pegmatite, 10 internal zones (in the probable sequence of formation) have been distinguished (Fig.3 and Table 1):

(I) graphic zone with a assemblage of quartz, feldspar and beryl;
(II) coarse grained sugary albite zone, Be (Nb) ore body;

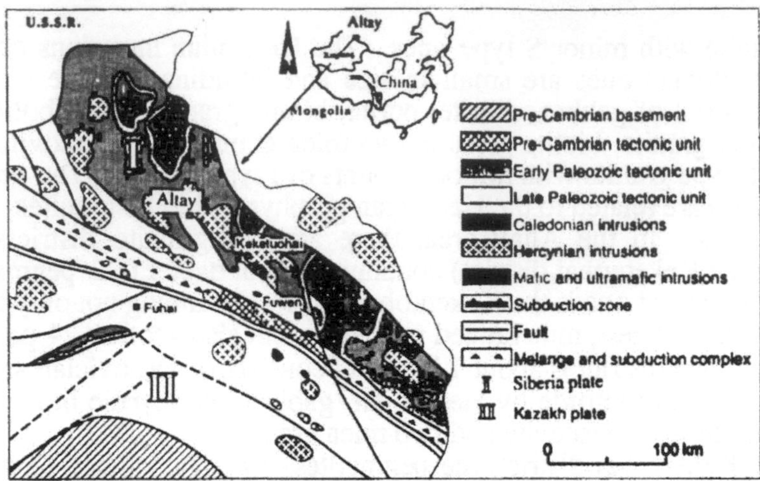

Fig. 1. Geological map of Altay region, Xinjiang province, China (modified from Zhang et al. 1990) Tectonic units of this region include Pre-Cambrian basement, Pre-Cambrian tectonic unit, Early and late Paleozoic tectonic units. The Caledonian and Hercynian intrusions intrude into the three tectonic units. In the subduction zone of Siberia and Kazakhs plates, mafic and ultramafic intrusions, and some melange and subduction complex occurred. The lithologic units are suffered various degree of metamorphism including schist, granulite, gneiss and phyllite.

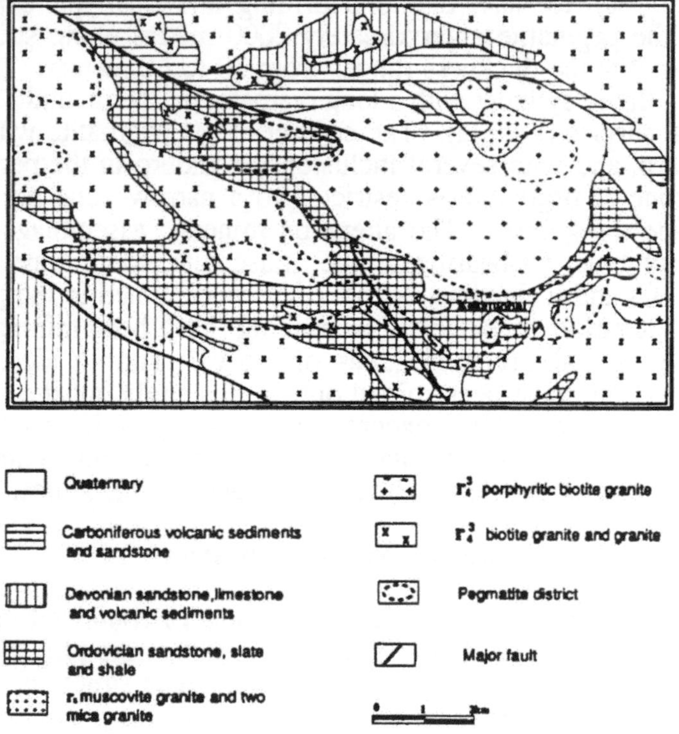

Fig. 2 Geological map of Keketuohai district (modified from Zhang et al. 1990)

(III) massive microcline zone, Be (Nb) ore body;
(IV) muscovite and quartz zone, with Be, Nb, and Ta mineralization;
(V) cleavelandite and spodumene zone, with Li, Nb and Ta mineralization;
(VI) quartz-spodumene zone, rich in Li, Nb and Ta;
(VII) muscovite and albite zone, with Ta and Cs mineralization;
(VIII) albite-lepidolite zone, Li, Ta, Rb, Cs ore body;
(IX) quartz-pollucite zone, with Cs mineralization; and
(X) massive microcline and quartz zone (core) rich in Rb and Cs.

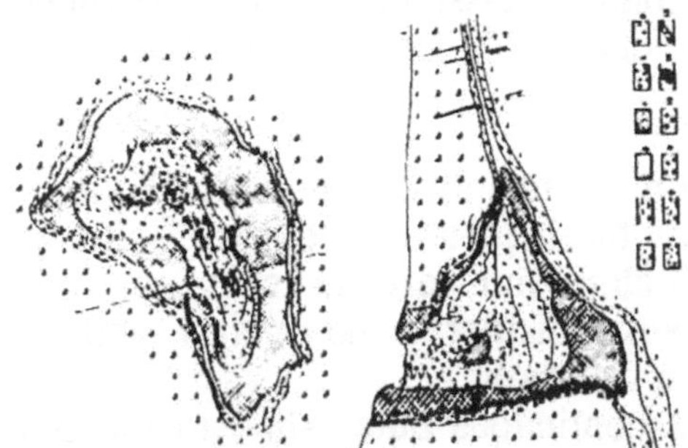

Fig. 3 Geological map of No.3 pegmatite in Keketuohai (after Xinjiang Nonferrous metal Co-op 1988) A: surface geological map; B: Cross section
1. gabbro; 2. graphic zone with a assemblage of quartz,feldspar and beryl; 3. sugary grained albite zone, Be (Nb) ore body; 4. massive microcline zone, with beryl,Be (Nb) ore body; 5. muscovite and quartz zone, with Be, Nb, and Ta mineralization; 6. cleavelandite and spodumene zone, with Li, Nb and Ta mineralization; 7. quartz-spodumene zone, rich in Li, Nb and Ta; 8. muscovite and albite zone, with Ta and Cs mineralization; 9. albite-lepidolite zone, Li, Ta, Rb, Cs ore body; 10. quartz-pollucite zone, with Cs mineralization; 11. massive microcline and quartz zone (core) rich in Rb and Cs and 12. granite dyke.

The calculated volume of different zones of the pegmatite, from zone I to zone VI occupied about 95 % of the total volume. A complete important mineral assemblage and zonation are given in Table 1 and Fig. 3. The mineralization consists of Li, Be, Nb, Ta, Cs and Rb; Be mineralization occur mainly in zones I to IV, that for Li is important in zone V to VIII, Nb and Ta mineralization are in zone I to VIII, and that for Rb and Cs in zone IX and X, but Rb is mostly substitute in muscovite, lepidolite and microcline, Cs is substitute in beryl, lepidolite and microcline, and mostly concentrated in pollucite. According to Zhang et al 1990, the bulk composition of minerals in No. 3 pegmatite are: muscovite 6.00%, Microcline, 32.10%; albite 21.28%; quartz, 35.00%; spodumene, 3.94%; biotite 0.50%; Beryl, 0.47%; lepidolite, 0.05%; pollucite, 0.005%; garnet, 0.17%, apatite,0.11%; tourmaline, 0.02%; columbite, 0.02% and amazonite, 0.02%. The major Be ore mineral is beryl, and the major Li ore minerals are spodumene and lepidolite, no petalite or eucryptite have been found. From the view of sequence of formation, these zones can be classified into two units: primary crystallization unit and metasomatic (replacement) units formed at the expense of pre-existing zones. Primary crystallization units differ with the replacement one in mineral assemblages, modes and textural styles (Table 1). With progressive crystallization from the margins inward, zones usually show increasing grain size, decreasing number of rock-forming minerals, and textural changes from granitic through graphic in intermediate zones to blocky and coarse grained monomineralic in the core [2]. In zones of I to VI both primary and metasomatic mineral assemblage are exist, but in zones VII to X only primary ones occurred, and the mineral assemblages are very similar to the metasomatic mineral assemblages in zones of I to VI. The metasomatic mineral assemblages are muscovite, albite, quartz, tourmaline, apatite, Li-mica and Rb-mica. The metasomatism is not as easily identified and defined as the primary units. Metasomatic effects range widely in extent at No.3 pegmatite including sodium, potassium and fluorine metasomatism. Details of the internal zones are discussed in the following section:

I. Graphic zone

The border zone consists mainly of fine grained quartz and albite with the accessory minerals being biotite, tourmaline, muscovite, garnet, apatite, beryl and columbite. Tourmaline and beryl crystals up to 5 cm in length are oriented more or less perpendicular to the contact, and sometime tourmaline continue growth into the wallrock.

This zone forms a continuous outer shell to the pegmatite with about 19.7% of volume of whole pegmatite and ranges in thickness from 3 to 10 meters (Fig.3). It is usually thicker along the footwall contact (Fig. 3b) and in lower part of the pegmatite. The albite and quartz form a beautiful graphic texture (Fig. 4c). In some parts of this zone, a braid albite and quartz zoning has been found (Fig 4b).

The contact between pegmatite and gabbro is filled with alteration assemblage of tourmaline and muscovite (Fig. 4a). The width of alteration zone varies from 0.5 to 3 m and usually is thicker in the upper part of the pegmatite.

II. Sugary albite zone

The sugary albite zone consists of sugar like crystals of albite and quartz with tourmaline, muscovite, beryl and columbite accessory minerals (Fig.4d). The zone forms a secondary outer shell of the pegmatite with about 16.1% volume of whole pegmatite and it ranges in thickness from 3 to 8 m. It is commonly thicker in the lower part than at the upper part (Fig. 3b). The grain size of minerals increases from the graphic zone inwards.

III. Massive microcline zone

The massive microcline zone is the thickest zone of this pegmatite. It ranges from 5 m. to 50 m.and occupies about 19.9% of the pegmatite's total volume. It consists of microcline, amazonite, albite, quartz (Fig. 4e) and muscovite, beryl and columbite accessory minerals. The crystal size of microcline ranges from 0.1 m. to several meters. The crystal size increases greatly from zone II to this zone.

IV. Muscovite quartz zone

The forth zone- muscovite quartz zone consists of muscovite, quartz, microcline with beryl, cleavelandite, tourmaline and columbite (Fig. 4f) accessory minerals. of It is an almost continuous zone with about 15.0% of volume of whole pegmatite. This is the Be, Nb, Ta ore body.

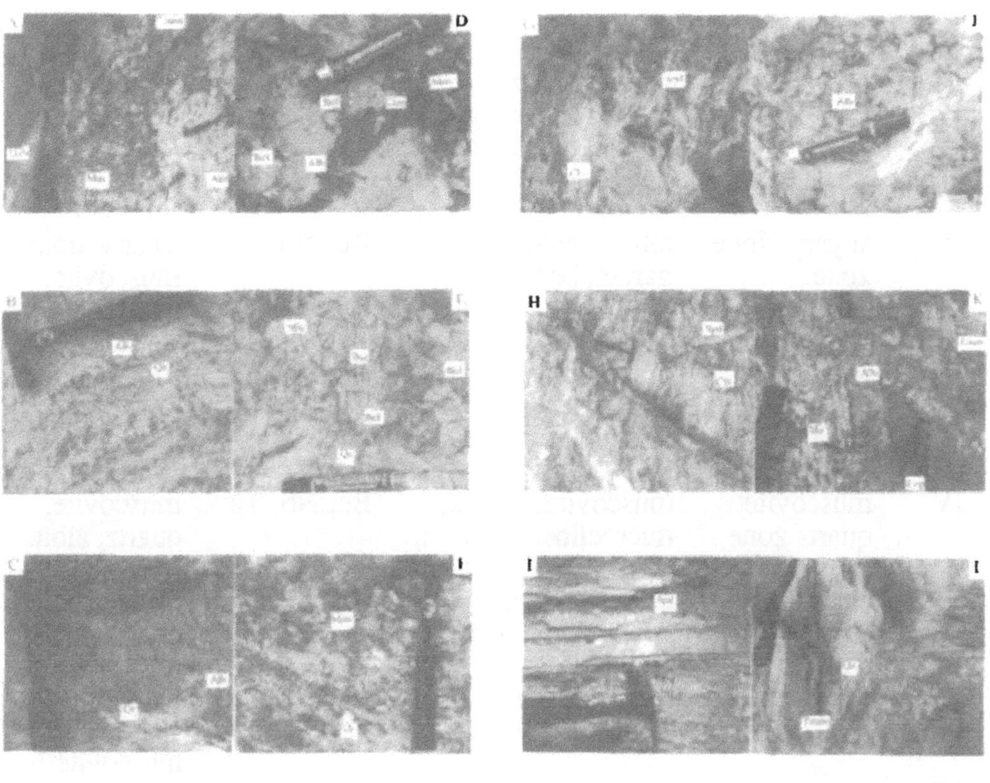

Fig. 4 Internal structural zones of No. 3 pegmatite

A: The contact between pegmatite and gabbro, developing an alteration zones consisting of tourmaline and muscovite; B: Graphic zone (I) with zoned quartz and albite; C: Graphic zone(I); D: Sugary albite zone (II) with albite, quartz, beryl and garnet; E: Massive microcline zone (III), with microcline, quartz, beryl and albite; F: Muscovite quartz zone (IV), with muscovite, beryl, quartz, and cleavelandite; G: cleavelandite and spodumene zone(V), consist of cleavelandite, spodumene, quartz, columbite and montebrasite; H: Spodumene and quartz zone(VI), with quartz, spodumene, cleavelandite and columbite; I: Big spodumene crystal in photo H; J: Muscovite and albite zone (VII); K: Albite lepidolite zone (VIII) with Albite, lepidolite and tourmaline; and L: Massive quartz microcline core (X), consist of massive quartz and microcline.

Spd=spodumene; Mic=microcline; Qz=quartz; Alb=albite; Toum=tourmaline; Bel=beryl ; Clv=cleavelandite; Mus=muscovite; Lep=lepidolite; Gab=gabbro; gan= garnet

Table 1. Internal structure zones of No.3 pegmatite.

Zone	Name	Primary mineral assemblage	Ore body	Metasomatic
I	graphic zone	albite, quartz, tourmaline, beryl biotite, muscovite garnet	Be, Nb	muscovite, quartz, sugary albite
II	sugary albite zone	albite, quartz, garnet, beryl	Be, Nb	sugary albite muscovite, quartz
III	massive microcline zone	microcline, quartz, albite, beryl, amazonite	Be, Nb, Ta	sugary albite
IV	muscovite quartz zone	muscovite, quartz, microcline, beryl	Be, Nb, Ta	muscovite, quartz, albite
V	cleavelandite and spodumene zone	cleavelandite, spodumene, quartz, columbite, montebrasite	Li, Nb, Ta	cleavelandite, quartz, muscovite, lepidolite, Li bearing muscovite
VI	quartz spodumene zone	quartz, spodumene, cleavelandite, columbite	Li, Ta	cleavelandite, Rb bearing muscovite
VII	muscovite albite zone	muscovite, albite	Ta, Cs	
VIII	albite lepidolite zone	albite, lepidolite, tourmaline	Ta, Li, Rb, Cs	
IX	quartz pollucite zone	quartz, pollucite, lepidolite, albite	Cs	
X	quartz microcline core	quartz, microcline	Rb, Cs	

V. Cleavelandite spodumene zone

This is the principal Li, Nb, Ta ore body. It consists of cleavelandite, spodumene, albite, microcline, muscovite and quartz with montebrasite, lepidolite, columbite beryl and Li-muscovite accessory minerals (Fig.4g). The thickness of this zone ranges from 3 to 30 m and in thicker in the lower part of the pegmatite (Fig. 3). The tabular spodumene crystals are from 0.2 to serval meters in length and form intergrowths with quartz and cleavelandite (Fig. 4H). The zone is commonly continuous and occupies about 14.8% of the total volume of the pegmatite. A low temperature alteration of primary minerals of this zone has produced a series of secondary assemblages such as albite, quartz, montebrasite, Rb bearing muscovite and K-feldspar, and apatite. Eucryptite and petalite have not been found in this pegmatite.

VI. Quartz spodumene zone

The quartz spodumene zone and Zone V are the major occurrence of spodumene. This zone consists of quartz, spodumene, cleavelandite with muscovite, columbite, montebrasite, and Rb-muscovite accessory minerals. It is a continuous inner shell of the pegmatite with about 8.7% volume of whole pegmatite and ranging 3 to 8 m in thickness. Large spodumene crystals are intergrown with quartz and cleavelandite (Fig. 4i). Zones V and VI are main Li ore body. Several crystal cavities with 1 to 5 cm in size containing albite, Rb-muscovite, quartz crystal and natrolite have been found in this zone.

Zone I to zone VI, it occupy of about 94.3% of the total volume of the No. 3 pegmatite.The rest consists of 5.7% of the total volume.

VII. Muscovite-albite zone

This zone consists of muscovite, albite, microcline, and quartz with columbite, beryl, microlite and lepidolite accessory minerals (Fig. 4j). It occupies about 3.2% volume of the pegmatite.

VIII. Albite lepidolite zone

In this zone, the major Li mineral is changes from spodumene to lepidolite (Fig. 4k). It contains albite, lepidolite, muscovite, quartz, and tourmaline with accessory apatite. This zone is not a shell and it is only located in the upper part of the pegmatite. It only occupied about 0.08 volume of whole pegmatite.

IX. Quartz pollucite zone

This zone contains quartz, pollucite with accessory albite, lepidolite and montebrasite. This is the Cs mineralization zone, and it is located in the upper part of the pegmatite only and with about 0.02% volume of whole pegmatite.

X. Quartz microcline core (zone)

This zone consists of quartz, microcline with accessory albite, montebrasite and tourmaline (Fig. 4l). It is located in the central part of the pegmatite and occupies about 2.4% of total volume of pegmatite.

Ore mineralization stages

Three mineralization stages can be found, in the probable order of formation, the Be, Nb, and Ta (beryl, columbite) mineralization stage; the early Li (spodumene) mineralization one and the late Li (lepidolite), Rb and Cs mineralization stage

Some of the characteristics of the No.3 pegmatite have already been established by earlier investigators [18,15]. There is little doubt that the pegmatite is of igneous origin, and is the product of advanced differentiation of granitic magma that was injected into a dilation zone crosscutting the wall rock [2,3]. There was no extensive interaction between the pegmatite and the surrounding amphibolic gabbro, except for minor introduction of Li, F, B, P and H_2O into the latter. Zones of I, II, III, IV, V, VI, and VII form symmetrically shells continuing from the outer shell to inner shell. The mineralogy of No. 3 pegmatite are similar to that of a granite the only difference being the grain size and mineralization. The mineral size in pegmatite gradually increases from the outer shell (zone I) to the inner shells with a decrease in mineral assemblage. The compositional and spatial relationships of the zones in the No. 3 pegmatite indicate that Zones I and III seem to be the products of early magmatic crystallization, and the Zones IV to VI could be the results of magma immiscibility by separation of a hydrothermal fluids from the magma. In other word, these zones are crystallized during the magma to fluid transition. In this process the cleavelandite was replaced by spodumene and quartz. Some spodumene occur in Zone V form a pseudomonrph of cleavelandite. From the zone VI to X metasomatic replacement is rare, but it exists from Zone I to Zone V. This indicates that the minerals in the zones of VII to X are crystallized from same system that is probably a super critical fluid system.

Lithium minerals occur as lithium aluminosilicates located in the central parts of the pegmatite. The occurrence of lithium aluminosilicates at Keketuohai is notable in that spodumene and lepidolite are found together with quartz. The Li mineralization can be divided into 2 stages: early spodumene, and later lepidolite. Textural evidence indicates that primary petalite was extensively replaced by two texturally different intergrowths of spodumene and quartz. Spodumene also occurs in very coarse grained laths (from one to several meters).

MELT AND FLUID INCLUSION STUDY

The fluid inclusion, and specially the magmatic and fluid-magmatic inclusion studies of No.3 Pegmatite have not been carried out. The one of the objective of this paper is to study the melt, fluid-melt and fluid inclusions in the ten zones of the No.3 pegmatite and to establish the character of the inclusions associated with the Pegmatite evolution. Then to determine the temperature, pressure of formation and composition of melt and fluid inclusions thus provide further evidence as to the Li, Be, Nb, Ta Rb and Cs mineralization. For realizing the objective the quartz, beryl, spodumene and tourmaline from ten zones have been chosen for study. The mineral beryl and spodumene are the ore minerals and primary minerals in the pegmatite, to study the inclusions in those minerals, we can directly obtain the information of Be, LI mineralization. Only primary quartz has been chosen for study, quartz is a mineral occur in all zones and trapping a lot of inclusions. The inclusions in tourmaline represent the metasomatic process since tourmaline only found in alteration assemblage of this pegmatite. The inclusion study have been carried on quartz from zones I to X, beryl from zone I to IV, spodumene from Zone V to VII, and tourmaline from alteration zone and zone VIII.

Fluid inclusion petrography

Fluid, fluid-melt and melt inclusion studies in spodumene, beryl, and coexisting quartz, and tourmaline from which show four types of inclusion: silicate melt inclusion, fluid-melt inclusion, H_2O-CO_2, halite bearing and aqueous inclusions (Fig. 5 and table 2). Melt inclusions are found in beryl and quartz of zones I, and III, and melt-fluid inclusions (consisting of melt and fluid, the fluid being mostly H_2O and CO_2, in a inclusion) in quartz, beryl and spodumene occur in zones I, III IV and V. Aqueous inclusions occur in all minerals that studied. Most H_2O-CO_2 inclusions are found from spodumene and quartz in Zone IV to X, the halite bearing inclusions occur in beryl, spodumene and quartz of zones IV to VII and coexisting with aqueous inclusions.

Silicate melt inclusions: These melt inclusions contains silicate melt, a vapor phase and several crystals which may be the same minerals that forming the pegmatite (Fig. 5a and 5b). The vapor phase of those inclusions dominated with or without CO_2. Halite is found in some those inclusions. These inclusions occur in zones I and III, may be in Zone II too, but since it has been intensely metasomatized. Silicate melt inclusions are difficult to find.

Fluid-melt inclusions: At room temperature, inclusions consist of silicate melt and fluid phase(liquid and gases) are called fluid melt inclusions. This type of inclusion represents the products of crystallization from silicate-rich hydrous fluids whose compositions lie between those of typical silicate magmas (silicate melt inclusion) and aqueous fluids (fluid inclusions). They consist of silicate melt + saline melt + H_2O (Fig. 5c, 5d and 5f).The aqueous phase can be present in the form of both saturated liquid and vapor. The ratio between the phases is highly variable, and in some inclusions, the saturated aqueous phase dominates over the silicate melt phase.

H_2O-CO_2 inclusion: These inclusions contain of variable amount of CO_2 and H_2O (Fig. 5h and 5i). Sometimes there are three phases (vapor and liquid CO_2, liquid H_2O) sometimes there are only two(vapor and liquid CO_2). A halite daughter mineral has been found in two H_2O-CO_2 inclusions of spodumene of zone V.

Halite bearing inclusions: These inclusions consist of vapor ,liquid phases and several daughter minerals, in most cases halite (Fig. 5e 5g and 5h). Inclusions with tube or negative in shape occur within the growth zones of beryl and spodumene crystals. This type occurs in zones from IV to VII and is associated with aqueous inclusions.

Aqueous inclusion: Aqueous inclusions occur in all zones(Fig.5h), with most being primary in zones V to X (specially in spodumene growth zones).

Where they occur in zones of I to IV, they are mostly secondary in origin, existing along sealed fractures.

The metasomatic alteration are associated with all types of fluid inclusions. In the tourmaline alteration zone, aqueous, saline and H_2O-CO_2 inclusions are found in tourmaline with negative crystal shape (mostly tubular).

In summary, three types inclusion: melt, fluid-melt and fluid inclusions have been found in No. 3 pegmatite. The fluid inclusions can be further subdivided into: aqueous, halite bearing and H_2O-CO_2 inclusions.

Fluid and melt inclusions in beryl: Most inclusions in beryl crystals are melt and fluid melt inclusions. Usually beryl crystals in zones I to III, containing melt inclusions (Fig. 5b), but these in zone IV, containing fluid melt inclusions (Fig. 5c). Besides these two types, fluid inclusions, mostly saline fluid inclusions with negative crystal shape also occur (Fig.5e). The fluid-melt and fluid inclusions are mostly negative crystal shape and located in the growth zones of beryl.

Fluid inclusions in spodumene: Spodumene mostly contains fluid melt inclusions (Fig.5d and 5f) and fluid inclusions including aqueous, saline and H_2O-CO_2 inclusions (Fig. 5g). Spodumene is abundant in zones V and VI, in zone V, spodumene as relatively medium grained laths (10 cm to tens cm) intergrown with quartz, cleavelandite, and in zone VI, as coarse grained crystals (up to 3 m) embedded in cleavelandite, quartz and sometimes with montebrasite. Both textural types of spodumene are densely packed with fluid-melt inclusions and fluid inclusions. These inclusions are relatively large, possess subhedral to euhedral negative crystal forms of spodumene (tubular in shape) and are located in the growth zones of spodumene crystal, or as isolated individuals, thus are of primary origin. The fluid-melt inclusions and fluid inclusions distribute either together or in different location.

Fluid inclusions in quartz: Quartz is mineral host to a lot of inclusions in the pegmatite. Fluid inclusions were examined in quartz from three associations: (1) from intergrowth with beryl in zones I and III, (2) from intergrowth with spodumene, (3) massive quartz in zones VII to X. The fluid inclusions in quartz of zones I and III are mostly melt inclusions (Fig. 5a) and fluid melt inclusions, similar to those in beryl in same zone. In zones IV to VI, it consists of fluid melt and fluid inclusions analogous to those in spodumene. The fluid inclusions in quartz of zone VII to X are aqueous and H_2O-CO_2 inclusions (Fig. 5h and 5i). The fluid inclusions in quartz are not as easily distinguished as primary or secondary as those in beryl and spodumene, but since the quartz occur as intergrowths with spodumene and beryl, and the type of fluid inclusions in quartz, beryl or spodumene are same, it may be assumed that these inclusions are primary ones as same as that in beryl and spodumene.

The fluid inclusions of quartz in zones VII to X are mostly aqueous inclusions with minor of H_2O-CO_2 inclusions. The aqueous ones are irregular in shape and distributed as cluster and sometimes along the fractures.

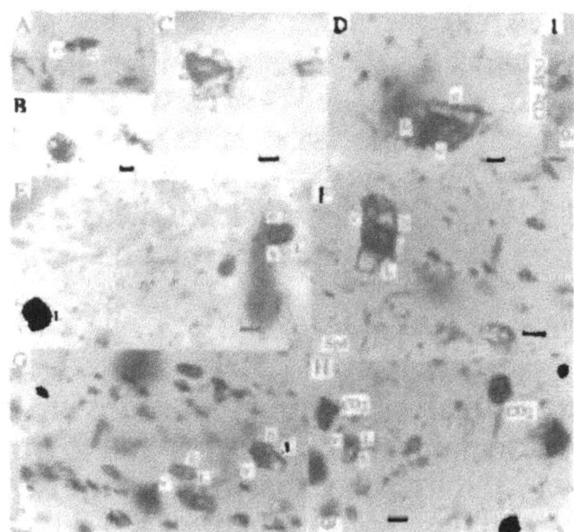

Fig. 5. Melt, Melt-fluid and fluid inclusions in No. 3 Pegmatite

A. Melt inclusions in quartz of zone No.1, it consists of several crystals (c: possible mineral: lithium tetraborate, cookeite, and albite) and a vapor phase. B. melt inclusion in beryl with negative shape of beryl and it consists of several crystals (possible lithium-tetraborate $Li_2B_4O_7$,cookeite,pollucite,quartz,microlite) and a vapor phase; C. fluid melt inclusion of beryl in zone III, it contains of a low salinity fluid phase (almost 50% in volume) and several crystals; D. fluid melt inclusion in spodumene of zone V, with several crystals and a fluid phase rich in CO_2; E. saline inclusions in beryl of zone III, it consists several daughter minerals and a vapor. These inclusions have a negative crystal in shape and with several daughter minerals in the two big ones and one halite in the small ones; F. fluid-melt inclusions in spodumene of zone V, it consists of halite and other daughter minerals with various size of negative inclusions occurred in the growth zone of spodumene; G. saline inclusions in spodumene of zone VI, it contains halite and other daughter minerals; H. aqueous, saline inclusions and CO_2-rich inclusions in quartz of zone X; and I. H_2O-CO_2 inclusions in quartz of zone VII. bar=20 μm; C=silicate crystal; V=vapor; L=liquid; h=halite.

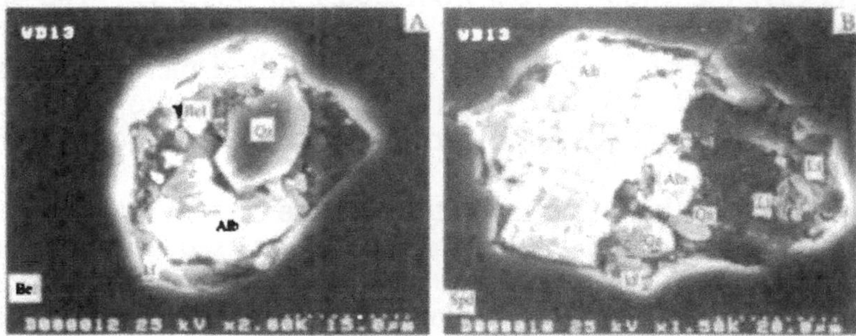

Fig. 6. Scanning Electron Microscope analysis of fluid melt inclusions in beryl and spodumene

A. Fluid-melt inclusion in beryl, it consists of albite, quartz, apatite, a possible beryllium mineral and unknown crystals, the dark part of the inclusion originally occupied by the fluid phase; B. fluid melt inclusion in spodumene with k-feldspar, quartz, albite and possible lithium silicate minerals: cookeite, lithium tetraborate, and other unknown minerals. The dark part of the inclusion originally occupied by the fluid phase.

alb=albite; ap=apatite; c=unknown crystal ; kf=K-feldspar; Qz=quartz; Li=possible lithium mineral.

Analytical methods: Fluid inclusion studies were carried out on doubly polished thinsections, using a Chaixmeca heating -freezing stage. The stage was calibrated at -56.6, -6.6, 0.0, +374.0°C using the synthetic fluid inclusions. Calibration indicates a precision at the standard reference points of ±0.1°C at 0°C and ±0.7°C at -56.5°C, in the heating, calibration gives a precision of ±6°C at 374°C. High temperature measurements were carried out on a Leitz 1350 heating stage and a quench method.

Electron microscope analysis

The electron microscope analysis of fluid inclusions are carried out mainly the fluid -melt inclusions (including 4 from beryl and 2 from spodumene). The results are as follows: through the analyses the most abundant mineral in inclusions are quartz, albite, halite, apatite, K-feldspar and some lithium silicate minerals possible cookeite, lithium tetraborate, one possible beryllium mineral, and other unknown minerals. From Fig.6, it may be seen that the volume of fluid inside the inclusions varies but most from 15 to 25 percent, the rest of the inclusion is occupied by the solid minerals mentioned above. The solid minerals inside the inclusions are very similar to the minerals that form the pegmatite, but the composition of the fluid phases in inclusions are H_2O, CO_2 and NaCl which is different to the composition of the solid part which formed from silicate melt. This indicates that immiscibility took place during the crystallization of pegmatite particularly in the formation of zones from IV to VI.

Microthermometric data

Microthermometric data (Fig.7 and table 3) show that homogenization temperature (Th) of the melt inclusions ranges from 700 to 850°C. Th of melt-fluid inclusions ranges from 299.1 to 465.2°C, and that for aqueous, halite bearing and H_2O-CO_2 inclusions ranges from 220.1 to 403.7°C. The salinity of aqueous inclusions range from 5 to 9 wt.% NaCl. equiv., which is relatively low. The salinity of halite bearing and some fluid melt inclusions ranges from 28.50 to 32.0 wt.% NaCl equivalent. The pressure calculated (using Flincor program) from H_2O-CO_2 and CO_2 inclusions is from 1800 to 2500 bars.

To measure the homogenization temperature of melt inclusions in quartz and beryl of zones I and III, a quench method was used. The melt inclusions in both minerals started to melt at 650° and homogenized from 700 to 850°C. Some bigger melt inclusions started to melt at 650 there is resudial solid rest but until 900°C. This could be the results of leakage of these inclusions.

The fluid-melt inclusions have been studied in beryl and spodumene. The homogenization temperature ranges from 390.7 to 465.2°C for beryl and 299.1 to 401.3°C for spodumene. Most of these inclusions decrepitated, when heated 50°C over their Th. The decrepitate temperatures for these inclusions range from 370 to 510.0°C. Some fluid melt inclusions have a halite daughter mineral which dissolves at temperatures from 147.9 to 201.6°C and correspond to a salinity of 29.40 to 32.0 wt.% NaCl.

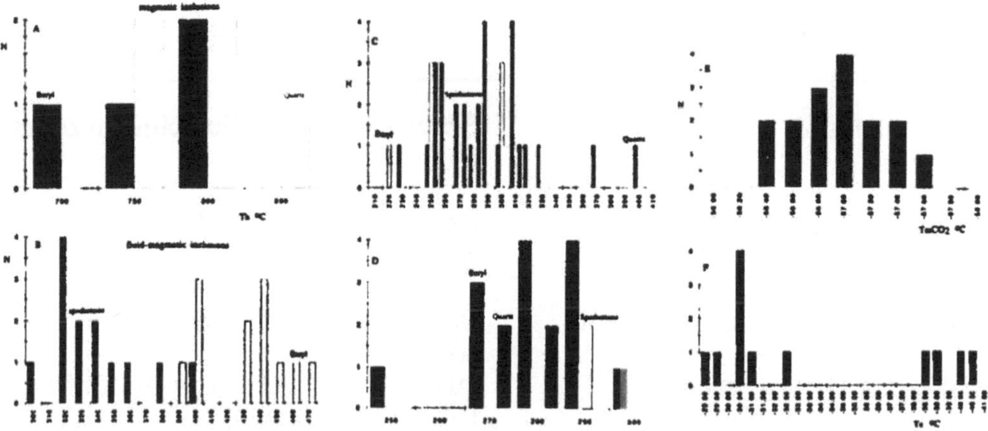

Fig. 7 Microthermometric data of fluid and melt inclusions in No.3 pegmatite
A. Temperatures of homogenization of melt inclusions in quartz and beryl of zones I and III; black: inclusions in beryl, white: inclusions in quartz. B. Temperatures of homogenization of fluid-melt inclusions in beryl and spodumene; Black; spodumene, white: beryl. These temperature representing the disappear of vapor phase only, not for the solid phase in the inclusions. Most of these inclusions when temperature over 50°C of their Th,they are decrepitated, C. Temperatures of homogenization of aqueous inclusions in beryl, spodumene and quartz from Zone III to zone X. Black: inclusions in quartz, white in beryl and grey : in spodumene, D. Temperatures of homogenization of halite bearing inclusions in beryl, spodumene and quartz, Black: inclusions in beryl, white: in spodumene and grey: in quartz. E. Temperature of melting in CO_2 and H_2O-CO_2 inclusions and F. Eutectic temperatures of aqueous inclusions in quartz.

The phenomenon of decrepitation, relative low Th and halite bearing are three distinguishing features between melt inclusion and fluid melt inclusions.

The Th of halite bearing inclusions in beryl, spodumene and quartz range from 253.0 to 298.0°C with a salinity from 28.50 to 31.50 wt.%. Comparing these inclusions with some halite bearing fluid melt inclusions, gave similar results for salinity, but the Th is lower than that in fluid melt ones.

The Tm of H_2O-CO_2 range from-56.40 to -57.60°C indicating close to the Tm of CO_2, but with some other gases, the Th range from 220.1 to 311.0°C and the pressure calculated (using Flincor program, [1]) from H_2O-CO_2 and CO_2 inclusions is from 1800 to 2500 bars.

Aqueous inclusions occurring in beryl, spodumene and quartz from Zone V to X have been studied. The Th of these inclusions ranges from 218.9 to 403.7°C, with the eutectic temperatures range from -29.0 to -32.5 and -38.5 to 40.5°C indicating there are other cations such as Mg, Ca and K in the fluid besides the NaCl.

Table 2 The occurrence of melt, fluid-melt and fluid inclusions in No.3 Pegmatite

Type of inclusions	Host mineral and zones
Melt inclusion	Beryl, quartz and microcline in zones I and III
Fluid-melt inclusion	Beryl, spodumene and quartz in zones III to VI
Fluid inclusion:	
saline multiple phases inclusion	Beryl, quartz and spodumene in zones II to VII
CO_2 or CO_2-H_2O inclusion	quartz and spodumene in zones from IV to VI and zone X
Aqueous inclusion	Beryl, quartz and spodumene in all zones

Table 3 Homogenization temperatures of melt, fluid-melt and fluid inclusions

	Host minerals		
Inclusion	beryl	spodumene	quartz
Melt inclusion	700-800 (4)*		750-850 (5)
Fluid-melt inclusion	390.7-465.2 (11)	299.1-401.3 (12)	
Halite bearing inclusion	253.0-298.0 (13)	265.0-89.8 (5)	268.9-278.6 (4)
Aqueous inclusion	218.9-311.2 (10)	245.3-307.1 (8)	250.5-403.7 (21)
H_2O-CO_2 inclusions			220.1-311.0 (12)

* number of measurement

DISCUSION AND INTERPRETION

There are ten textural and internal mineral zones in the Keketuohai No. 3 pegmatite, which represent a crystallization process from a granitic magma to a Li, Be, Nb, Ta and Cs pegmatite. These ten zones are distinguished in mineral assemblage, texture and location, but they are occur in a single continuous pegmatite body, and the boundary between zones is a transition and not a knife shape. Secondly from petrographic studies show that there are primary minerals and metasomatic mineral assemblages. Although these two groups are difficult to distinguish, may in the zones from I to V and partly in zone of VI, both groups exist, but from zone VI to X only one group mineral assemblage occur. In other words, the primary mineral in zones from I to V have been partly replaced by a later metasomatic process. The minerals formed during the metasomatic process are muscovite, quartz, albite, cleavelandite, Li and Rb bearing micas which are the same mineral assemblage as occurs in the later zones (Zone VI to X, see table 1). This indicates that two main stages: magmatic stage and metasomatic stage occur in the No. 3 pegmatite.

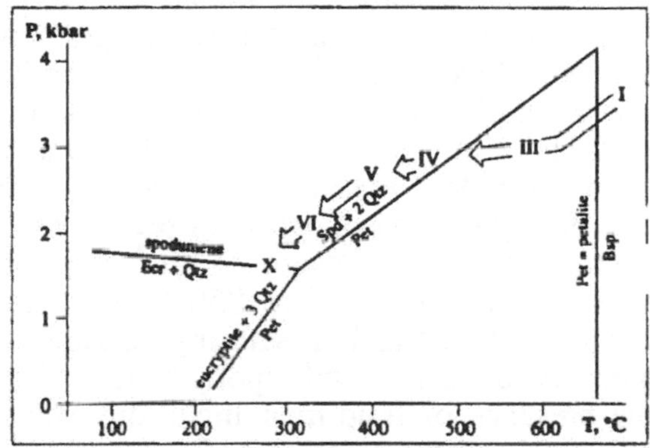

Fig. 8 The cooling path and internal evolution of the No.3 pegmatite, Keketuohai (Phase diagram of Lithium alummosilicate from London 1986)
The P-T condition starting graphic zone (I) at temperature of 850°C, then gradually cooling to form zones II and III. At temperature of 470°C spodumene crystalized and forming the zone V. Continually cooling forming the zone of VI. After that the pegmatite continually crystallization, but since we have found eucryptite and petalite in this pegmatite, the cooling path is not showing in this diagram.

Experimental P-T phase diagram for the bulk composition lithium minerals have been carried out by several authors [6]. London (1986) provide a P-T phase diagram for most common lithium minerals (spodumene, petalite, eucryptite) with quartz saturating. The P-T phase diagram is directly applicable to natural pegmatite systems and explains many of the observed reaction relations among the lithium aluminosilicates. The field of spodumene and petalite is bounded at T > 700°C by a reaction that produce the high temperature, stuffed-silica derivatives virgilite and beta -spodumene [7]. Natural beta-spodumene has not been reported, and virgilite is known only from one locality, therefore, lithium aluminosilicates saturation occurs at temperatures below 700°C at any pressure. Spodumene and petalite are the stable lithium aluminosilicates at magmatic temperatures, with spodumene at high pressure and petalite at low pressures. The major lithium mineral in Keketuohai No.3 pegmatite are spodumene and lepidolite, and no petalite and eucryptite have been found, therefore, the dominance of spodumene in pegmatite places constrains on the regional pressure and temperatures at the time of pegmatite consolidation. In the No.3 pegmatite, major industry minerals are beryl and spodumene which are intergrown with quartz (or saturated with quartz). The beryl occurs in the outer shell (from zone I to IV) and the spodumene exists in the central shell (from V to VI). The paragenesis indicates that the beryl is formed early than spodumene. After forming the spodumene, another Li mineral, lepidolite deposited in the inner zones (from VII to IX). Since no phase diagram of beryl with quartz and other silicate minerals is available mean time, but it is known that beryl formed earlier than spodumene and is in the same system, it may be presumed that the temperature (possible pressure) of formation of beryl is higher than that of spodumene. According to London (1986), spodumene + quartz is stable over a temperature range from 150 to 700°C and pressure ranges from 1800 to 6000 bars. The No. 3 pegmatite formed at relatively high pressure and temperature, it also can be provided by the source granites for this pegmatite shown at deeper levels of exposure (Fig. 3b). The beryl in No. 3 pegmatite is a primary mineral and intergrown with feldspar and quartz, the formation conditions of beryl can be indicated by melt and fluid inclusion in this study.

Fluid inclusions provide a means of evaluation of P-T conditions and fluid compositions during the consolidation of pegmatite. In No. 3 pegmatite, homogenization temperatures of magmatic inclusions in quartz and beryl from Zones I and III range from 700 to 850°C. This temperature is higher than that of spodumene + quartz P-T stability conditions, but is coincident with the paragenesis of beryl and spodumene. The most important occurrence is the presence of fluid-melt inclusions whose composition lies between those of typical silicate magmas and aqueous fluids. At Keketuohai No.3 pegmatite, the fluid-melt inclusions in spodumene beryl and quartz intergrowth have been studied. The Th of melt-fluid inclusions ranges from 299.1 to 465.2°C, which is in the range of spodumene and quartz P-T condition stability. The Scanning Electron Microscope analysis for these inclusions show that it consists of albite, quartz, K-feldspar, Lithium minerals, apatite and some unknown minerals. The fluid phase in these inclusions is rich in NaCl with salinity about 29 to 32 wt.% NaCl. The fluid in some fluid-melt inclusions is rich in CO_2 in addition tothe saline one. This indicate the fluid derived from the magma is a dense saline fluid.

On the basis of fluid and melt inclusion data from Keketuohai No. 3 pegmatite, there is no indication that an exsolved aqueous fluid existed in the early stages of pegmatite consolidation, especially in the zones I, II and III. However, phase separation of an aqueous fluid from a pegmatite melt took place after the solidification of the above zones. This study presents evidences that there is a dense, hydrous, alkali silicate fluid which represented by the fluid-melt inclusions, are compositionally intermediate between silicate melt and aqueous fluids that led to crystallization of Zones IV, V, VI, and VII. The presence of primary miarolitic cavities (in the zones of V and IV) indicate vapor saturation. It shows clear evidence of various stages of silicate melt - hydrous saline melt - aqueous fluid - CO_2 fluid immiscibility. The immiscible processes occurred during the evolution of the pegmatite and silicate melt/hydrous saline melt and silicate/CO_2 and silicate melt/H_2O indicates heterogeneous trapping of immiscible phases during the pegmatite crystallization and mineralization stages (Fig. 8).

The data derived from fluid inclusions: aqueous, H_2O-CO_2 and halite bearing inclusions in No. 3 pegmatite indicate that after silicate-fluid immiscibility, separating a fluid which is a saline and CO_2 rich. Roedder (1992) point out that the pegmatite melt separated a dense H_2O, salts, CO_2 fluid and this fluid further separated a CO_2 phase. Lu (1990) and Lu et al. (1993) studied the fluid-melt inclusion and show an evidence of a hydrothermal fluid derived from a magma [9,10]. It is not clear that the magma separated one fluid with the composition of NaCl-H_2O-CO_2 or separated into two fluids with the composition of NaCl-H_2O and H_2O-CO_2. In the first case, the NaCl-H_2O-CO_2 fluids then separated into two fluids: NaCl-H_2O and H_2O-CO_2. In No.3 pegmatite, halite occurs as a daughter mineral in two H_2O-CO_2 inclusions, and halite bearing inclusions are associated with aqueous ones, not with H_2O-CO_2 ones. It is possible at first separation of a NaCl-H_2O-CO_2 fluid and took place further later by separation into two fluids: NaCl-H_2O-CO_2. The P-T conditions of aqueous fluids for No. 3 pegmatite as follows: The microthermometric data of three types of inclusions are the Th of aqueous, halite bearing and H_2O-CO_2 inclusions range from 220.1 to 403.7°C. The salinity in aqueous inclusions ranges from 5 to 9 wt.% NaCl. equiv., the salinity of halite bearing ranges from 28.50 to 31.50 wt.% NaCl equivalent. The Th of H_2O-CO_2 inclusions ranges from 220.1 to 311.0°C and the pressure calculated from H_2O-CO_2 and CO_2 inclusions ranges from 1800 to 2500 bars. This is similar to the fluid in charge of metasomatic process in No. 3 pegmatite. The metasomatic process occur in zones I to VI formed second stage minerals such as muscovite, quartz, albite, cleavelandite, Li-mica and Rb-mica. In the zones VII to X, the metasomatic feature is very weak, but the mineral assemblage are the same as the metasomatic minerals in the zones from I to VI. This indicates these saline dense and CO_2 bearing fluids are responsible for the zones of VII to X and the metasomatism of the zones form I to VI.

The history of fluid evolution of No. 3 pegmatite involved a magmatic, silicate melt/hydrous saline melt and silicate/CO_2 and silicate melt/H_2O immiscible fluids, and fluids. The magmatic process formed zones I, II and III, the silicate melt-fluid immiscibility process took place during formation of zones IV, V, VI (starting partly in Zone III), and the dense saline fluid are in charge to form the rest of zones. Although three stages can be recognized, but actually they took place in a same and continually system.

The available fluid inclusion data from the No.3 and other pegmatite in Keketuohai, couples with the mineral assemblage of spodumene + quartz, spodumene + cleavelandite, constrain the conditions of pegmatite formation to the approximate range of 300 to 850°C and over 1900 bars (Fig. 8). This P-T range for Li mineralization stages (300 - 470°C and > 1900 bars) is comparable to that of Tanco pegmatite [6], but it is different other fluid inclusion studies on pegmatites[5, 13].

The results indicate that the No.3 pegmatite was formed during the transition from magmatic to subsolidus hydrothermal fluid conditions, occurring probably in three stages. Stage 1 was magmatic; stage 2 required a processes of immiscibility to produce both melts and fluids; the final stage was dominated hydrothermal fluids. Li, Be, Nb, Ta and Cs mineralization occurred throughout this transition.

Acknowledgements

This research has been financial supported by the K.C. Wong Education Foundation, Hong Kong and a Grant to H.Z. Lu from Natural Sciences and Engineering Research Council of Canada. We would like to thank the mine staff at the Keketuohai mining Company for their generous assistance to the authors during the field work. Thoughtful review of the early version of the manuscript by Drs. E. Chown, London and J. Guha is much appreciated.

REFERENCES

1. Brown, P., and Hagemann, S., 1995 MacFlincor and its application to fluids in Archean lode-gold deposits, Geochimica et Cosmochimica Acta.v.59, pp. 3943-3952.
2. Cerny, P. 1991, Rare-element granitic pegmatites. Part I: Anatomy and internal evolution of pegmatite deposits, Geoscience Canada, V. 18, No.2 pp. 49-81.
3. Cerny, P. 1982, Granitic pegmatites in science and industry; MAC short course handbook, V. 8, 555 p.
4. Gresens, R.L., 1967,Tectonic-hydrothermal pegmatites; 1, The model: Contributions to Mineralogy and petrology, v. 15, pp. 345-355.
5. Linnen, R.L., and Williams-Jones, A.E., 1994, The evolution of pegmatite-hosted Sn-W mineralization at Nong Sua, Thailand: evidence from fluid inclusions and stable isotopes. Geochimica et Cosmochimica Acta.v.58, pp. 735-747.
6. London, D., 1986, Magmatic-hydrothermal transition in the Tanco rare-element pegmatite: Evidence from fluid inclusions and phase-equilibrium experiments. Am. Mineralogist, V. 71, pp. 376-395.
7. London, D., 1990, Internal differentiation of rare elements pegmatite; a synthesis of recent research, GSA special paper 246, pp. 35-50.
8. London, D. 1986, Formation of tourmaline -rich gem pockets in miarolitic pegmatites, Am. Mineralogist, V. 71, pp. 396-405.
9. Lu, H.-Z., and Wang, Zhonggang, 1993, Mineralization, fluid and melt inclusions studies on No.3 pegmatite of Keketuohai, Xinjiang province, China in the abstract volume of GAC| MAC Annual meeting.
10. Lu, H.-Z., 1990, Fluid-melt inclusion: an evidence of hydrothermal fluid derived from magma? Geochemica, 3: pp. 225-229.
11. Roedder,E.1992, Fluid inclusion evidence for immiscibility in magmatic differentiation. Geochimica et Cosmochimica Acta.v.56, pp. 5-20.
12. Stewart, D.B. 1978, Petrogenesis of lithium-rich pegmatites: Am. Mineralogist, v.63, pp. 970-980.

13. Turnbull, R.B., 1995, A fluid inclusion study of the Sinceni rare-element pegmatites of Swaziland, Mineralogy and Petrology. v. 55, pp. 85-102.
14. Wang, Z.G. 1993, Granites and Pegmatite in Altay area. in Chinese, 305 project report and a book, Science Press, in press.
15. Xinjiang Nonferrous metal Co-op 1988, The No. 3 pegmatite
16. Zhang, L., Liu, D., and Tang, Y. 1990, Mineral resources of Xinjinag, published by Xinjiang Peoples Publisher and Hong Kong Cultural and education publisher, 328 p. in Chinese.
17. Zhou, T., Zhang, X., Jia,F., Wang,R., Cao, H., and Wu, B., 1986, The genesis of No.3 pegmatite of Keketuohai, Mineral deposits v.5, No. 4., pp. 34-48. in Chinese.

Milton Keynes UK
Ingram Content Group UK Ltd.
UKHW020314111024
449327UK00040B/1083

9 780367 447953